T0295199

WINDS OF STARS AND EXOPLANETS

IAU SYMPOSIUM 370

COVER ILLUSTRATION:

Three-dimensional simulation of the escaping atmosphere of a close-in exoplanet interacting with the wind of its host star. The red field lines represent the planetary magnetic field, while the contours represent the total density (horizontal plane) and the neutral hydrogen density (vertical plane). Figure from Carolan et al 2021.

IAU SYMPOSIUM PROCEEDINGS SERIES

INTERNATIONAL ASTRONOMICAL UNION

UNION ASTRONOMIQUE INTERNATIONALE

WINDS OF STARS AND EXOPLANETS

PROCEEDINGS OF THE 370th SYMPOSIUM OF THE INTERNATIONAL ASTRONOMICAL UNION BUSAN, KOREA, REP OF 8–11 AUGUST 2022

Edited by

ALINE A. VIDOTTO

Leiden Observatory, Leiden University, the Netherlands

LUCA FOSSATI

Space Research Institute, Austrian Academy of Sciences, Austria

and

JORICK S. VINK

Armagh Observatory and Planetarium, Armagh, N. Ireland, UK

CAMBRIDGE UNIVERSITY PRESS
University Printing House, Cambridge CB2 8BS, United Kingdom
1 Liberty Plaza, Floor 20, New York, NY 10006, USA
10 Stamford Road, Oakleigh, Melbourne 3166, Australia

First published 2023

Printed in Great Britain by Henry Ling Limited, The Dorset Press, Dorchester, DT1 1HD

Typeset in System LaTeX 2ε

A catalogue record for this book is available from the British Library Library of Congress Cataloguing in Publication data

This journal issue has been printed on FSC™-certified paper and cover board. FSC is an independent, non-governmental, not-for-profit organization established to promote the responsible management of the world's forests. Please see www.fsc.org for information.

ISBN 9781009352789 hardback
ISSN 1743-9213

Table of Contents

Part 3: Physical ingredients of winds

Part 4: Flow-flow interactions

Part 5: Relevance of winds on stellar/planetary evolution

Preface

Winds form an integral part of astronomy – from regulating rotation of stars through enriching galaxies with fresh materials, outflowing winds persist during the entire lives of stars and play a key role in shaping the observed exoplanet demographics. In the case of massive stars, their winds are a vital ingredient of their evolution, from the main sequence to the pre-supernova stage, determining black hole masses as measured from gravitational waves. In the case of low-mass stars, their winds dictate rotational evolution, which affect angular momentum distribution within the stellar interior and thus affect generation of magnetic fields. Finally, in the case of planets, winds take the form of atmospheric escape, which can strongly affect their atmospheric evolution. Strong escape of highly irradiated exoplanets have now been observed in several close-in exoplanets during transits and are indirectly detected in the observed exoplanet radius distribution.

Although the only astrophysical wind that we are able to directly probe is that of the Sun, the past decades have seen great progress in observing winds of other astrophysical objects. In particular, in recent years, several observing programmes and space missions have focused on studying winds from our Sun, other stars and exoplanets.

On the solar side, two new space missions, Parker Solar Probe and Solar Orbiter, are dedicated to studying the physics of the solar wind. By traveling much closer to the Sun than any other spacecraft has ever been, these new missions allow direct measurements of the solar wind at an unprecedented close distance. Data from these missions might provide interesting implications for the variability of the plasma environment at the orbits of close-in exoplanets.

On the stellar side, winds of low-mass stars are magnetically driven, and magnetism has been either directly (through Zeeman effects) or indirectly (through activity proxies) observed in these stars. Recently, many new magnetospheres were detected around massive stars as well. In spite of similarities, there is a major difference between winds of low- and high-mass stars: their mass-loss rates are orders of magnitude different, due to different physical processes driving their winds. Even with substantially lower mass loss rates, winds of low mass-stars play a fundamental role in removing angular momentum, and thus, shaping the rotational evolution of these stars.

On the planetary side, missions like Kepler, TESS and Plato (will) provide the statistics for planet population studies and hence infer the indirect presence of outflowing planetary winds in shaping the distribution of sizes of close-in exoplanets. HST has been fundamental in detecting strong atmospheric escape of close-in giant planets through ultraviolet transmission spectroscopy. Recent observations have also opened the possibility to detect escaping planetary winds from the ground.

In order to gain insight in the physics and the modelling tools used by different communities, and to foster communication between communities that do not usually interact with each other, we brought together researchers working on winds of close-in exoplanets (atmospheric escape), winds of low- and high-mass stars and the solar wind in a symposium dedicated to "winds". The IAU Symposium "S370: Winds of stars and exoplanets", that took place in Busan, Republic of Korea, from 8 to 11 August 2022.

In this book, you will find contributions from most of the symposium presenters. The first Part presents an overview of winds of stars and planets, introducing its similarities and differences. The remaining parts contain the four main themes discussed in the symposium

- Observational evidence of winds
- Physical ingredients of winds

- Flow-flow interactions
- Relevance of winds on stellar/planetary evolution

The IAU Symposium 370 took place during the XXXI General Assembly meeting in Busan, in the midst of the Covid-19 global pandemic. We wish to thank the Local Organisation of the General Assembly meeting for their substantial efforts to make it possible that we all could meet in person again. We also warmly thank all the participants of the meeting for their cooperation and understanding during these difficult times.

Aline A. Vidotto
Luca Fossati
Jorick S. Vink

Editors

Aline A. Vidotto
Leiden Observatory, Leiden University, the Netherlands

Luca Fossati
Space Research Institute, Austrian Academy of Sciences, Austria

Jorick S. Vink
Armagh Observatory and Planetarium, Armagh, N. Ireland, UK

Organizing Committee

Scientific Organizing Committee

Aline A. Vidotto (Leiden Observatory, Leiden University, the Netherlands) – **Chair**
Luca Fossati (Space Research Institute, Austrian Academy of Sciences, Austria) – **Co-chair**
Jorick S. Vink (Armagh Observatory and Planetarium, Armagh, UK) – **Co-Chair**
Steve Cranmer (University of Colorado Boulder, US)
Richard Ignace (East Tennessee State University, US)
Moira Jardine (University of St Andrews, UK)
Kristina Kislyakova (University of Vienna, Austria)
Shazrene Mohamed (University of Cape Town and South African Astronomical Observatory, South Africa)
Takeru Suzuki (University of Tokyo, Japan)

National and Local Organizing Committees of the XXXIth General Assembly of the IAU

Hyesung Kang (Chair), Pusan National University, Republic of Korea
Byeong-Gon Park (Vice-Chair), Korea Astronomy & Space Science Institute, Republic of Korea
Deokkeun An, Ewha Womans University, Republic of Korea
Jungyeon Cho, Chungnam National University, Republic of Korea
Joon-Young Choi, Busan National Science Museum, Republic of Korea
Aeree Chung, Yonsei University, Republic of Korea
Junga Hwang, Korea Astronomy & Space Science Institute, Republic of Korea
Ho-Seong Hwang, Seoul National University, Republic of Korea
Chunglee Kim, Ewha Womans University, Republic of Korea
Dohyeong Kim, Pusan National University, Republic of Korea
Ji-hoon Kim, Seoul National University, Republic of Korea
Jongsoo Kim, Korea Astronomy & Space Science Institute, Republic of Korea
Minjin Kim, Kyungpook National University, Republic of Korea
Sungsoo S. Kim, Kyung Hee University, Republic of Korea
Woong-Tae Kim, Seoul National University, Republic of Korea
Woojin Kwon, Seoul National University, Republic of Korea
Jeong-Eun Lee, Kyung Hee University, Republic of Korea
Kang Hwan Lee, Institut Pasteur Korea, Republic of Korea
Sang-Sung Lee, Korea Astronomy & Space Science Institute, Republic of Korea
Seo-gu Lee, Korea Astronomy & Space Science Institute, Republic of Korea
Soo-Chang Rey, Chungnam National University, Republic of Korea
Hyunjin Shim, Kyungpook National University, Republic of Korea
In-Ok Song, Korea Science Academy of KAIST, Republic of Korea
Hong-Jin Yang, Korea Astronomy & Space Science Institute, Republic of Korea
Suk-Jin Yoon, Yonsei University, Republic of Korea
Sung-Chul Yoon, Seoul National University, Republic of Korea

List of Contributors

N. Alameri
A. Allan
M. Alqasimi
J. Alvarado-Gómez
K. Amada
A. Anand
I. Araya
C. Arcos
B. Arora
M. Bernini Peron
A. S. Brun
V. Brunn
T. Cang
S.-J. Chang
J. Chebly
E. Costa-Almeida
M. Cure
S. Daley-Yates
A. Danehkar
J.-M. Desert
L. Dos Santos
F. Driessen
V. Elbakyan
A. Finley
L. Fossati
G. González-Torà
P.-G. Gu
C. Hawcroft
G. Hazra
A. S. Hojaev
A Ibraimova
H. Imai
A. Javadi
R. Kavanagh
E. Keles
Z. Keszthelyi
H. Kim
K. Kislyakova
V. Kocharovsky
D. Koh
T. Konings
A. Kosherbayeva
D. Kubyshkina
M. Kunitomo
X. Leng
J. Mackey
A. Manchado
A. Manousakis
A. Mehner
C. Meskini
H. Mitani
D. Modirrousta-Galian
N. Moens
T. Moriya
D. Nandi
A. Nanni
D. O Fionnagain
A. Okazaki
R. Osten
S. Owocki
A. Pai Asnodkar
S. Parenti
G. Pinzon
S. Popov
K. Pukitis
V. Ramachandran
A. Ray
M. Saberi
G. N. Sabhahit
A. Sander
I. Shaikhislamov
M. Shoda
A. Strugarek
T. Su
T. Suzuki
A. Taani
G. Telford
O. Verhamme
A. Vidotto
J. Vink
R. Waugh
S.-C. Yoon
S. Yun

IAUGA 2022

Part 1:
Overview of Winds of stars and exoplanets

Winds of Stars and Exoplanets
Proceedings IAU Symposium No. 370, 2023
A. A. Vidotto, L. Fossati & J. S. Vink, eds.
doi:10.1017/S1743921322003489

Winds and magnetospheres from stars and planets: similarities and differences

Stan Owocki

Department of Physics & Astronomy, Bartol Research Institute, University of Delaware, Newark, DE 19716 USA
email: owocki@udel.edu

Abstract. Both stars and planets can lose mass through an expansive wind outflow, often constrained or channeled by magnetic fields that form a surrounding magnetosphere. The very strong winds of massive stars are understood to be driven by line-scattering of the star's radiative momentum, while in the Sun and even lower-mass stars a much weaker mass loss arises from the thermal expansion of a mechanically heated corona. In exoplanets around such low-mass stars, the radiative heating and wind interaction can lead to thermal expansion or mechanical ablation of their atmospheres. Stellar magnetospheres result from the internal trapping of the wind outflow, while planetary magnetospheres are typically shaped by the external impact from the star's wind. But in both cases the stressing can drive magnetic reconnection that results in observable signatures such as X-ray flares and radio outbursts. This review will aim to give an overview of the underlying physics of these processes with emphasis on their similarities and distinctions for stars vs. planets.

Keywords. Sun: solar wind; stars: early-type; stars: mass loss; stars: planetary systems

---

1. Introduction

To set the stage for this symposium's exploration of "The Winds from Stars and Exoplanets", I have been asked to review the similarities and differences in the physical processes that drive wind outflows from stars vs. planets, including the distinct roles that magnetic fields play in trapping and diverting plasma flows in an associated *magnetosphere*.

As illlustrated in figure 1, one can identify three broad classes of stellar wind:

1. the pressure-driven coronal wind of the Sun and other cool stars;
2. radiatively driven winds from OB stars;
3. the slow "overflow" mass loss from highly evolved giant stars.

For the last a key is the greatly reduced gravity, which allows even surface convection or pulsations to eject outer layers to escape, somewhat like Roche-lobe overflow in binary systems. The resulting mass loss can be irregular and difficult to quantify, and since it has little overlap with winds from planets, we set this aside to focus on the former two.

We first review (§2) the radiative driving of OB winds, which occurs through a kind of line-driven "suction" effect. We then discuss (§3) how the winds of the Sun and other cool stars are, in contrast, driven by the gas pressure associated with hot corona, with mass loss representing now an escape valve analogous to that of pressure cooker.

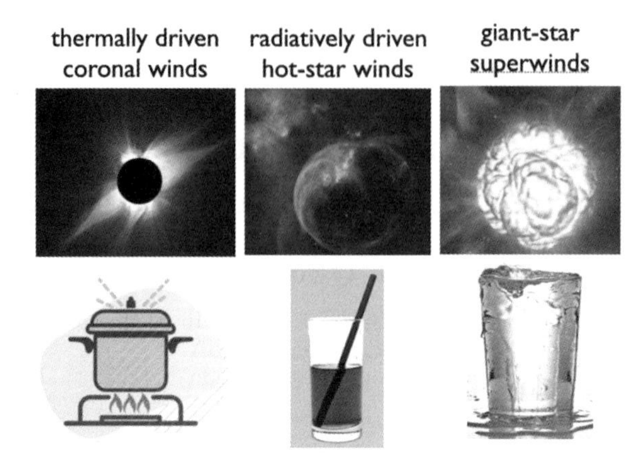

Figure 1. Icons to represent analogies for processes inducing the three different types of steady stellar wind outflow.

This provides a basis for discussion (§4) of planetary outflows, which are likewise largely driven by gas-pressure expansion, now powered by the UV and X-ray heating from the chromospheric and coronal emission from the underlying cool star.

We conclude with a review (§5) of stellar magnetospheres, contrasting their inside-out, internal filling by the stellar wind with the outside-in, external stress imposed on planetary magnetospheres by the wind of their host star.

2. Radiatively driven winds from OB stars

2.1. *Radiative acceleration and Eddington parameter*

In hot stars with a high luminosity, the outward force from scattering of stellar radiation can overcome gravity and so drive a stellar wind outflow. For opacity κ_ν at a frequency ν with radiative flux F_ν, the total radiative acceleration depends on the frequency integral,

$$g_{rad} = \int_0^\infty d\nu \, \frac{\kappa_\nu F_\nu}{c} \equiv \frac{\bar{\kappa}_F F}{c} \tag{2.1}$$

where the last equality defines the flux-weighted mean opacity, $\bar{\kappa}_F$, with F the bolometric flux.

In the idealized case of continuum scattering by free electrons, $\bar{\kappa}_F$ is just equal to the electron scattering opacity $\kappa_e = (1+X)0.2 = 0.34\,\mathrm{cm}^2/\mathrm{g}$, where the latter value applies to a fully ionized gas with solar hydrogen mass fraction $X = 0.72$. The ratio of the associated radiative acceleration to gravity defines the classical Eddington parameter,

$$\Gamma_e \equiv \frac{\kappa_e F/c}{g} = \frac{\kappa_e L}{4\pi G M c} = 2.6 \times 10^{-5} \frac{L/L_\odot}{M/M_\odot} \sim 0.26 \left(\frac{M}{100\, M_\odot} \right)^2 , \tag{2.2}$$

wherein the inverse-square radial dependence of both the radiative flux $F = L/4\pi r^2$ and gravity $g = GM/r^2$ cancels, showing this Eddington parameter depends only on the ratio L/M of luminosity to mass. The third equality shows that Γ_e is very small for stars like the Sun; but if one applies the standard main-sequence mass-luminosity scaling $L \sim M^3$, the last equality shows that massive stars can have Eddington parameters that approach unity. This provides a basic rationale for the upper limit to stellar mass, which is empirically found to be around $200M_\odot$, remarkably close to the mass for which (2.2) gives $\Gamma_e \approx 1$. Stars that approach or exceed this classical Eddington limit can have strong eruptive mass loss, as thought to occur in eruptive Luminous Blue Variable stars like η Carinae.

But since generally $\bar{\kappa}_F \gg \kappa_e$, even stars with $\Gamma_e \ll 1$ can have a total $\Gamma > 1$, and drive a more steady-state *stellar wind* mass loss. Ignoring the small gas-pressure acceleration, the associated steady-state acceleration has the scaling,

$$v\frac{dv}{dr} = -\frac{GM}{r^2} + \frac{\kappa L}{4\pi r^2} = (\Gamma - 1)\frac{GM}{r^2}\,, \tag{2.3}$$

where to simplify the notation, we have set $\bar{\kappa}_F = \kappa$. For constant Γ, this can be trivially integrated to give the variation of wind velocity with radius r,

$$v(r) = v_\infty\,(1 - R/r)^{1/2} \quad ; \quad v_\infty = v_{esc}\sqrt{\Gamma - 1}\,, \tag{2.4}$$

which shows that the wind terminal speed v_∞ just scales with the escape speed $v_{esc} \equiv \sqrt{2GM/R}$ from the wind initiation surface radius R. We thus see that Γ represents an effective *anti-gravity*, which for the simple case $\Gamma = 2$ gives a direct gravitational *reversal*, with material flying away from the star, asymptotically reaching the escape speed.

Moreover, if we multiply (2.3) by $4\pi\rho r^2 dr$, and note the standard definition of mass loss rate $\dot{M} \equiv 4\rho v r^2$, we find upon integration a relationship between the wind momentum $\dot{M}v_\infty$ and its optical depth $\tau \equiv \int_R^\infty \kappa\rho dr$,

$$\dot{M}v_\infty = \frac{\tau L}{c}\left(\frac{\Gamma - 1}{\Gamma}\right)\,. \tag{2.5}$$

OB stars generally have $\tau < 1$ and so fall within the single-scattering limit, $\dot{M}v_\infty < L/c$. In contrast, Wolf-Rayet (WR) stars can have $\tau \approx 1 - 10$ and so require multi-line scattering to explain their optically thick winds with $\dot{M}v_\infty > L/c$.

2.2. *The CAK model for line-driven stellar winds*

In practice, the enhancement of the flux-weighted mean opacity above the simple electron scattering value results mainly from the *bound-bound resonance* of radiation with electrons bound into heavy ions ranging from CNO to Fe and Ni. As illustrated in the left panel of figure 2, the resonance nature of such bound-bound line-scattering greatly enhances the opacity, typically by a factor $\bar{Q} \gtrsim 10^3$ (Gayley 1995). This means any star with $\Gamma_e \gtrsim 1/\bar{Q} \approx 10^{-3}$ can have a line-force that overcomes gravity and so drive a wind outflow.

In practice this maximal line acceleration from optically *thin* scattering is reduced by the saturation of the reduced flux within an optically *thick* line. But as illustrated in the right panel of figure 2, the Doppler shift associated with the wind acceleration acts to *desaturate* this line absorption, effectively sweeping the absorption through a broad frequency band, extending out to the frequency associated with the Doppler shift from the wind terminal speed, v_∞. This wind Doppler shift of line resonance concentrates the interaction of continuum photons into a narrow resonance layer with width set by the Sobolev length, $\ell \equiv v_{th}/(dv/dr)$ (Sobolev 1960), associated with acceleration through the ion thermal speed v_{th} that broadens the line profile. In the outer wind where $\ell \ll r$, the line acceleration for optically thick lines is reduced by $1/\tau$, where the Sobolev optical depth $\tau \equiv \bar{Q}\kappa_e\rho\ell$, giving then a line acceleration $\Gamma_{thick} \sim (1/\rho)(dv/dr)$ that itself scales with the wind acceleration.

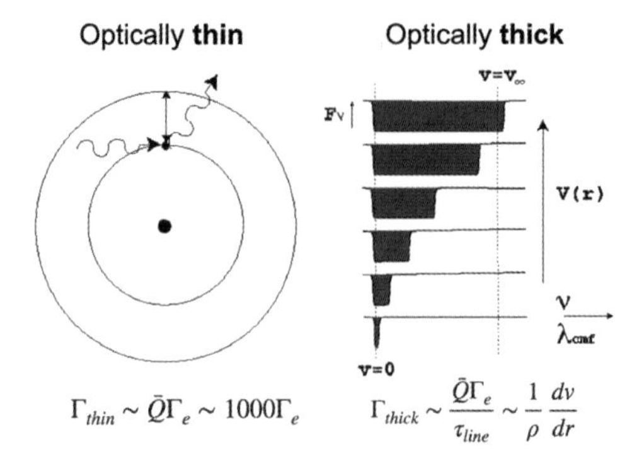

Figure 2. *Left*: Illustration of the resonance nature of line opacity for an optically thin line, for which the spectral average over the resonance quality ($\bar{Q}$) results in an optically thin line acceleration that is of order a thousand times the gravitationally scaled acceleration from electron scattering Γ_e. *Right*: The corresponding acceleration for an optically thick line, which is reduced by the line optical thickness τ_{line}, which in an accelerating wind scales with the ratio of local density to velocity gradient. This line desaturation is a consequence of the Doppler shift of the line absorption from the increasing wind velocity v, resulting in a net line force that scales as $\Gamma_{thick} \sim (1/\rho)(dv/dr)$.

Within this Sobolev approximation, Castor et al. (1975, hereafter CAK)) developed a formalism that accounts for the cumulative radiative acceleration from a *power-law ensemble* of both optically thick and thin lines,

$$\Gamma_{CAK} \approx \frac{\bar{Q}\Gamma_e}{(\bar{Q}t)^\alpha} \gtrsim 1 \;\; ; \;\; t \equiv \kappa_e c \frac{\rho}{dv/dr} \,, \tag{2.6}$$

where the CAK power index $\alpha \approx 0.5 - 0.7$ characterizes the relative number of strong vs. weak lines. Applying (2.6) into the equation of motion (2.3) and using the fact that the critical solution requires $v(dv/dr) \sim GM/r^2$, we find the maximal CAK mass loss rate has the scaling

$$\dot{M} \sim \frac{L}{c^2} \left(\frac{\bar{Q}\Gamma_e}{1 - \Gamma_e} \right)^{-1+1/\alpha} . \tag{2.7}$$

For canonical values $\alpha = 1/2$, $\bar{Q} = 2000$, and $L = 10^5 L_\odot$, this CAK scaling gives $\dot{M} \approx 10^{-5} M_\odot/\mathrm{yr}$, which is indeed a billion times the mass loss rate of the solar wind! Much as in the solar wind, the terminal wind speeds scale with the surface escape speed $v_{esc} \equiv \sqrt{2GM/R}$, with values up to $v_\infty \approx 2000\,\mathrm{km/s}$.

An overall point is that in such models the onset of line-driving near the sonic point represents an effective *line-driven suction*, which draws up mass from the underlying hydrostatic equilibrium of the subsonic region. The reduction in pressure from the outer line-driving induces the underlying subsonic region to expand upward, much as the suction on a straw draws up liquid from a glass (see figure 1). This outside-in suction contrasts with the inside-out thermal expansion of a pressure cooker, and of the analogous gas pressure-driven solar wind discussed below in §3.

2.3. *The Conti mechanism*

A longstanding question is whether such line-driven winds might, over an O-star's main sequence lifetime, lead to loss of the star's Hydrogen envelope, representing the so-called "Conti mechanism" for producing the strong depletion of Hydrogen inferred for Wolf-Rayet (WR) stars. Using the scaling $L \sim M^3$ to estimate the star's main sequence

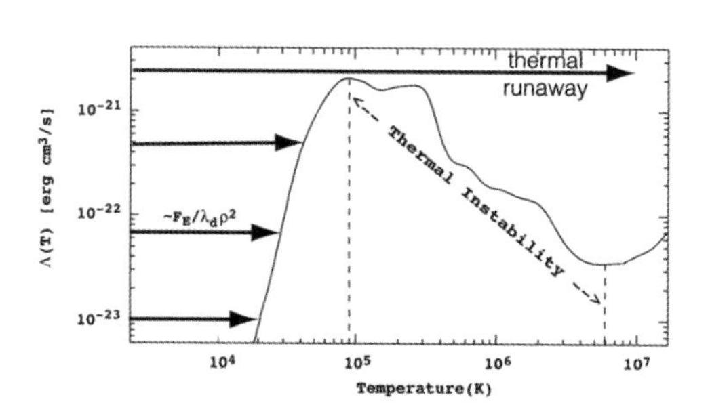

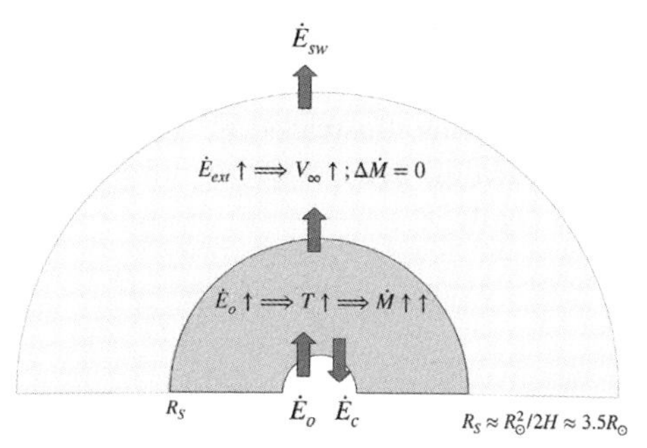

Figure 3. Left: Illustration of mechanical energy flux F_E deposited over a damping is balance optically thin cooling $n_e n_i \Lambda(T)$; the exponential decrease in densit ρ with height leads to high temperature and eventual thermal run above the peak in cooling function Λ. Right: Schematic to illustrate energy input into the nearly hydrostatic, subsonic coronal base vs. that into the supersonic wind above the Parker sonic radius R_s. For the former, the net input vs. loss by conduction back into the underlying atmosphere leads to coronal heating that sets the coronal temperature and location and density at the sonic radius, thus fixing the associated wind mass loss rate. For the latter, any further energy addition increases the wind flow speed, with the wind expansion providing the primary mechanism to carry out the total net amount of coronal heating.

lifetime $t_{\rm ms} \approx 1\,{\rm Myr}\,(100M_\odot/M)^2$, we find for the canonical line opacity factor $\bar{Q}=2000$ that the cumulative mass loss fraction follows the scalings,

$$
\begin{aligned}
\frac{\dot{M}t_{\rm ms}}{M} &\approx 0.18\left(\frac{M}{100M_\odot}\right)^2 \;;\; \alpha=1/2 \\
&\approx 0.016\left(\frac{M}{100M_\odot}\right)^1 \;;\; \alpha=2/3
\end{aligned}
\tag{2.8}
$$

which shows a very sensitive dependence on the CAK power index α. When corrected for wind clumping, empirically inferred mass loss rates agree better with predictions for higher $\alpha \approx 0.6$, and so seem to disfavor the Conti mechanism for H-envelope stripping.

As discussed in the review by A. Sander in these proceedings, quantitative models of OB-wind mass loss endeavor to derive the line-driving opacity self-consistently from NLTE solution of excitation and ionization of key driving ions.

3. Pressure-driven coronal winds

3.1. *Runaway heating of the solar corona and its pressure extension*

In the Sun and other cool stars with surface temperatures below 10^4 K, the recombination of Hydrogen leads to strong subsurface convection and an associated magnetic turbulence that drives mechanical heating of the upper atmosphere. As illustrated in the left panel of figure 3, deposition of mechanical energy flux F_E over a damping length λ_d must be balanced by radiative cooling. In the upper atmosphere this requires excitation of ions by electrons, so the associated cooling rate per unit volume scales with density-squared, $\rho^2 \sim n_e n_p$, multiplied by an optically thin cooling function $\Lambda(T)$ that reflects the excitation and ionization of the emitting ions.

As the density decreases exponentially with height, balancing the local heating F_E/λ occurs at progressively higher temperatures, reflecting the initially steep increase in $\Lambda(T)$ from more energetic collisional excitation. However, for temperatures $T \gtrsim 10^5$ K, such collisions begin to ionize away the bound electrons, reducing the efficiency of radiative cooling, and causing the $\Lambda(T)$ to decline with increasing T. Since radiative cooling can no longer balance the heating, this leads to a *thermal runaway* to coronal temperatures $T >$MK, limited now by inward thermal conduction from the corona to underlying atmosphere, as illustrated in right panel of figure 3.

Maintaining hydrostatic equilibrium at the coronal base radius R implies a pressure scale height,

$$H \equiv \frac{P}{|dP/dr|} = \frac{kT}{\mu g} = \frac{2a^2}{v_{esc}^2} R \equiv \frac{T}{T_{esc}} R \, , \tag{3.1}$$

where $g \equiv GM/R^2$ and $v_{esc} \equiv \sqrt{2GM/R}$ are the surface gravity and escape speed, and for a gas with molecular weight μ, the isothermal sound scales with temperature as $a \equiv \sqrt{kT/\mu}$. The last equality defines an "escape temperature" at which $H = R$, which for solar parameters has a value $T_{esc} \approx 13.8\,\mathrm{MK}$.

In the solar photosphere the low temperature $T \ll T_{esc}$ implies $H \ll R$ and thus a fixed gravity g, with pressure declining exponentially with scale height H. But in the corona, the higher temperature gives $T/T_{esc} = H/R \lesssim 1$, implying one now has to account for the radial decline in gravity in the hydrostatic balance.

For a simple isothermal model, the pressure stratification now takes the form,

$$P(r) = P(R) e^{(R/H)(R/r-1)} \, . \tag{3.2}$$

In contrast to the exponential decline in the photosphere, the coronal pressure at large radii now asymptotically approaches a finite value, P_∞. Relative to the initial pressure $P_o \equiv P(R)$ at the coronal base, the total drop in pressure for a hydrostatic, isothermal corona is given by

$$\frac{P_o}{P_\infty} \approx e^{13.8\mathrm{MK}/T} \quad ; \quad \log \frac{P_o}{P_\infty} \approx \frac{6}{T/\mathrm{MK}} \, . \tag{3.3}$$

The latter equality shows the pressure drops by 6 decades for $T = 1\,\mathrm{MK}$, and only 3 decades for $T = 2\,\mathrm{MK}$.

By comparison, from the solar transition region at the coronal base to the interstellar medium, the pressure drop is actually much greater, $\log(P_{tr}/P_{ism}) \approx 12$. The upshot is that an extended, hot corona can *not* be maintained in hydrostatic equilibrium; instead, as shown by the outward streamers from the eclipse image in figure 3, it must undergo an outward, supersonic *expansion* known as the solar wind.

As illustrated by the right panel of figure 3, this solar wind expansion can be thought of as analogous to the release valve of a pressure cooker, driven fundamentally by mechanical heating generated by magnetic turbulence in the underlying solar atmosphere. Some of this upward energy flux is lost back to the solar atmosphere through thermal conduction, but the net effect leads to a thermal runaway that raises the coronal temperature to temperature within a factor ten of T_{esc}.

3.2. *Isothermal Solar Wind*

The imbalance between outward pressure and inward gravity gives rise to an outward acceleration, which for a steady-state outflow takes the form

$$v \frac{dv}{dr} = -\frac{GM}{r^2} - \frac{1}{\rho} \frac{dP}{dr} \, . \tag{3.4}$$

For an isothermal case with $P = \rho a^2$, one can use mass conservation of the spherical outflow, $\rho v r^2 = \dot{M}/4\pi$=constant, to eliminate the density ρ in favor of the radius r and flow speed v. This allows one to split the pressure gradient force into terms that scale

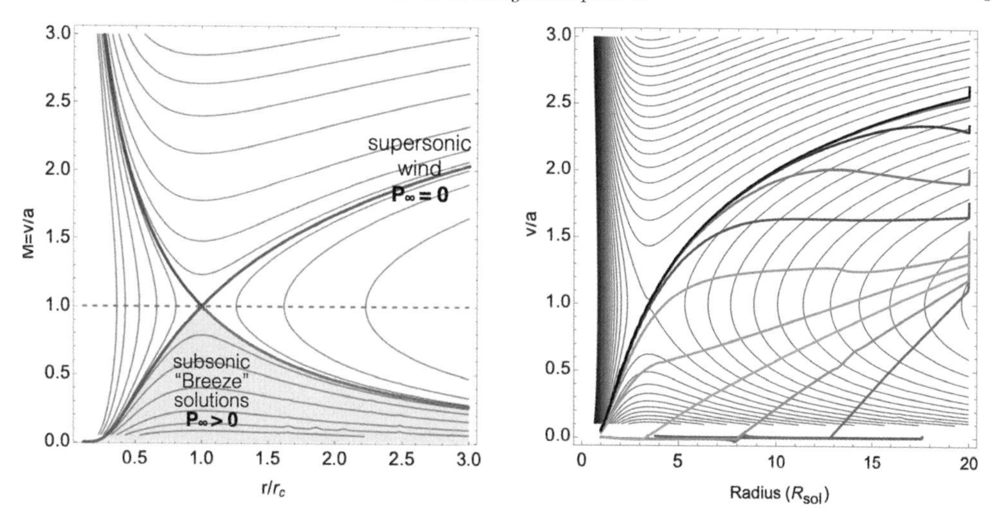

Figure 4. Left: Solution topology for Mach number v/a vs. scaled radius r/r_c for an isothermal corona. Solutions with initially low-speed at the wind base include subsonic "breeze" solutions, which however again have a large terminal pressure, P_∞. The transonic wind solution is the only with low enough P_∞ match the low pressure of the interstellar medium. Right: Steady-state velocity topology overplotted with snapshots at $\Delta t = 20$ ks intervals of a time-depedent hydrodynamical simulations for evolution away from an initially hydrostatic corona. As the high pressure reacts to the lower pressure outer boundary, an expansion develops that eventually evolves to the supersonic solar wind solution.

with the velocity and the sphericity, giving

$$\left(v - \frac{a^2}{v}\right)\frac{dv}{dr} = -\frac{GM}{r^2} + \frac{2a^2}{r} \, . \tag{3.5}$$

In the subsonic region $v \ll a$, this reduces to the condition for hydrostatic equilibrium.

But at a critical (a.k.a. "Parker") radius,

$$r_c = \frac{GM}{2a^2} = \frac{R^2}{2H} = R\left(\frac{T_{esc}}{2T}\right) \, , \tag{3.6}$$

the spatial component of pressure balances gravity, so making the RHS vanish. The LHS can likewise vanish if either $dv/dr = 0$ or $v = a$.

Figure 4 plots Mach number $M \equiv v/a$ vs. scaled radius r/r_c for the overall topology of solutions. We can eliminate all the bi-valued solutions, as well as those that start at supersonic speeds near the coronal base. There is a class of subsonic solutions that have a velocity peak, with thus $dv/dr = 0$, at the critical radius $r = r_c$; but, like the hydrostatic case, all such "breeze" solutions asymptote to a finite pressure at large radii, and thus again cannot match the required low pressure of the interstellar medium.

The only solution that matches both the requirement of low, subsonic speed at the coronal base $v(R) \ll a$, and vanishing pressure at large distances, $P_\infty \to 0$, is the critical transonic solution with $v(r_c) = a$.

3.3. *Energy balance and mass loss rate*

While such isothermal models provide a basic rationale for the supersonic outflow of the solar wind, they effectively ignore the central physics of what energy mechanisms keep the corona hot against the adiabatic cooling of the wind expansion. Moreover, because the density scales out of the equation of motion (3.5), they do not provide any basis for determining the associated wind mass loss rate, $\dot{M} \equiv 4\pi\rho v r^2$. For that one must consider

Figure 5. Left: White light eclipse picture of the solar corona, showing how coronal magnetic field confinement in loops extends into outward steamers by the wind expansion, while open field coronal holes allow direct wind outflow. Right: Corresponding X-ray corona from emission by the MK plasma. Image credits: NASA.

the overall wind *energy* balance, which we cast here in terms of integration from some stellar base radius R to an outer radius r,

$$\dot{M}\left[\frac{v^2}{2} - \frac{GM}{r} + \frac{5}{2}\frac{P}{\rho}\right]_R^r = 4\pi \int_R^r r'^2 q_{net}\, dr' + 4\pi[R^2 F_{c*} - r^2 F_c(r)]\,. \tag{3.7}$$

The terms in parenthesis on the LHS represent the wind kinetic energy, potential energy, and gas enthalpy, while the RHS include the integral of the local net heating per unit volume, and the net difference in conductive heat flux.

Since the change in gas enthalpy is generally relatively small, we can write the energy balance over the full wind by

$$\dot{M}\left(\frac{v_\infty^2}{2} + \frac{v_{esc}^2}{2}\right) \approx Q_{heat} - Q_{rad} - Q_{cond}\,, \tag{3.8}$$

where Q_{rad} and Q_{cond} are the energy losses from radiation and conduction, and the total heating Q_{heat} is associated with MHD waves and/or magnetic reconnection. Since the wind terminal speed is generally comparable to the escape speed, $v_\infty \approx v_{esc}$, we see that the mass loss rate scales directly with the net heating vs. cooling,

$$\dot{M} \approx \frac{Q_{net}}{v_{esc}^2}\,. \tag{3.9}$$

Measurements by interplanetary spacecraft give a typical solar wind speed of $v_\infty \approx 400\,\mathrm{km/s}$, ranging up to $v_\infty \approx 700\,\mathrm{km/s}$ in high-speed streams thought to originate from open-field coronal holes; these values thus do indeed straddle the solar escape speed, $v_{esc} \approx 618\,\mathrm{km/s}$.

The observed mass flux implies to a quite small global mass loss rate, $\dot{M} \approx 10^{-14} M_\odot/\mathrm{yr}$, which is even less than the mass loss associated with the Sun's radiative luminosity, $\dot{M}_{rad} \equiv L_\odot/c^2 \approx 5 \times 10^{-14} M_\odot/\mathrm{yr}$. Over the Sun's entire $\sim 10\,\mathrm{Gyr}$ lifetime, it will thus lose only 0.01% of its mass via the solar wind. The wind kinetic luminosity is $\dot{M} v_\infty^2/2 \approx 10^{-7} L_\odot$, which turns out quite comparable the observed total coronal emission in X-ray and XUV, L_{xuv}.

As discussed in section 4, such coronal emission from cool stars with exoplanets can play an important role in heating planetary atmospheres and inducing their own wind outflows.

3.4. *Angular momentum loss and spindown: the Skumanich law*

While the mass loss from coronal winds from the Sun and other cool stars is negligible, the coronal magnetic field provides an extended moment arm that can make the associated loss of angular momentum in the magnetized wind have a significant effect in causing an evolutionary spindown in the star's rotation. This is indeed thought to be the key cause of the relatively slow rotation period of the present-day Sun, $P_{rot} \approx 27\,\mathrm{d}$.

Following pioneering analysis of Weber and Davis (1967), for a star of rotation frequency Ω, the loss of stellar angular momentum $\dot{J}$ from a wind with mass loss rate $\dot{M}$ is given by

$$\dot{J} = (2/3)\dot{M}\Omega R_A^2 \,, \tag{3.10}$$

where the effective moment arm is set by the Alfvén radius R_A, defined to be where the wind outflow speed equals the Alfvén speed, $V(R_A) \equiv V_A \equiv B/\sqrt{4\pi\rho}$. For the present-day solar wind and magnetic field, one finds $R_A \approx 20R_\odot$, which when applied in (3.10) gives a spindown time,

$$t_s \equiv \frac{J}{\dot{J}} \approx 10\,\mathrm{Gyr} \,. \tag{3.11}$$

The fact that this comparable to the Sun's main sequence lifetime is consistent with the notion that Sun's present-day slow rotation period of 27 d is the result of wind spindown.

More quantitatively, observations of a large population of solar-type stars shows their rotation follow a simple relation, known as the "Skumanich law" (Skumanich 2019). From an initial period P_o, the period P increases with the square root of the age t,

$$P(t) \approx P_o \left(\frac{2t}{t_{so}} + 1 \right)^{1/2} \,. \tag{3.12}$$

If one assumes that the magnetic field scales with $B \sim \Omega \sim 1/P$, then application of the Weber and Davis (1967) analysis gives a simple scaling for the initial spindown time with the initial field strength B_o,

$$t_{so} \approx 0.09\,\frac{MV_w}{R^2 B_o^2} = 0.38\frac{\rho R V_w}{B_o^2} \approx 11.7\mathrm{Gyr} \left(\frac{B_o}{2\,G} \right)^{-2} \,, \tag{3.13}$$

where ρ and R are the star's mean density and radius, and V_w is its wind speed. The last equality provides a numerical evaluation for solar parameters. Remarkably, the dependence on wind mass loss rate scales out, through cancellation of the $\dot{M}$ in (3.10) with the inverse dependence in the Alfvén radius, $R_A^2 \sim 1/\dot{M}$.

For example, taking the Sun's initial rotation period to be $P_o \approx 1\,\mathrm{d}$, achieving the present $P \approx 27\,\mathrm{d}$ for the Sun's current age $t \approx 4.6\,\mathrm{Gyr}$ requires an initial field $B_o \approx 61\,\mathrm{G}$ and initial spindown time of just $t_s \approx 12.6\,\mathrm{Myr}$. This gives for the present global field $B_{now} = 61/27 = 2.3\,\mathrm{G}$.

The upshot is that the young Sun was likely much more rapidly rotating, with stronger magnetic activity, a stronger wind, and greater coronal EUV emission. These likely played a role in stripping the initially much denser atmosphere of Mars, implying that younger, more active stars might have similar effects on the atmospheres of the exoplanets, as we next discuss.

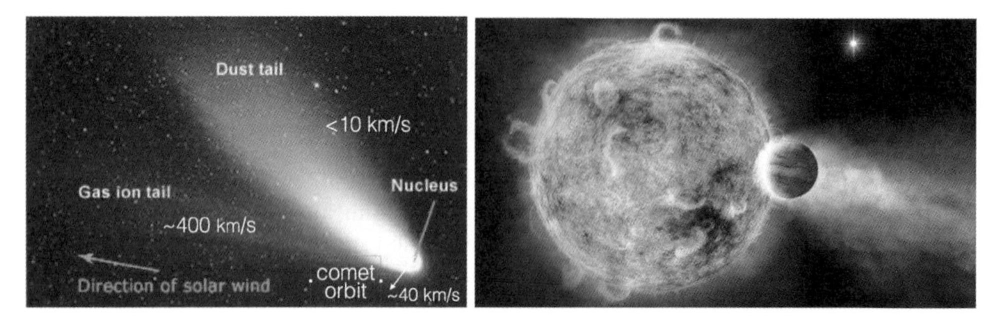

Figure 6. Left: Illustration of dust and ion tails from comets. The former is driven away slowly by solar radiation, thus trailing behind the comet's orbital motion. The latter is driven by the much faster solar wind, and so is oriented nearly in an anti-solar direction. Right: illustration of mass loss from a hot-Jupiter exoplanet can be similarly driven away in a comet tail by interaction with the star's light and wind. Image credits: NASA.

4. Planetary winds and mass loss

The above discussion of solar and stellar wind mass loss from expansion of a hot corona provides a good basis for exploring planetary winds and the associated depletion of their atmospheres. One key difference is that solar and stellar coronae form from inside-out heating from the star, in contrast to the mainly external heating and ablation of a planetary atmosphere from the parent star and its wind. As illustrated in figure 6, this can be expected to cause any planetary outflow to take the form of a cometary tail, much as has been observed extensively for comets orbiting the Sun.

Unlike the more spherical, solar coronal expansion, which must become supersonic to match the outer boundary condition of a low interstellar pressure, the cometary tail mass loss from planets could, in principal, be quite slow and subsonic, even representing the outer layer loss of an otherwise nearly hydrostatically stratified atmosphere. But even if the kinetic energy of the outflow thus remains small and negligible, there remains a key energy requirement for some source of heating to power any mass loss against the gravitational binding from the planet.

4.1. *Planetary escape temperatures*

In this context, it helpful to consider the general scaling and values of the *escape temperatures* of planetary bodies as a function of their mass M and radius R, and the molecular weight μ (in units of proton mass m_p) of their atmospheres,

$$T_{esc} \equiv \frac{GM/R}{k/\mu m_p} = 131\,\text{kK}\,\frac{M/M_J}{R/R_J} = 131\,\text{kK}\,\frac{\rho}{\rho_J}\left(\frac{R}{R_J}\right)^2 . \tag{4.1}$$

Here the last two equalities give scalings relative to Jupiter, assuming $\mu = 0.6$ that applies for a fully ionized mix of H and He at solar abundances. For gas giants, the mean density ρ is (like the Sun) typically of order unity CGS (e.g., $\rho_J \approx 1.3\,\text{g/cm}^3$), and so the last equality shows that the escape temperature mainly scales quadratically with the planetary radius.

For smaller, rocky, terrestrial planets like the Earth, the density is a bit higher ($\rho_E \approx 5.5\,\text{g/cm}^3 \approx 4.1\rho_J$,) but the radius is much smaller ($R_E \approx R_J/10$), giving the Earth an escape temperature of about 4500 K.

By comparison, if we approximate planetary absorption and emission as a simple blackbody, then for a planet at a distance d from a star with effective temperature T_* and radius R_*, the equilibrium temperature is given by

$$T_{eq} = 290\,\mathrm{K}\,\frac{T_*}{T_\odot}\sqrt{\frac{R_*/R_\odot}{d/\mathrm{au}}} = 4250\,\mathrm{K}\,\frac{T_*}{T_\odot}\sqrt{\frac{R_*}{d}}\,. \tag{4.2}$$

The first equality is cast in terms of interplanetary distances like that of the Earth at 1 au; it actually matches quite well the case of Earth's actual mean temperature, but this is due to a rather fortuitous cancelation between the effect of the high albedo of clouds in reducing the absorption of solar radiation, and the greenhouse effect in trapping Earth's own cooling radiation.

The latter equality frames this in terms of very close-in planets, like the so-called "hot Jupiters". A key point here is that even for close-in planets, the direct heating from the star's thermal radiation is almost never sufficient to bring a planet anywhere near the typical escape temperature associated with the thermal wind expansion from the planet.

Instead, it is the much harder, XUV radiation from stellar coronae that can lead such escape-level heating, as we next discuss.

4.2. *Planetary mass loss from heating by XUV coronal radiation*

For photons (γ) of a given wavelength λ and thus energy $E = h\nu = hc/\lambda$, we can cast this energy in terms of an associated temperature,

$$T_\gamma \equiv \frac{h\nu}{k} = 11.6\,\mathrm{kK}\,\frac{h\nu}{eV} = 14\,\mathrm{kK}\,\frac{1\mu m}{\lambda}\,. \tag{4.3}$$

For photons above the Hydrogen ionization limit with $h\nu > 13.6\,\mathrm{eV}$ and $\lambda < 91.2\,\mathrm{nm}$, the associated escape temperature is $T_{esc} > 156\,\mathrm{kK}$. Coronal emission at EUV energies can thus readily ionize H with a sufficient excess energy to heat the gas to temperature well above typical escape temperature.

In analogy with the scaling in eqn. (3.9) that was derived for mass loss from stellar coronae, the associated planetary mass loss depends on the net XUV heating of its outer atmosphere. For a planet of mass M_p and radius R_p at a distance d from a star with coronal XUV luminosity that is a fraction of $f_{xuv} \equiv L_{xuv}/L_*$ of the stellar luminosity, the energy balance to drive a planetary mass loss $\dot{M}$ is

$$\dot{M}\frac{GM_p}{R_p} \approx \epsilon\, f_{xuv}\,\frac{L_*}{4\pi d^2}\,\pi R_p^2\,, \tag{4.4}$$

where ϵ is an order-unity efficiency factor for this XUV heating. Solving for the mass loss shows that it scales with the ratio of XUV flux to the planet's mean density,

$$\dot{M} = 1.5\epsilon f_{xuv}\,\frac{L_*/4\pi d^2}{G\rho_p}\,. \tag{4.5}$$

For example, assuming $\epsilon \approx 1$ and a solar value $f_{xuv} \approx 10^{-7}$ for XUV fraction, we find a hot Jupiter at distance $d = 2R_\odot$ from a solar-type star should have mass loss rate $\dot{M} \approx 10^{10}\,\mathrm{g/s}$. The associated mass loss time is $t_{atm} \equiv M/(dM/dt) \approx 4 \times 10^{12}\,\mathrm{yr}$. A central conclusion is thus that even for such an extreme case, the total mass loss from such gas giants is small even over the evolutionary lifetime of the system.

In contrast, for terrestrial planets like Earth, the atmosphere is a tiny fraction $f_{atm} \approx 10^{-6}$ of the planet's mass. Even at the greater distance $d = 1\,\mathrm{au}$, the associated time for Earth to loss a Hydrogen atmosphere is $t_{atm} \approx 670\,\mathrm{Myr}$, and even shorter if one accounts for the likely stronger coronal emission from the early Sun.

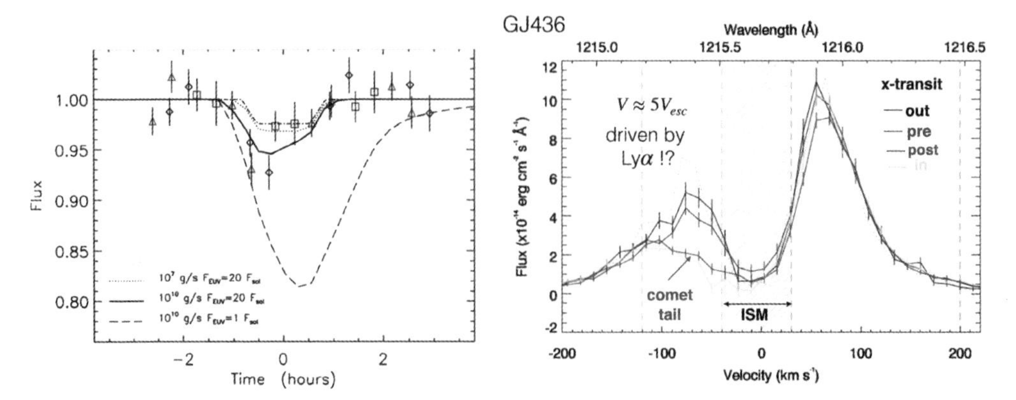

Figure 7. Left: Lyα transit light for hot Jupiter HD1897333b, comparing data points with values models with various assumed mass loss rates and EUV fluxes. Figure from Lecavelier Des Etangs et al. (2010). Right: Lyα transit spectrum for warm Neptune GJ436, overplotted for 4 phases in which the planet was pre, post, in and out of transit. Note, in particular, that the post-transit spectrum shows evidence for a comet tail, with a speed that likely reflects either radiative acceleration by stellar Lyα or entrainment in the outflowing stellar wind, rather than acceleration by thermal expansion from the planet. Adapted from Ehrenreich (2015).

By contrast, the present-day atmosphere of Nitrogen and Oxygen has been mostly retained, because of its higher molecular weight, although there is evidence of a weak "polar wind" mass loss from polar regions of open magnetic fiield.

As illustrated in figure 7, mass loss from exoplanets can be observationally diagnosed through transit light curves and spectra, often centered as here on the UV Lyman-α transition of Hydrogen at $\lambda \approx 91.2$ nm. The results here are for two hot giants, with the light curve in the left panel providing constraints on the planets mass loss and star EUV flux. The overplot of the Lyα spectral line at four phases of the transit provides information on the outflow speed and its orientation in a comet tail extending away from the star.

5. Magnetospheres of stars and planets

Mass loss from both stars and planets can be strongly affected by magnetic fields. Figure 5 illustrates vividly the key role of fields in structuring the solar coronal X-ray emission and the solar wind outflow. The white-light coronal image on the left shows how gas trapped by closed magnetic loops near the Sun gives way to a radially pointed streamer structures. As illustrated in left panel of figure 8, the basic mechanism was first demonstrated in a pioneering analysis by Pneuman & Kopp (1971), who examined how an initially closed magnetic dipole in the solar corona is forced open by gas-pressure expansion. The opening of opposite polarity field lines induces a current-sheet in the outer, expanding wind, where they are a central feature of *in situ* spacecraft measurements. Analogous magnetic effects likely occur other cool stars with hot coronal winds, but the complexity of these dynamo-generated fields leads to significant cancellation in disk-integrated light, making them difficult to diagnose remotely through spectro-polaritmetry.

5.1. *Massive-star magnetospheres*

In contrast, although massive, luminous, hot stars lack the hydrogen recombination convection zone that induces the magnetic dynamo cycle of cooler, solar-type stars, modern spectropolarimetry has nonetheless revealed that about 10% of O, B and

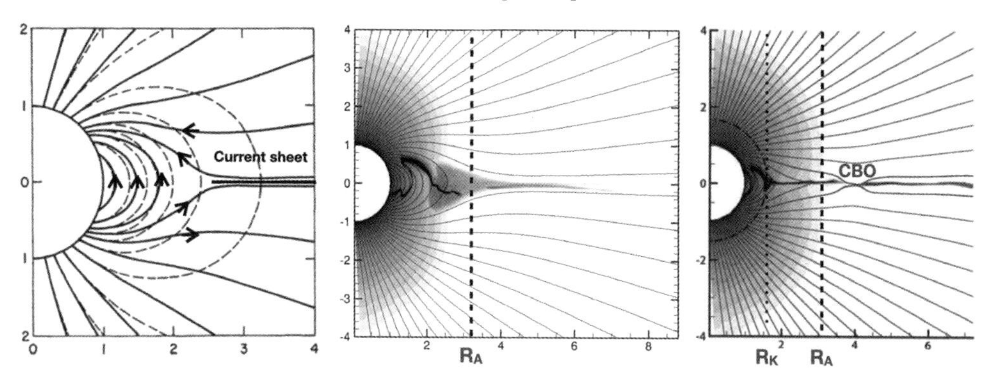

Figure 8. Left: Iterative solution from Pneuman & Kopp (1971) for how an initial dipole field (dashed lines) in the solar corona is by altered by gas pressure and solar wind expansion into a stretched and eventually open form, with lines of opposite polarity (denoted by arrows) separated by a current sheet. Middle: Snapshot of MHD simulation of analogous distortion of dipole field from a magnetic hot star, this time by its strong, radiatively driven wind, with the Alfvén radius R_A marking the transition from inner closed loops to radially stretched field lines in the outer wind. Right: Analogous MHD snapshot for a hot star with rapid rotation, giving a Kepler co-rotation radius $R_K \ll R_A$, forming a wind-fed CM in the region between, punctuated by episodic CBO events, with associated magnetic reconnection.

A-type stars harbor large-scale, organized (often predominantly dipolar) magnetic fields ranging in dipolar strength from a few hundred to tens of thousand Gauss. These fields, which are likely fossils of an earlier epoch, channel and trap the strong, radiatively driven winds of such stars, feeding a circumstellar magnetosphere.

The inside-out building of these wind-fed magnetospheres is in some way complementary to the outside-in nature of the planetary magnetospheres impacted the star's wind. But there are also some interesting similarities in the role of the characteristic magnetospheric and corotation radii.

5.1.1. Alfvén radius and Kepler co-rotation radius

MHD simulation studies (e.g., ud-Doula & Owocki 2002, Ud-Doula, Owocki, & Townsend 2008 show that the overall net effect of a large-scale, dipole magnetic field in diverting such a hot-star wind can be well characterized by a single *wind magnetic confinement parameter* and its associated *Alfvén radius*,

$$\eta_* \equiv \frac{B_{eq}^2 R_*^2}{\dot{M}\, v_\infty} \quad ; \quad \frac{R_{\rm A}}{R_*} \approx 0.3 + (\eta_* + 0.25)^{1/4} \; , \tag{5.1}$$

where $B_{\rm eq} = B_{\rm p}/2$ is the field strength at the magnetic equatorial surface radius R_*, and $\dot{M}$ and v_∞ are the fiducial mass-loss rate and terminal speed that the star *would have* in the *absence* of any magnetic field. This confinement parameter sets the scaling for the ratio of the magnetic to wind kinetic energy density. For a dipole field, the r^{-6} radial decline of magnetic energy density is much steeper than the r^{-2} decline of the wind's mass and energy density; this means the wind always dominates beyond the Alfvén radius, which scales as $R_{\rm A} \sim \eta_*^{1/4}$ in the limit $\eta_* \gg 1$ of strong confinement.

For a model with $\eta_* = 100$, the middle panel of figure 8 shows magnetic loops extending above $R_{\rm A} \approx 3.2 R_*$ are drawn open by the wind, while those with an apex below $R_{\rm A}$ remain closed, trapping wind upflow from opposite footpoints of closed magnetic loops. Once this material cools back to near the stellar effective temperature, it falls back onto the star over a dynamical timescale.

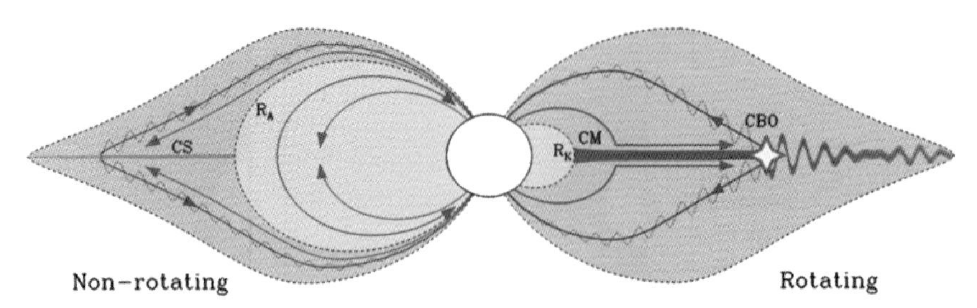

Figure 9. Schematic of stellar magnetospheres, contrasting those from non-rotating (or slowly rotating) stars (left) from rapidly rotating stars (right). On the left R_A denotes the Alfvén radius, separating the region of closed magnetic loops from the magnetized outflowing wind, which stretches the field into open lines of opposite polarity, separated by a current sheet (CS). On the right R_K denotes the Kepler co-rotation radius, above which the centrifugal force exceeds gravity, trapping wind material in a centrifugal magnetosphere (CM), until the mass build-up overwhelms the magnetic field confinement, leading to centrifugal breakout (CBO) events.

The dynamical effects of rotation can be analogously parameterized (Ud-Doula, Owocki, & Townsend 2008) in terms of the *orbital rotation fraction*, and its associated *Kepler corotation radius*,

$$W \equiv \frac{V_{\rm rot}}{V_{\rm orb}} = \frac{V_{\rm rot}}{\sqrt{GM_*/R_*}} \quad ; \quad R_{\rm K} = W^{-2/3}\, R_* \tag{5.2}$$

which depends on the ratio of the star's equatorial rotation speed to the speed to reach orbit near the equatorial surface radius R_*. Insofar as the field within the Alfvén radius is strong enough to maintain *rigid-body rotation*, the Kepler corotation radius $R_{\rm K}$ identifies where the centrifugal force for rigid-body rotation exactly balances the gravity in the equatorial plane. Figure 9 contrasts the roles of magnetic confinement and centrifugal support in stellar magnetospheres.

5.1.2. Dynamical vs. Centrifugal Magnetosphere

If $R_{\rm A} < R_{\rm K}$, then material trapped in closed loops will again eventually fall back to the surface on a dynamical timescale, thus forming what's known as a *dynamical magnetosphere* (DM).

But if $R_{\rm A} > R_{\rm K}$, then wind material located between $R_{\rm K}$ and $R_{\rm A}$ can remain in static equilibrium, forming a *centrifugal magnetosphere* (CM) that is supported against gravity by the magnetically enforced co-rotation. As illustrated in the upper left schematic in figure 10, the much longer confinement time allows material in this CM region to build up to a much higher density than in a DM region.

Eventually the centrifugal force of this material overwhelms the confining effect of magnetic tension, leading to *centrifugal breakout* (CBO) events, as illustrated by the right panel of figure 8 for the case $\eta_* = 100$ and $W = 0.5$. (See also the right side of figure 9.)

For general 2D MHD simulations in the axisymmetric case of a rotation-axis aligned dipole, the mosaic of greyscale plots in figure 10 shows the time vs. height variation of the equatorial mass distribution $\Delta m/\Delta r$ for various combinations of rotation fraction W and wind confinement η_* that respectively increase upward and to the right. This illustrates vividly the DM infall for material trapped below $R_{\rm K}$ and $R_{\rm A}$, vs. the dense accumulation of a CM from confined material near and above $R_{\rm K}$, but below $R_{\rm A}$. For such CM stars, note also the episodic CBO events.

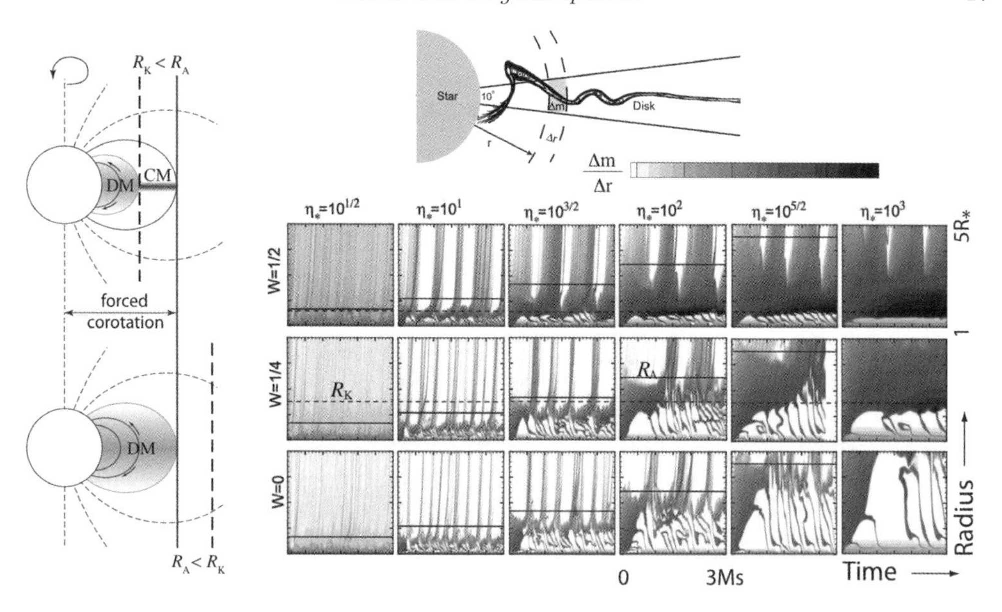

Figure 10. *Left:* Sketch of the regimes for a dynamical vs. centrifugal magnetosphere (DM vs. CM). The lower panel illustrates the case of a slowly rotating star with Kepler radius beyond the Alfvén radius ($R_\mathrm{K} > R_\mathrm{A}$); the lack of centrifugal support means that trapped material falls back to the star on a dynamical timescale, forming a DM, with shading illustrating the time-averaged distribution of density. The upper panel is for more rapid rotation with $R_\mathrm{K} < R_\mathrm{A}$, leading then to a region between these radii where a net outward centrifugal force against gravity is balanced by the magnetic tension of closed loops; this allows material to build up to the much higher density of CM. *Right, Upper:* Contour plot for density at arbitrary snapshot of an isothermal 2D MHD simulation with magnetic confinement parameter $\eta_* = 100$ and critical rotation factor $W = 1/2$. The overlay illustrates the definition of radial mass distribution, $\Delta m/\Delta r$, within 10° of the equator. *Right, Lower:* Density plots for log of $\Delta m/\Delta r$, plotted versus radius (1-5 R_*) and time (0-3 Msec), for a mosaic of 2D MHD models with a wide range of magnetic confinement parameters η_*, and 3 orbital rotation fractions W. The horizontal solid lines indicate the Alfvén radius R_A (solid) and the horizontal dashed lines show Kepler radius R_K (dashed).

5.1.3. Comparison with Observations of Confirmed Magnetic Hot-stars

For observationally confirmed magnetic hot-stars with $T_\mathrm{eff} \gtrsim 16\,\mathrm{kK}$, figure 11 plots positions in a log-log plane of R_K vs. R_A (Petit et al. 2013). The vertical solid line representing $\eta_* = 1$ separates the domain of non-magnetized or weakly magnetized winds to left, from the domain of stellar magnetospheres to the right. The diagonal line representing $R_\mathrm{K} = R_\mathrm{A}$ divides the domain of centrifugal magnetospheres (CM) to the upper right from that for dynamical magnetospheres (DM) to the lower left.

For strong CM stars with $R_\mathrm{K} \gg R_\mathrm{A}$ one can apply a semi-analytic, rigidly rotating magnetosphere (RRM) model developed by Townsend & Owocki (2005), which assumes the wind-fed mass accumulates along surfaces of minimum combined gravitational-centrifugal potential. The strong confinement leads to a high enough density to make the CM have a net emission in the Hydrogen Balmer line, Hα. In the common case that the magnetic dipole axis has a non-zero titl angle β with the rotation axis, the highest density and strongest emission comes from circumstellar clouds near the common rotational and magnetic equator. As rotation brings these clouds from in front of the star to off the limb, the Hα line profile shows corresponding variations from absorption near line center to emission in the blue and red wings. As shown in figure 12, the overall line-profile variation can be characterized by a *dynamic spectrum* that renders on a grayscale the rotational phase variations of doppler-shifted emission and absorption.

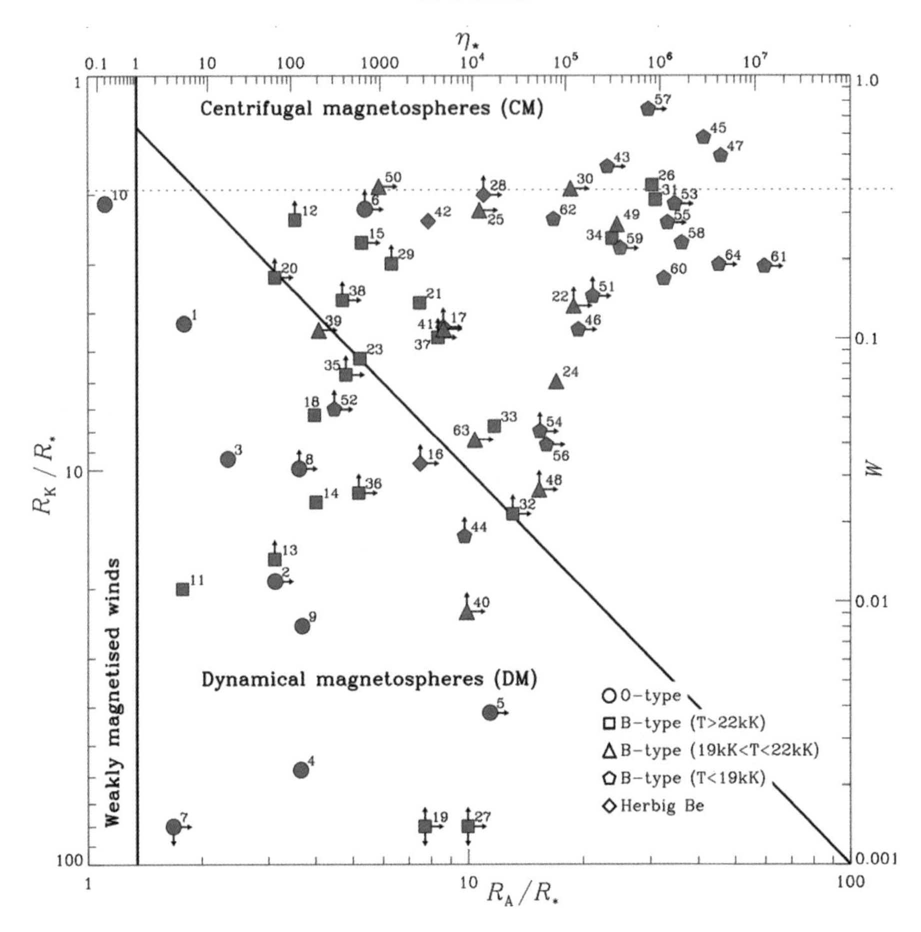

Figure 11. Classification of 64 observationally confirmed magnetic massive stars in terms of magnetic confinement vs. rotation fraction, characterized here by a log-log plot of Kepler radius $R_{\rm K}$ increasing downward vs. Alfvén radius $R_{\rm A}$ increasing to the right. The labeled ID numbers are sorted in order of decreasing effective temperature $T_{\rm eff}$, with stellar identities given in Table 1 of Petit et al. (2013). Stars to the left of the vertical solid line have only weakly magnetized winds (with $\eta_* < 1$), Star below and left of the diagonal solid line have dynamical magnetospheres (DM) with $R_{\rm A} < R_{\rm K}$, while those above and right of this line have centrifugal magnetospheres (CM) with $R_{\rm A} > R_{\rm K}$.

For example, the strong CM star σ Ori E has a nearly dipole field of polar strength $B_p > 10^4$ G and strong tilt $\beta > 60^o$, and rotation fraction $W \approx 0.25 - 0.5$. The corresponding confinement parameter $\eta_* > 10^6$ implies an Alfvén radius $R_{\rm A} \approx 30 R_*$ well beyond the Kepler radius $R_{\rm K} \approx 2 R_*$. Figure 12 compares dynamical spectra from a RRM model (left) with that obtained from actual Hα observations of σ Ori E (right). The good overall agreement provides strong evidence for the overall CM paradigm, as well as the basic RRM analysis method.

5.1.4. Role of centrifugal breakout in Hα and radio emission

Despite this success of the RRM model in characterizing the *relative* density of wind-fed material along potential minima, the lack of a clear mechanism for emptying this build-up had precluded any prediction of the *absolute* density and associated emission. But recent theoretical analyses and numerical MHD simulations (Owocki et al. 2020) have

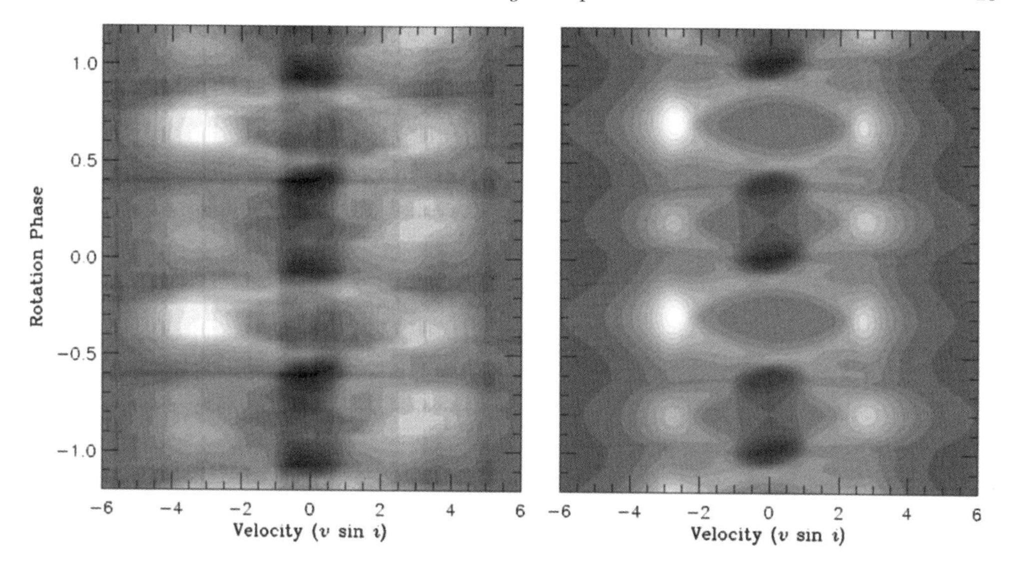

Figure 12. Greyscale renditions of dynamic spectrum for variation of doppler-shifted emission and absorption with rotation phase. The right panel shows results from an RRM model assuming rotation and parameters quoted in the text for the strong CM star σ Ori E. This is in good overall agreement with the left panel showing observed variations in Hα from this star. Adopted from Townsend, Owocki, & Groote (2005).

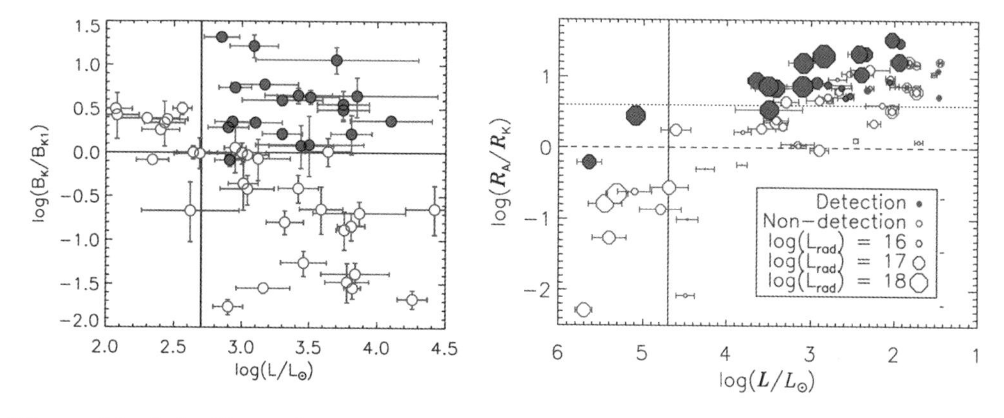

Figure 13. Left: Observed magnetic B-stars plotted in the $\log B_K/B_{K1}$ vs. $\log L/L_\odot$ plane, with stars showing Hα in filled circles, and those without in open circles (Shultz et al. 2020). Here B_K is the observationally inferred field strength at the Kepler co-rotation radius $R_{\rm K}$, while is B_{K1} is the strength the CBO model predicts is needed to make the CM have unit optical thickness in Hα at $R_{\rm K}$. Above a threshold luminosity, the onset of Hα emission occurs right at this CBO value, independent of luminosity; this indicates the associated wind feeding of the CM overwhelms any leakage, to build the density to its breakout values. Right: Occurrence of radio emission for magnetic B stars in a log-log plane of the ratio $R_{\rm A}/R_{\rm K}$ vs. luminosity (now increasing from right to left). Filled circles with detected radio lie the upper diagram, indicating rapid rotation again plays a key role in radio emission; the lack of clear dependence on luminosity again indicates the wind feeding rate is relatively unimportant, as expected for emission tied to CBO events and their associated magnetic reconnection.

provided strong evidence that this wind-fed density build-up in CM's proceeds up to a level wherein the outward centrifugal force exceeds the inward confinement of magnetic, whereupon material escapes in centrifugal breakout (CBO) events. This CBO analysis gives a specific, threshold value of the magnetic field strength at the Kepler radius, B_{K1},

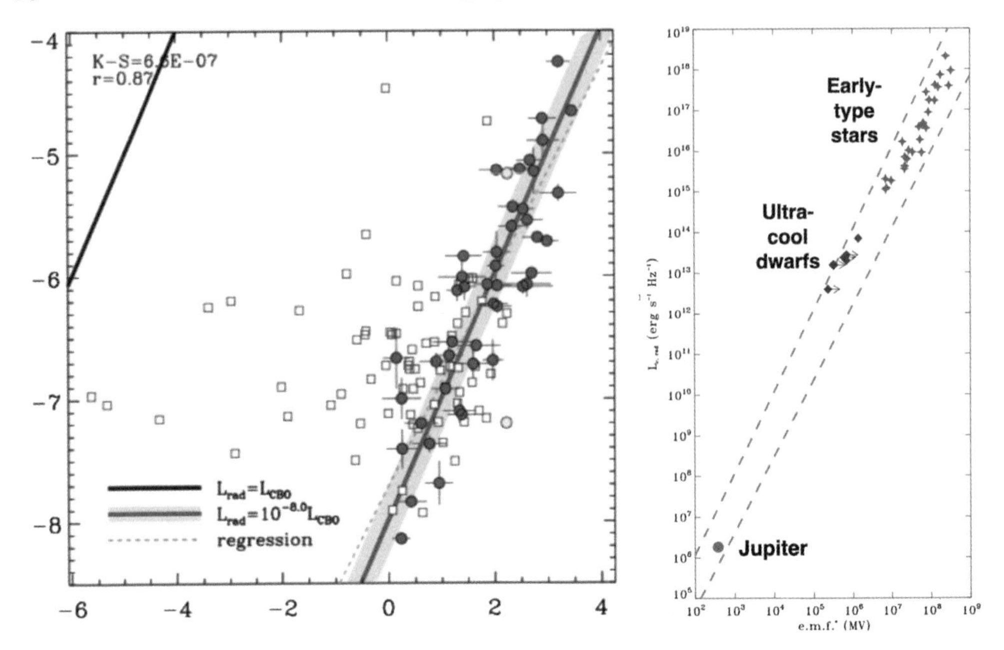

Figure 14. Left: Comparison of observed non-thermal radio luminosity with the luminosity scaling for CBO model, showing a strong correlation with an efficiency factor $\epsilon \approx 10^{-8}$ for conversion to radio emission (Owocki et al. 2022). Right: Empirical scaling law for radio emission, comparing early-type stars with ultra-cool dwarfs, and even extending to radio emission from Jupiter (Leto et al. 2021).

that is needed to give a sufficient density for Hα to become optically thick, and thus produce a net emission within the Hα line profile. The left panel of figure 13 shows a log-log plot of inferred ratio of B_K/B_{K1} vs. stellar luminosity. Remarkably, the onset of Hα emission, shown by the filled circles, occurs right at the line for unit ratio, with no apparent luminosity trend for the onset of emission. The provides strong evidence that CBO is the mechanism for controlling density and mass balance in the CM.

Moreover, the right panel of figure 13 shows a similar strong dependence on rotation for the observed incoherent *radio* emission from such magnetic B-stars (Shultz et al. 2022). As illustrated in the left side of figure 9, the previously favored model by Trigilio et al. (2004) proposed that electrons accelerated at the base of the wind-induced current sheet could be spiral along closed magnetic loops, producing the radio by gyro-synchrotron emission. Since this includes no role for stellar rotation, this model is now strongly disfavored.

A promising alternative, illustrated by the right side of figure 9, proposes instead that electrons are accelerated near sites magnetic reconnection associated with CBO events. An analysis by Owocki et al. (2022) show that, under the assumption that the Alfvén radius follows the split-monopole scaling $R_A \sim \sqrt{\eta_*}$, the luminosity available from such CBO events scales as

$$L_{CBO} \approx W\Omega B_*^2 R_*^3 \,. \tag{5.3}$$

While this seems to suggest the full volume energy of the magnetic field is dissipated on a rotation timescale, in practice the field and its breakout effectively just act as a *conduit* for the large energy reservoir associated with the stellar rotation.

The left panel of figure 14 compares the observed radio luminosity L_{rad} against this predicted L_{CBO} scaling (Owocki et al. 2022). The results show a nearly linear correlation, with however a small efficiency factor $\sim 10^{-8}$ for conversion of available CBO energy into

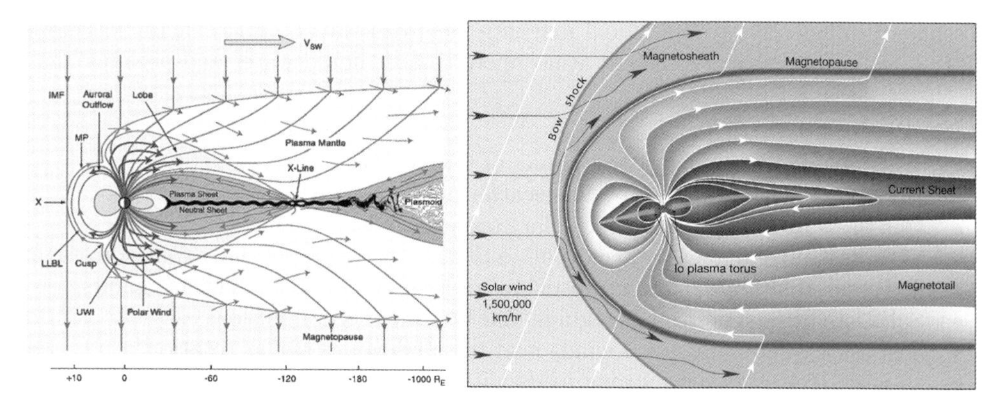

Figure 15. Schematics sketches for the magnetospheres of Earth (left) and Jupiter (right). Both are externally stressed by the solar wind into elongated tail, leading to episodes of magnetotail reconnection that drive particle acceleration manifest as auroral substorms. But Jupiter is also stressed internally from volcanic ejection of material from the active moon Io, which upon ionization by solar UV becomes trapped in Jupiters rotating magnetosphere. Jupiter's decametric radio emission is thought link to this Io plasma torus, possibly through CBO-driven reconnection and gyro-emission from associated energized electrons. Image credits: NASA.

radio. Moreover, when extended downward by several decades, the right panel shows that this empirical scaling for radio luminosity from early-type stars also matches well that observed from ultra-cool dwarfs (Leto et al. 2021). Such UCDs are also inferred to have both strong magnetic fields and rapid rotation, and thus are also strong candidates for CBO events and associated reconnection.

Even more remarkably, further downward extrapolation shows that the radio emission from Jupiter follows this same scaling relation, thus suggesting a potential link in physical process producing radio emission in stellar and planetary magnetospheres.

5.2. *Outside vs. inside stressing of planetary magnetospheres*

In contrast to the inside-out stressing of stellar magnetospheres, figure 15 illustrates how the impact of an external solar or stellar wind exerts an outside-in stress on planetary magnetospheres, stretching them into a long magnetotail away from the star. As seen for the Earth's magnetosphere on the left, thermal expansion now forms a "polar wind" outflow, which is then channeled along field lines toward the tail. The closed-field region near the equator traps material into a plasma sheet that extends beyond the Earth's co-rotation radius ($R_K \approx 6 R_E$), where the combination of centrifugal and external stresses from the solar wind drive X line magnetic reconnection and associated plasma ejection along the magnetotail, a process quite analogous to CBO events from stellar magnetospheres.

In the case (right panel) of the much larger magnetosphere of Jupiter, the volcanic moon Io provides an additional plasma source, at an orbital radius well beyond Jupiter's co-rotation radius, $r_{Io} \approx 2.6 R_K$. This can combine with other plasmas to drive CBO reconnection, accelerating electrons whose gyro-emission in magnetic loops produce the decametric radio emission thought linked to Io, again a process similar to the CBO-linked radio emission from stellar magnetospheres.

But a key overriding effect of such planetary magnetospheres is to *shield* the planet from the *atmospheric ablation* that can result from direct impact of coronal wind ions. Such ablation by the solar wind is thought to be a key factor in depleting the initially dense atmosphere of Mars, which lacks a strong global magnetic field. The presence or absence of sufficiently strong, global magnetic field can thus be a key factor in whether rocky exoplanets retain a sufficient atmosphere to host liquid water and perhaps life.

6. Summary

Let us conclude with a list of key points of this review:

- Hot-star winds are driven by line-scattering of the star's radiation.
- Gas pressure drives supersonic solar coronal winds.
- XUV heating of planetary atmospheres drives analogous thermal expansion/escape.
- This and ablation deplete atmospheres of rocky planets, but not gas giants.
- Solar/stellar magnetospheres guide and trap their wind outflow.
- The associated magnetized winds lead to spindown of stellar rotation.
- 10% of OBA stars have strong fields with wind-fed magnetospheres...
  - characterized by the Alfvén radius $R_{\rm A}$ and Kepler co-rotation radius $R_{\rm K}$.
  - If $R_{\rm A} < R_{\rm K}$, forms a *dyanmical magnetosphere* (DM).
  - If $R_{\rm K} < R_{\rm A}$, forms a *centrifugal magnetosphere* (CM).
- For CM's, centrifugal breakout (CBO) controls Hα and radio emission.
- Planetary magnetospheres both shield atmospheres from winds, and trap outflows.

Acknowledgments

I thank the IAUS370 SOC, and particularly its chair, Aline Vidotto, for the invitation to give this opening review. I acknowledge numerous helpful conversations with B. Das, M. Shultz, J. Sundqvist, R. Townsend, A. ud-Doula, and other members of the Magnetism in Massive Stars (MiMeS) collaboration. This work was supported in part by NASA ATP grant 80NSSC22K0628.

References

Castor, J. I., Abbott, D. C., & Klein, R. I. 1975, *ApJ*, 105, 157.
Ehrenreich et al. 2015, *Nature*, 522 ,459.
Gayley, K. G. 1995, *ApJ*, 454, 410.
Lecavelier Des Etangs, Ehrenreich, D.. Vidal-Madjar, A., Ballester. G. E., Désert, J.-M., Ferlet, R., Hébrard, G., Sing, D. K., Tchakoumegni, K.-O. and Udry, S. 2010, *A&A*, 514, A72.
Leto P., Trigilio C., Krtička J., Fossati L., Ignace R., Shultz M. E., Buemi C. S., et al., 2021, MNRAS, 507, 1979.
Owocki S. P., Shultz M. E., ud-Doula A., Sundqvist J. O., Townsend R. H. D., Cranmer S. R., 2020, MNRAS, 499, 5366.
Owocki S. P., Shultz M. E., ud-Doula A., Chandra P., Das B., Leto P., 2022, MNRAS, 513, 1449.
Petit V., Owocki S. P., Wade G. A., Cohen D. H., Sundqvist J. O., Gagné M., Maíz Apellániz J., et al., 2013, MNRAS, 429, 398.
Pneuman G. W., Kopp R. A., 1971, SoPh, 18, 258.
Shultz M. E., Owocki S., Rivinius T., Wade G. A., Neiner C., Alecian E., Kochukhov O., et al., 2020, MNRAS, 499, 5379.
Shultz M. E., Owocki S. P., ud-Doula A., Biswas A., Bohlender D., Chandra P., Das B., et al., 2022, MNRAS, 513, 1429.
Skumanich, A. 2019, *ApJ*, 878, 35.
Sobolev, V. V. 1960, Cambridge: Harvard University Press, 1960.
Townsend R. H. D., Owocki S. P., 2005, MNRAS, 357, 251.
Townsend R. H. D., Owocki S. P., Groote D., 2005, ApJL, 630, L81.
Trigilio C., Leto P., Umana G., Leone F., Buemi C. S., 2004, A&A, 418, 593.
ud-Doula A., Owocki S. P., 2002, ApJ, 576, 413.
Ud-Doula A., Owocki S. P., Townsend R. H. D., 2008, MNRAS, 385, 97.
Weber, E. J. and Davis, L. 1967, *ApJ*, 148, 217.

Part 2:
Observational evidence of winds

Winds of Stars and Exoplanets
Proceedings IAU Symposium No. 370, 2023
A. A. Vidotto, L. Fossati & J. S. Vink, eds.
doi:10.1017/S1743921322003714

Observations of Winds and CMEs of Low-Mass Stars

Rachel A. Osten[1,2]

[1]Space Telescope Science Institute, Baltimore MD USA

[2]Center for Astrophysical Sciences, Department of Physics and Astronomy, Johns Hopkins University, Baltimore MD USA

Abstract. In this invited review talk I summarize some of the recent observational advances in understanding mass loss from low-mass stars. This can take the form of a relatively steady wind, or stochastically occurring coronal mass ejections (CMEs). In recent years, there has been an expansion of observational signatures used to probe mass loss in low-mass stars. These observational tools span the electromagnetic spectrum. There has also been a resurgence of interest in this topic because of its potential impact on exoplanet space weather and habitability. The numerous recent observational and theoretical results also point to the complexities involved, rather than using simple scalings from solar understanding. This underscores the need to understand reconnection and eruption processes on magnetically active stars as a tool to putting our Sun in context.

Keywords. stars: low-mass, stars: mass loss, stars: flare

---

1. Introduction

One of the exciting parts of attending a symposium like this one is the opportunity for cross-disciplinary interactions. The solar/stellar "connection" can often feel more like a great rift, when one compares and contrasts studies of the Sun with those of low-mass stars. Observations of the Sun will always win compared to stars in terms of spatial, spectral, and temporal resolution, as it's impossible to beat the harsh reality of a $1/d^2$ sensitivity function. Because of its centrality to life on Earth, the Sun is studied in exquisite detail, with daily global monitoring of its surface and near-solar environment, as well as a multi-wavelength context for interpreting many events on the Sun thanks to the constellation of ground- and space-based heliophysics measurements that exist. On the other hand, it is only one star studied at one point in its evolutionary sequence. Given recent progress in understanding the Sun in the context of other Sun-like stars, the Sun may even be unusual in where it sits in activity space, with evidence of the Sun sitting in transition between two cycle types (Metcalfe & van Saders 2017). G stars additionally are not representative of most of the stars in our galaxy. Given the wealth of detail we have in observations and understanding of the Sun and particularly magnetic activity-related phenomena, it is a worthwhile question to ask how universal are the processes occurring in this one star, and whether they might manifest differently in stars of differing characteristics. In contrast, low-mass M dwarfs are the most type of star in our Galaxy. While we cannot as yet spatially resolve these stars, they more than make up for this in the number which can be studied, and the span of ages, rotation rates and evolutionary histories that provide a broad range of parameter space. Admittedly, observations are typically fairly sparse, with little multi-wavelength context. Stellar astronomers also need to compete with the rest of the universe for observing time and funding!

Stellar mass-loss affects the interaction between a planet and its host star. On a larger scale, stellar mass-loss shapes the interaction between the heliosphere and the interstellar medium. In the last several years, with the rise in number of exoplanets detected outside of our solar system, it has become increasingly more topical to examine the likely impacts of space weather experienced by these exoplanets in their varieties of configurations.

The Sun currently has a feeble mass-loss rate, $\approx 2\times10^{-14}$ $M_{\odot}$ yr^{-1}, especially in comparison to the high rates experienced by massive O and B stars, which can be up to 10^9 times stronger. Because of this it has been difficult to make progress in observational constraints on measuring the mass-loss rates of nearby and solar-like stars.

In this review I'll deal with observational signatures of winds separately from those of transient coronal mass ejections. I recognize that the limit of CMEs occurring frequently enough will provide an integrated signature that should mimic that of a fast stellar wind, and address this when necessary. Because of the dynamic relationship between open and closed field lines during an eruptive, and the connection of CMEs with flares, I will also take a bit of an interlude to discuss connections between stellar winds, stellar flares, and CMEs.

2. Observational signatures of stellar winds in low-mass stars

Noting that the first detection of the solar wind was from in-situ particle measurements (Neugebauer & Snyder 1962), it is clear that astronomers need to be creative in searching for observational signatures of these winds. There are currently three main methods used to provide such observational constraints: excess absorption in the blueward wing of the Hydrogen Lyman α line; radio bremsstrahlung emission from the escaping material; and X-ray scattering from charge exchange emission. The use of exoplanets as test particles to diagnose stellar flows is also now possible with the rise of detections of exoplanets close in to their parent star, such as in Villarreal D'Angelo et al. (2021).

2.1. *Hydrogen Lyman alpha*

Hydrogen is the most abundant element in the universe and easily experiences absorption due to intervening material between us and astronomical objects. Pioneering work using high-resolution spectroscopic measurements of the Hydrogen Lyman alpha transition towards a sample of nearby stars (Wood 2004) revealed evidence for excess absorption on the short wavelength side of the Hydrogen Lyman alpha transition at 1216 Å, after accounting for the impact of absorption on this transition by passage through the interstellar medium. Figure 1 illustrates the geometry, including extra redward absorption of the Lyman alpha transition by material in our heliosphere's hydrogen wall. This method is currently the main method used to produce constraints on steady stellar winds in the cool half of the main sequence. It can only be done with high resolution ultraviolet spectroscopy from space. The Space Telescope Imaging Spectrograph on the Hubble Space Telescope currently corners the market on such observations, a situation that will remain until a true successor to HST's high resolution UV spectroscopic capabilities is launched sometime in the next two decades.

Wood and his collaborators use the word "astrosphere" to describe these detections, in parallel with the heliosphere in our solar system. The method does not detect the stellar wind itself, but rather the bow shock created when the wind interacts with the local interstellar medium. Detection of astrospheric absorption by itself does not provide a measurement of a stellar wind; this must be combined with models which describe the local flow of interstellar medium and assumptions about the gas in the astrosphere in order to arrive at a mass-loss rate. While there have been a range of

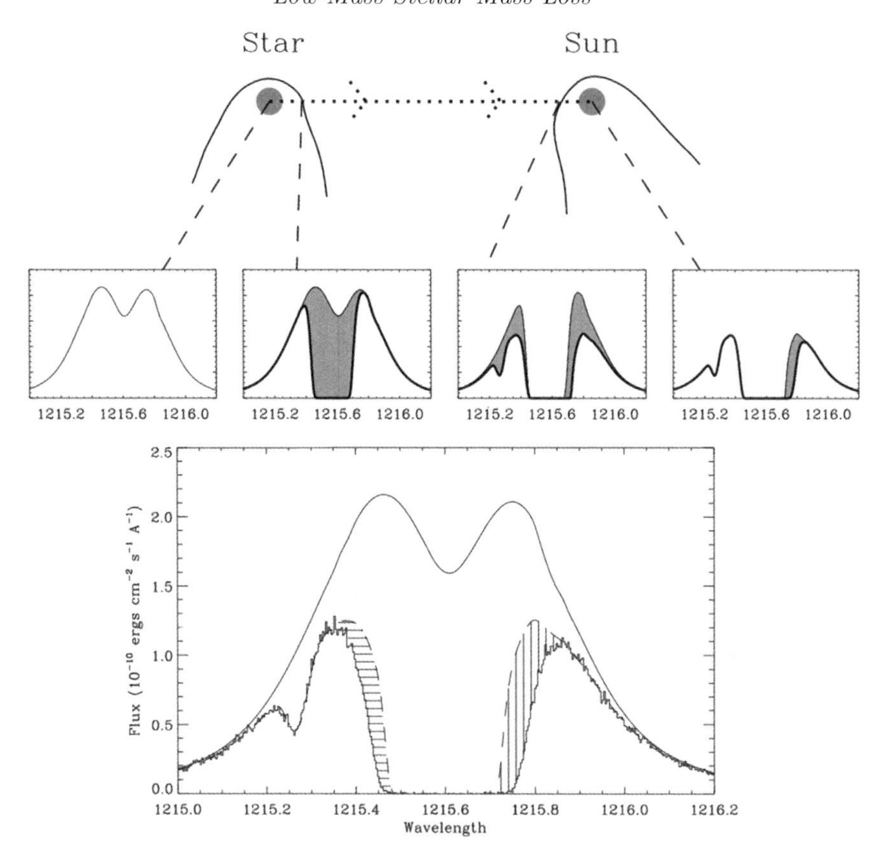

Figure 1. Figure 6 from Wood (2004) illustrating the geometry of astrospheric absorption and its impact on various parts of the Hydrogen Lyman α line profile. The top panel indicates sources of emission and absorption: Lyman α emission from the star, absorption first by its astrosphere, then from passage through the interstellar medium, and finally passage through the heliosphere on its way to being detected around the Earth. The middle panel delineates the spectral signature of each of these, while the profile on the bottom combines all of these effects. In the bottom plot, the thin black line shows the reconstructed intrinsic chromospheric emission from the star; the dashed line indicates the attenuation due to the ISM. The horizontally lined fill area to the left of line center indicates excess absorption from the astrosphere, and vertically lined fill area to the right of line center represents redward absorption from our own heliosphere. The underlying histogram is the resultant observed spectrum.

results, the most surprising thing about these measurements has been a general lack of evidence for significantly increased mass-loss rates in active stars.

A compilation of recent results from Wood et al. (2021) shows a good correlation between the surface X-ray flux of a sample of stars with detected astrospheres and the implied mass loss per unit surface area (Figure 2). X-ray coronal emission derives from closed magnetic field regions, whereas the stellar wind derives from open field regions. Given the large spread in the trend, it is clear that coronal activity and spectral type alone do not determine wind properties. Coronal mass ejections (described in more detail below) are one specific example of stochastically occurring transient stellar mass loss; for a star with a high enough rate of flares and associated coronal mass ejections, the integrated signature of such a CME-dominated wind should appear as a fast, dense stellar wind. That there is an inconsistency between what would be expected from extrapolating from the solar flare/CME occurrence relations and the winds observed on stars with known high rates of flaring suggests more complexity to the problem.

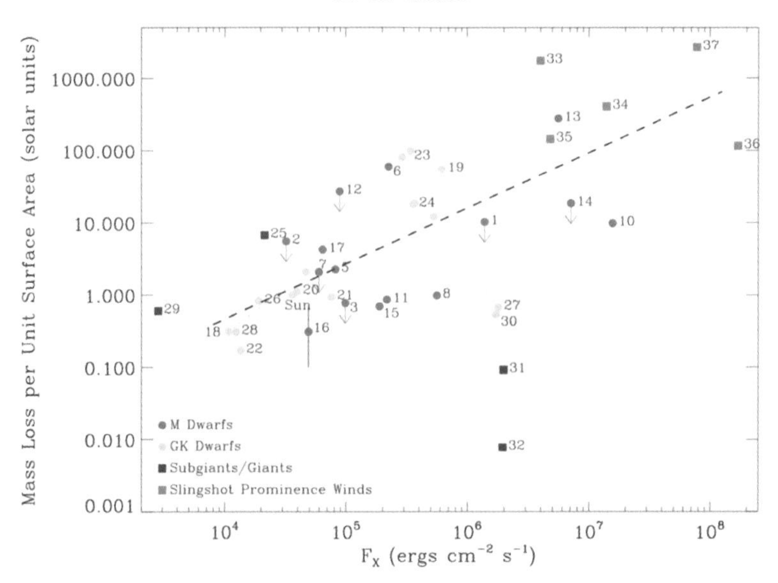

Figure 2. Figure 10 from Wood et al. (2021) summarizing recent measurements of steady mass loss from a variety of cool stars. Red and green circles, as well as blue squares indicate mass loss inferences from astrospheric detections, while fuschia squares come from slingshot prominence winds. The inferred mass loss rate per unit surface area appears to increase with surface X-ray flux, albeit with a large scatter.

2.2. *Radio Bremsstrahlung*

For a homogeneous plasma outflowing from a central star, incoherent radio emission can be detected via bremsstrahlung processes, with well-known flux-frequency $F_\nu \propto \nu^{-2}$ behavior in the optically thin regime, and a flat spectral dependence in the optically thick regime (Dulk 1985). Due to the orders of magnitude higher amounts of mass being lost in hot, massive stars, this detection technique has been most successful at providing constraints in the upper left part of the HR diagram, although recent investigations have probed the sensitivity of mass-loss from solar analogs under a range of assumptions about the directivity of the stellar wind (Fichtinger et al. 2017). This method can return constraints on mass-loss given a flux detection and assumptions about the temperature of the stellar wind; upper limits can provide useful constraints in contrast to the Lyman α method. The expected increase in sensitivity in future radio telescopes such as the next generation Very Large Array, along with an emphasis on frequencies higher than microwave regions, will enable constraints or detections of a number of the closest solar-neighborhood M dwarfs (Figure 3).

2.3. *X-ray scattering from charge exchange emission*

Wargelin & Drake (2001) proposed the use of X-ray scattering from charge exchange emission of astrospheric material with the interstellar medium as a method to provide constraints on stellar mass loss. For the nearest star outside our solar system, Proxima, they could provide limits on the mass loss rate via constraints on the amount of extended X-ray emission from the point source. For stars which are nearby and have a high enough inferred mass loss rate, detecting extended X-ray emission which has the spectral energy distribution expected from charge-exchange emission (which peaks at low X-ray energies) is a potentially viable tool for these limited numbers of objects.

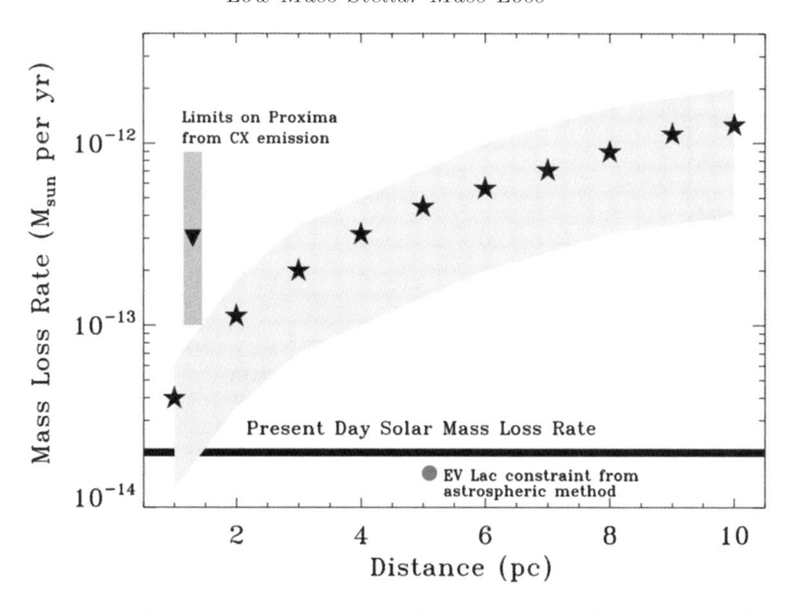

Figure 3. Grasp of the proposed next generation Very Large Array (ngVLA) for studying mass loss from nearby M dwarfs; Figure 3 from a community studies report by Osten & Crosley (2017). Constraints on mass loss rate are determined from the sensitivity imparted from a 12 hour observation, observing at 28 GHz, for a coronal wind with a range of wind speeds from 200–1000 km/s. The limits on mass loss from Proxima due to upper limits on charge exchange emission are indicated with the downward facing arrow and green rectangle. The red circle indicates the constraint on mass loss from the nearby flare star EV Lac obtained using the astrospheric method.

3. Connections between winds, flares, CMEs

As noted earlier, the source region on the disk of the Sun for wind emission is generally open magnetic fields, which allow mass and angular momentum to be lost to the system. X-ray emission and most observational signatures of stellar magnetic activity originates largely from closed-field regions. Magnetic reconnection flares, including coronal mass ejections, are a cross-over between these two regimes, as flares and CMEs involve the temporary opening of field lines as material is expelled away from the star. With the advent of high precision long timescale stellar photometry for the purpose of identifying transiting planets, it is relatively easy to identify and study stellar flares. Because of the positive correlation between many markers of magnetic activity, including flares, and the solar trend for a positive correlation between flares and CMEs, it is important to consider flare events in this discussion of mass loss. There are a few questions that arise in this context that must be considered to advance understanding of stellar mass loss in low-mass stars.

3.1. *How do flares in different wavelength regions relate to each other?*

The standard picture of magnetic reconnection flares (Benz & Güdel 2010) involves all layers of the stellar atmosphere, from the rarefied corona to the dense photosphere, and includes a variety of different physical processes, from plasma heating to particle acceleration and mass motions. The general assumption is that the emission mechanisms produced in separate layers of the stellar atmosphere are related to each other via the flare mechanism. However, even on the Sun correlations are not universal. As pointed out in Osten & Wolk (2015) the optical and X-ray flare frequency distributions for several well-studied M dwarfs show a general agreement in the index used to characterize the

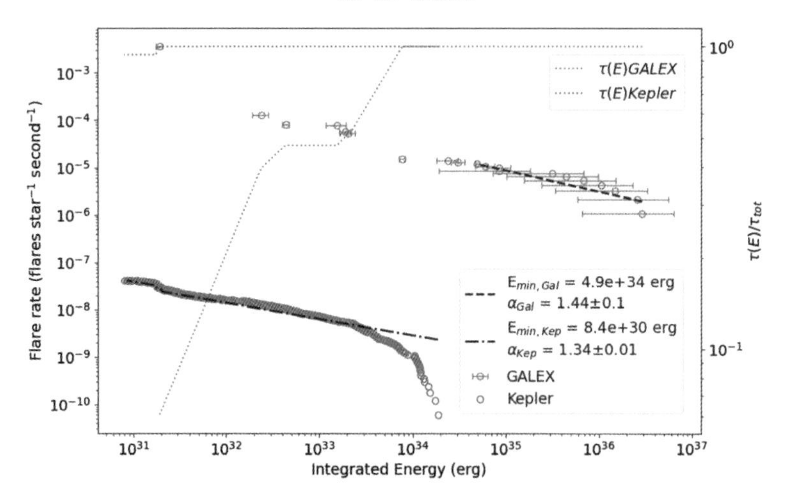

Figure 4. Figure from Brasseur et al. (submitted) illustrating the flare frequency distributions for a sample of twelve stars observed to flare in both Kepler (red circles) white-light and GALEX (blue circles) Near Ultraviolet bandpasses. The integrated energy on the abscissa is in the relevant bandpass. The dashed and dash-dotted lines are fits to the flare frequency distribution parameterized as a power-law. The flares are not simultaneous, yet the index of each distribution is remarkably similar, and suggestive of a common flare energy partition. The dotted lines and axis on the right side indicate the completeness of each dataset, namely the percentage of time where flares of energy E could have been detected based on the minimum detectable energy for that object.

flare frequency, assuming the cumulative flare rate decays as flare energy to the power $-\alpha$. The ultraviolet spectral region is especially important to characterize due to the huge impulsive flare increases seen in this portion of the electromagnetic spectrum, along with the impact on biological systems.

Recent work by Brasseur et al. (2023) uses a unique dataset of constraints from white-light flare measurements via the Kepler spacecraft, and near ultraviolet measurements with the GALEX spacecraft, to put constraints on the energy partition between these two bandpasses. From a sample of 12 stars observed to flare in both wavelength regions, but not simultaneously, we see the same conclusion as noted above from Osten & Wolk: the index of the flare frequency distributions of the two bandpasses are within the errors (Figure 4). This suggests that the offset between the two distributions could potentially be aligned with an common energy partition to convert from energy in a given bandpass to a bolometric flare energy partition.

The more tantalizing result in Brasseur et al. (2023) lies in the strictly simultaneous flare measurements. For over 1500 flares observed in the near UV, there are measurements obtained with the Kepler spacecraft. Curiously, there is no evidence for flare enhancements in the white-light portion of the spectrum to accompany the impulsive NUV flares, whose properties were largely studied in Brasseur et al. (2019). Figure 5 plots the NUV flare energy on the abscissa, with the ordinate displaying the upper limit of the Kepler to GALEX flare energy. That is, the ratio shows the energy of the largest Kepler band flare that would not have been detected at the time of the GALEX flare, to the observed NUV GALEX flare energy. This ratio shows a large scatter over four orders of magnitude, with evidence of a systematic trend with NUV flare energy. The few targetted multi-wavelength flare campaigns of nearby M dwarfs, shown with reddish-purple symbols in the figure, display a much narrower range of bandpass ratios which are more in line with what is expected from state-of-the-art radiative hydrodynamic models to describe the response of the atmosphere to the input of a range of electron beams.

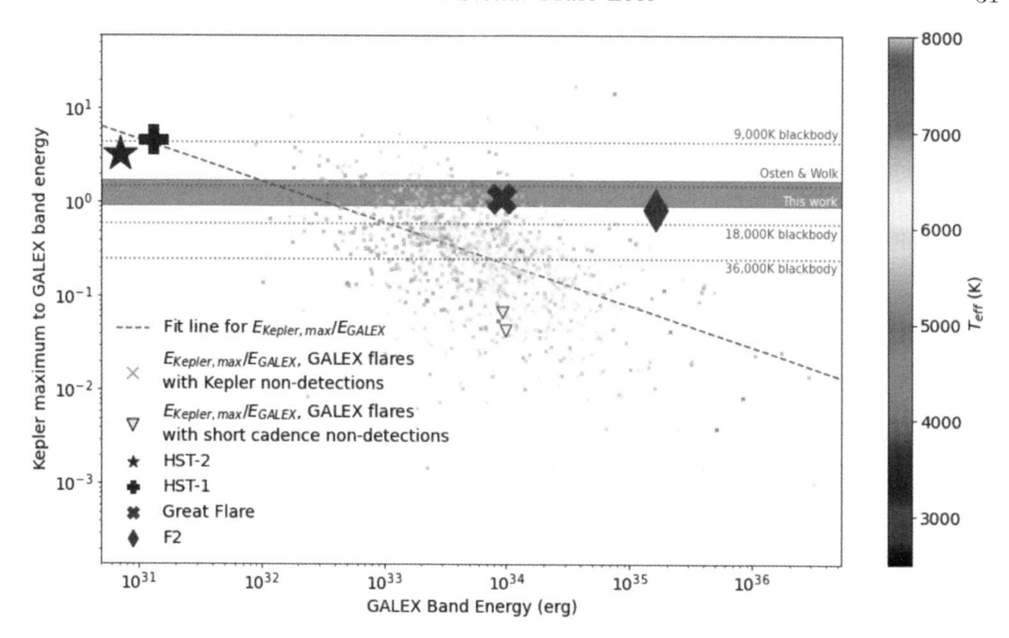

Figure 5. Figure from Brasseur et al. (submitted) indicating the spread of constraints on observed flare energy fractionation for a sample of 1559 datasets with simultaneous Kepler and GALEX measurements. The ordinate is the limit between the maximum undetected optical flare energy from Kepler data and the corresponding measured NUV flare energy. The abscissa is the measured NUV flare energy. The vast majority of the points come from long cadence Kepler data, while the two downward pointing triangles provide deeper constraints from short-cadence Kepler data. The dashed line is a fit to the data, indicating systematic dependence on increasing flare energy. The points are color-coded by the stellar effective temperature. Reddish-purple points are taken from literature studies of multi-wavelength flare campaigns on M dwarfs which place constraints on the NUV to optical flare energy ratios. Horizontal dotted lines are energy fractionations implied by a series of blackbody curves of given temperatures, or the fractionation described in Osten & Wolk (2015). The grey shaded area indicates the range of expected flare energy fractionations using radiative hydrodynamic models of the response of a flaring atmosphere to the input of energy from accelerated particles. These models tend to work better for M dwarf flares than for the range of superflares seen here on G and K stars.

3.2. *Are all flares part of eruptive events?*

As noted previously, on the Sun there is a good correspondence between the most energetic flare events and large coronal mass ejections. These relations have been developed by Emslie et al. (2012) relating the kinetic and potential energy of the CME to the bolometric radiated flare energy, and demonstrated by Drake et al. (2013) between the CME kinetic energy and an assumed partition between X-ray flare energies and bolometric flare energies. For both, the energy of the CME is 2–3 times that the total bolometric radiated flare energy. This led to a number of papers exploring the influence of flare-associated transient mass loss (e.g. Aarnio et al. 2013; Drake et al. 2013; Osten & Wolk 2015; Odert et al. 2017). The general result of these studies was an astoundingly high value of stellar mass loss for active stars, much higher than values inferred from models or from astrospheric detections of time-integrated mass loss.

So while the Sun shows good evidence for such correlations, extrapolating this result to the much higher energy flares observed on active stars produced results seemingly in contradiction with other techniques. There are a few supporting examples from the Sun itself, which occurred when a large active region produced a barrage of highly energetic X-ray flares with no accompanying eruptions. The explanations put forward by Zuccarello et al. (2014) invoking the torus instability have a nice potential to explain the stellar cases. Several papers by now have investigated a potential work-around to this, recognizing that magnetically active stars often have large magnetic fields overlying active regions. Alvarado-Gómez et al. (2018) demonstrated through modelling that such large field strengths can confine the plasma and prevent an eruption from happening.

4. Observational signatures of CMEs in low-mass stars

Coronal mass ejections are now recognized as one part of a solar eruptive event, which comprises a flare, the CME, and possibly solar energetic particles. While flares on stars outside our solar system have been seen since the early days of the 20th century (Hertzsprung 1924), the search for signatures of transient stellar mass loss have often relied on interpretations of odd-looking flares. It has only been in the last several years, with the rise of interest in exoplanetary space weather and impact of stars' magnetic activity on planetary habitability, that many concerted efforts have converged to explore systematic behavior of potential CME signatures in stars. Similarly to the case for wind studies, here too astronomers need to be creative. The workhorse observational technique for studying solar CMEs is the coronagraph, which blots out the main disk of the Sun and enables observation of Thomson scattering of photospheric photons off coronal electrons to probe the structures and dynamics of these eruptions. The parameter space of current astronomical coronagraphs, vis. the requisite sensitivities and angular scales achievable, do not allow for such searches to take place as of yet. There have been several lines of study to demonstrate the existence of stellar CMEs, as well as provide a probe of the systematic behavior of such events.

4.1. *Type II Radio Bursts*

The lynchpin for a CME detection method is one that demonstrates the presence of the eruption and is not dependent on the presence of a flare. Type II radio bursts nicely fit this bill. This is a unique radio signature formed from a super-Alfvenic shock as the CME propagates through the stellar atmosphere. Because of the stratified nature of the atmosphere, observing frequency traces density in the atmosphere, and the observed drift rate of the signal in frequency and time depends on the exciter speed, observing frequency, and coronal scale height. Thus measurements of a drift rate can be used to constrain the propagation speed of the eruption. In a series of papers, Crosley et al. (2016); Crosley & Osten (2018a,b) explored numerous low-frequency radio observations of nearby highly active stars, to look for the existence of these bursts. After a total of 64 hours on a binary star with a high flaring rate (exceeding one flare per hour at and above large solar flare energies where there is a good correlation between solar flares and CMEs), there were no detections of bursts that resembled the expected properties of type II bursts. Simultaneous optical observations occurring with some of the radio data show the lack of correlation between optical flares and any radio activity in the dynamic spectrum (Figure 6). This breakdown suggests that either CMEs occurring with flares on active M dwarfs are a rare occurrence, or that the conditions needed to create the super-Alfvenic shocks are not present at the observing frequencies and distances from the star being probed. Thus the type II radio bursts are an unfulfilled as of yet potential for diagnosing stellar mass loss.

4.2. *High velocity outflows in optical lines and coronal lines*

Flaring regions have many different types of flows associated with them. Observations of large blue shifts in emission (generally larger than the escape velocity) have been

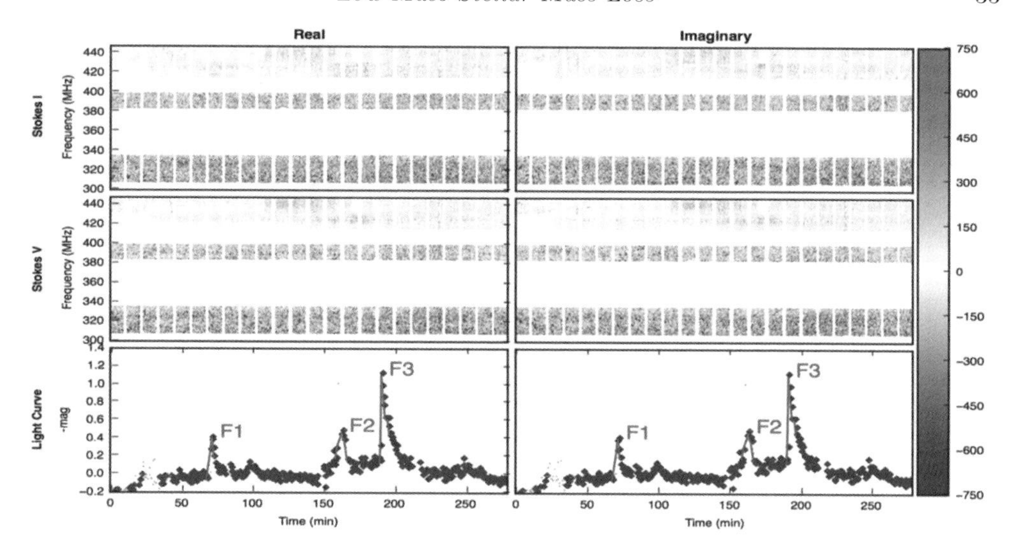

Figure 6. Figure 3 from Crosley & Osten (2018a) showing the radio dynamic spectrum (real and imaginary flux density as a function of frequency and time) of the binary EQ Peg, during a period of time when multiple optical flares were occurring (U-band light curve is the blue curve at the bottom). There is no evidence for radio bursts with signatures expected from passage of a super-Alfvenic shock produced by a CME travelling outward through the atmosphere. The total of 64 hours of observations, together with the flaring rate implying more than one flare per hour above an energy similar to solar flares where there is a nearly one to one correspondence between flares and CMEs, suggests either a breakdown in the flare-CME occurrence for highly active M dwarfs, or the inability of the CME to produce a shock at these observing frequencies sufficient to be detected.

interpreted as evidence of high velocity outflows attributable to an eruption or ejection. Argiroffi et al. (2019) detected an outflow in the lowest temperatures of coronal emission during a flare event on the cool giant HR9024 observed with the Chandra X-ray Observatory. Because this star is evolved, its escape velocity is lower than solar, and the blueshift of $\sim$90 km/s was seen only in emission lines of Oxygen, and only during the decay phase of the flare. They interpreted this as plausible evidence of a coronal mass ejection. More recently, Chen et al. (2022) reported on X-ray observations of the nearby flare star EV Lac, finding evidence for blue-shifted emission in lines of helium-like Oxygen (but not at higher temperatures), however below the escape velocity. They hesitated to call this evidence for a coronal mass ejection.

While X-ray observations of stars with sufficient spectral resolution and sensitivity do not occur often enough to be able to place limits on the occurrence rate of coronal flows associated with flares, optical observations of chromospheric lines enable longer term monitoring and more constraints. Maehara et al. (2021) reported on a sample of M dwarf flares, finding that only one out of 4 Hα-observed flares were associated with blue-ward asymmetries. of them exhibited blue shifts of some magnitude. Namekata et al. (2021) detected blueshifted emission as well as Hα absorption during a superflare on the solar-type star EK Dra. They interpreted this as an eruptive filament (Figure 7).

4.3. *Mass-loss dimming signatures*

Mason et al. (2014) described the possibility of a few types of dimming measurements that might accompany CMEs. The first, mass-loss dimming, would occur after the ejection of mass from the star leads to a reduction in emission compared to before the eruptive event. Veronig et al. (2021) explored dimming signatures in X-ray and Extreme

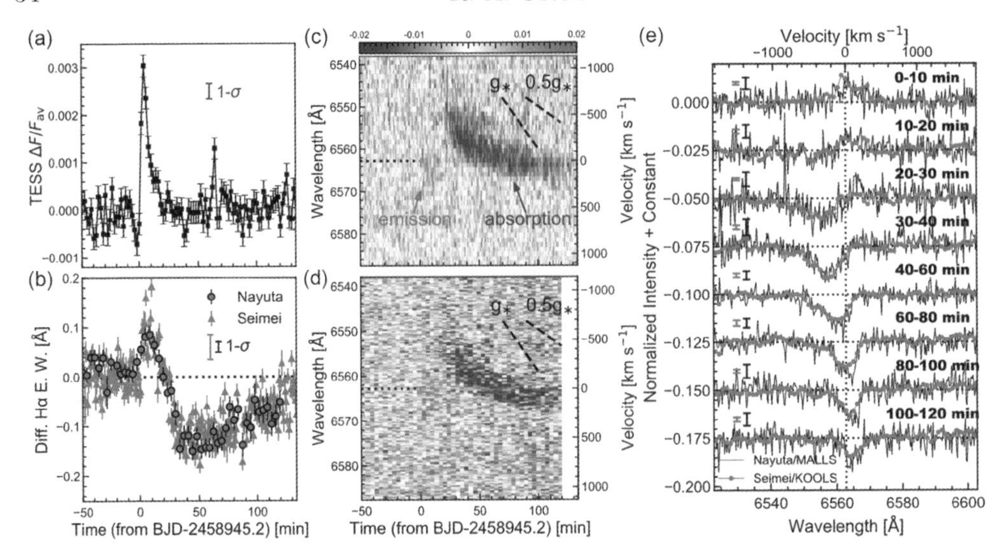

Figure 7. Figure 1 from Namekata et al. (2021) showing detection of an eruptive filament during a superflare on the solar-type star EK Dra. Left panels show TESS light curve at top, with differential Hα equivalent width variations during the time of the flare at bottom. The middle panel displays relative intensity as a function of time and wavelength. The right panel shows the Hα line profile integrated over the indicated time ranges, expressed in velocity space and wavelength. The vertical dashed line indicates line center. Blue-shifted absorption appears in the line profile immediately after the apparent cessation of the flare event.

Ultraviolet (EUV) light curves of magnetically active stars. They found several instances of large flares which were followed by a level of quiescent emission lower than that found immediately pre-flare. These observations were benchmarked by Sun-as-a-star EUV measurements. Because the quiescent emission levels of these stars show large variations, only strong dimming events can be identified. This technique does not enable determination of key parameters of the putative coronal mass emission like mass, velocity, height versus time. These have been seen in spatially resolved images of solar CMEs, and are also demonstrated in Sun-as-a-star high energy light curves tracing temperatures characteristic of the quiescent solar corona rather than flaring plasma (Harra et al. 2016). The geometry is illustrated schematically in Figure 8 left.

4.4. *X-ray absorption dimming*

Mason et al. (2014) also described the absorption dimming (Figure 8 right) phenomena as a possible signature of mass ejections. In this scenario, there is a temporary increase in absorption of the flaring region as absorbing material (here interpreted to be the expanding coronal mass ejection) moves across the line of sight. There have been a few detections of transient increases in absorbing column during large X-ray flares, most notably by Favata & Schmitt (1999). Moschou et al. (2017) used the results of the original fits to time-resolved spectroscopy from the flare on Algol described in Favata & Schmitt and modelled the trend of absorbing column with time as due to a self-similar expansion at constant velocity of a CME. This nicely explained the $N_H(t) \propto t^{-2}$ trend observed in the data. More recently, Osten et al. (in prep.) have been re-examining a sample of the largest stellar flares yet observed at X-ray wavelengths, and finding that many of these events are consistent with Hydrogen column density initially increasing, then decreasing with time during the decay of the flare. More work needs to be done to determine whether

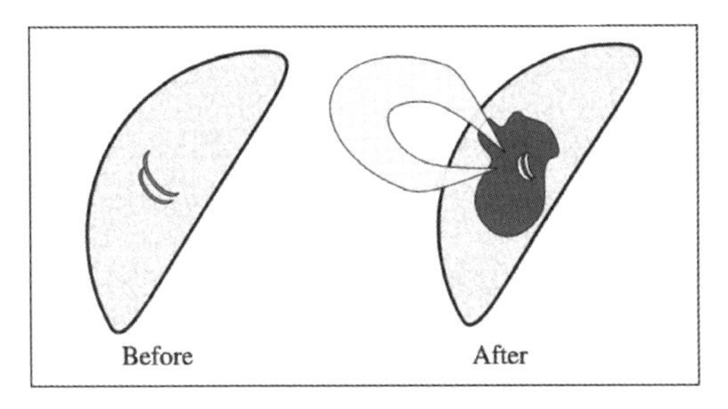

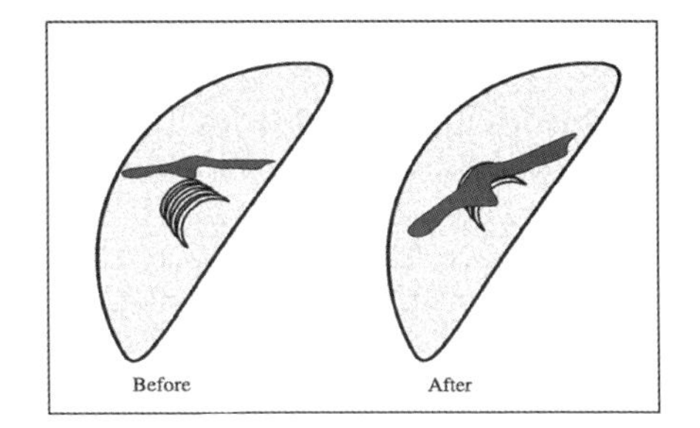

Figure 8. Figures 1 and 4 from Mason et al. (2014) pictorially explaining the circumstances surrounding two types of dimming potentially associated with coronal mass ejections. The left two panels indicate the situation before and after a mass-loss dimming event. Because of the evacuation of mass after the ejection event, there is less emitting material in the region compared to before the event, and differential measurements will indicate this as a decrease in emission. The right two panels indicate an absorption dimming event. Here, there is an arcade of flaring loops. As the ejected material expands it covers the lower-lying loop emission and provides a transient increase in the amount of absorption seen towards the line of sight to the flaring loops. With further expansion and associated decrease in column density, the amount of absorption will decrease back to a characteristic level. mass-loss dimming and absorption dimming.

these variations are in fact consistent with what would be expected from an ejection of mass from the corona.

5. Conclusions

The last several years have seen an expansion of observational signatures used to probe mass loss in low-mass stars. These techniques straddle the electromagnetic spectrum, and are beginning to return interesting insights into mass loss in cool stars. This activity will hopefully continue with the current suite of facilities as well as future facilities currently in development. There has been a concomitant expansion of interest in probing mass loss in low-mass stars. Part of this stems from a desire to deepen understanding of stellar astrophysics. With the rapid increase in the number of exoplanet detections, and the realization that close-in planets around low-mass stars will provide the first opportunities to study potentially habitable planets in the near future, it is more important than ever to understand the impact of stars on exoplanet space weather. Occurring along with these is an expansion of understanding of the complexities in extrapolating simple scalings from a solar understanding. This motivates the need to understand reconnection and eruption processes on magnetically active stars as a tool to putting our Sun in context.

References

Aarnio, A. N., Matt, S. P., & Stassun, K. G. 2013, Astronomische Nachrichten, 334, 77
Alvarado-Gómez, J. D., Drake, J. J., Cohen, O., Moschou, S. P., & Garraffo, C. 2018, ApJ, 862, 93
Argiroffi, C., Reale, F., Drake, J. J., Ciaravella, A., Testa, P., Bonito, R., Miceli, M., Orlando, S., & Peres, G. 2019, Nature Astronomy, 3, 742
Benz, A. O. & Güdel, M. 2010, ARA&A, 48, 241
Brasseur, C. E., Osten, R. A., & Fleming, S. W. 2019, ApJ, 883, 88
Brasseur et al. 2023, ApJ, 944, 5
Chen, H., Tian, H., Li, H., Wang, J., Lu, H., Xu, Y., Hou, Z., & Wu, Y. 2022, ApJ, 933, 92
Crosley, M. K. & Osten, R. A. 2018a, ApJ, 856, 39
—. 2018b, ApJ, 862, 113
Crosley, M. K., Osten, R. A., Broderick, J. W., Corbel, S., Eislöffel, J., Grießmeier, J. M., van Leeuwen, J., Rowlinson, A., Zarka, P., & Norman, C. 2016, ApJ, 830, 24
Drake, J. J., Cohen, O., Yashiro, S., & Gopalswamy, N. 2013, ApJ, 764, 170

Dulk, G. A. 1985, ARA&A, 23, 169
Emslie, A. G., Dennis, B. R., Shih, A. Y., Chamberlin, P. C., Mewaldt, R. A., Moore, C. S., Share, G. H., Vourlidas, A., & Welsch, B. T. 2012, ApJ, 759, 71
Favata, F. & Schmitt, J. H. M. M. 1999, A&A, 350, 900
Fichtinger, B., Güdel, M., Mutel, R. L., Hallinan, G., Gaidos, E., Skinner, S. L., Lynch, C., & Gayley, K. G. 2017, A&A, 599, A127
Harra, L. K., Schrijver, C. J., Janvier, M., Toriumi, S., Hudson, H., Matthews, S., Woods, M. M., Hara, H., Guedel, M., Kowalski, A., Osten, R., Kusano, K., & Lueftinger, T. 2016, Solar Physics, 291, 1761
Hertzsprung, E. 1924, BAIN, 2, 87
Maehara, H., Notsu, Y., Namekata, K., Honda, S., Kowalski, A. F., Katoh, N., Ohshima, T., Iida, K., Oeda, M., Murata, K. L., Yamanaka, M., Takagi, K., Sasada, M., Akitaya, H., Ikuta, K., Okamoto, S., Nogami, D., & Shibata, K. 2021, PASJ, 73, 44
Mason, J. P., Woods, T. N., Caspi, A., Thompson, B. J., & Hock, R. A. 2014, ApJ, 789, 61
Metcalfe, T. S. & van Saders, J. 2017, Solar Physics, 292, 126
Moschou, S.-P., Drake, J. J., Cohen, O., Alvarado-Gomez, J. D., & Garraffo, C. 2017, ApJ, 850, 191
Namekata, K., Maehara, H., Honda, S., Notsu, Y., Okamoto, S., Takahashi, J., Takayama, M., Ohshima, T., Saito, T., Katoh, N., Tozuka, M., Murata, K. L., Ogawa, F., Niwano, M., Adachi, R., Oeda, M., Shiraishi, K., Isogai, K., Seki, D., Ishii, T. T., Ichimoto, K., Nogami, D., & Shibata, K. 2021, Nature Astronomy, 6, 241
Neugebauer, M. & Snyder, C. W. 1962, Science, 138, 1095
Odert, P., Leitzinger, M., Hanslmeier, A., & Lammer, H. 2017, MNRAS, 472, 876
Osten, R. A. & Crosley, M. K. 2017, arXiv e-prints, arXiv:1711.05113
Osten, R. A. & Wolk, S. J. 2015, ApJ, 809, 79
Veronig, A. M., Odert, P., Leitzinger, M., Dissauer, K., Fleck, N. C., & Hudson, H. S. 2021, Nature Astronomy, 5, 697
Villarreal D'Angelo, C., Vidotto, A. A., Esquivel, A., Hazra, G., & Youngblood, A. 2021, MNRAS, 501, 4383
Wargelin, B. J. & Drake, J. J. 2001, ApJ Letters, 546, L57
Wood, B. E. 2004, Living Reviews in Solar Physics, 1, 2
Wood, B. E., Müller, H.-R., Redfield, S., Konow, F., Vannier, H., Linsky, J. L., Youngblood, A., Vidotto, A. A., Jardine, M., Alvarado-Gómez, J. D., & Drake, J. J. 2021, ApJ, 915, 37
Zuccarello, F. P., Seaton, D. B., Mierla, M., Poedts, S., Rachmeler, L. A., Romano, P., & Zuccarello, F. 2014, ApJ, 785, 88

Winds of Stars and Exoplanets
Proceedings IAU Symposium No. 370, 2023
A. A. Vidotto, L. Fossati & J. S. Vink, eds.
doi:10.1017/S1743921323000108

Observations of outflows of massive stars

Andrea Mehner

ESO - European Organisation for Astronomical Research in the Southern Hemisphere, Alonso de Cordova 3107, Vitacura, Santiago de Chile, Chile

Abstract. Mass loss plays a key role in the evolution of massive stars and their environment. High mass-loss events are traced by complex circumstellar ejecta and intricate line profiles across the upper Hertzsprung-Russell diagram for massive stars in different evolutionary stages. The basic physics of radiation-driven stellar wind for hot stars is well understood. However, the driving mechanisms and related instabilities for their enhanced mass-loss episodes and the driving mechanisms for the mass loss of cool stars are still debated. In this review, the mass-loss characteristics and the possible mechanisms will be surveyed for an observational set of prominent massive stellar populations that experience outflows, strong stellar winds, and periods of enhanced and eruptive mass loss; massive young stellar objects, OB-type stars, red supergiants, warm hypergiants, luminous blue variables, and Wolf-Rayet stars.

Keywords. stars: emission-line, Be, stars: evolution, (stars:) circumstellar matter, stars: winds, outflows, stars: variables: other

1. Introduction

The outflows of massive stars play a crucial role in the dynamics and chemical evolution of galaxies. They are key contributors to the turbulence budget and chemical enrichment of the interstellar medium, influencing directly and indirectly the star formation rate and efficiency. The mass-loss rates of massive stars determine their evolution, death, and end-product, yielding the observed diversity of evolved massive stars and of supernovae (SNe).

Mass loss of massive stars includes steady stellar winds and enhanced and eruptive mass-loss episodes. This is manifested in their complex circumstellar environments. Material in the form of molecular, atomic, and ionized gas, and dust can be seen in a variety of geometries, e.g., as arcs, collimated outflows, latitude dependent mass-loss geometries with equatorial density enhancements ("disks"), and large-scale circumstellar nebulae. Current observational efforts to characterize massive star outflows focus on their driving mechanisms, outflow rates, underlying fundamental physical and chemical processes that include the instabilities and triggers that can result in enhanced or eruptive mass-loss behavior, and the dependency on metallicity, rotation, and binarity.

At high effective temperatures, corresponding to OB-type stars, the dominant physical process resulting in steady mass loss is radiative acceleration (Kudritzki & Puls 2000; Puls et al. 2008; Vink 2021). Metal ions, such as Fe, are efficient photon absorbers at specific line frequencies, resulting in radiative acceleration (Lucy & Solomon 1970; Castor et al. 1975). Line driven winds in lower metallicity (Z) environments are predicted to be much weaker, with lower mass-loss rates. Bi-stability jumps may exist at specific temperatures where changes in the wind ionization occur (Vink et al. 2001; Petrov et al. 2016; Björklund et al. 2021). Some of the observed very strong mass-loss events may only be explained by continuum-driven winds or explosions. The mass-loss mechanism operational among the cooler post-main sequence super- and hypergiants, which can also exhibit enhanced and eruptive mass-loss behavior, is debated. Radiation pressure on dust grains or molecules, aided by pulsation or convection and additionally enhanced by magnetic phenomena can all be invoked.

Measured mass-loss rates are subject to systematic uncertainties and each mass-loss diagnostic has its own limitation. For example, Hα recombination emission (Puls et al. 1996) is sensitive to wind clumping (Fullerton et al. 2006) and limited to relatively high mass-loss rates. Ultraviolet (UV) P Cygni resonance lines (Lamers et al. 1995) are less sensitive to clumping but require access to UV satellites, while X-ray light curves and spectroscopy provide excellent information but only for close binary systems. Huge progress has been made in the past decade by applying non-LTE stellar atmosphere codes such as CMFGEN (Hillier & Miller 1998), PoWR (Gräfener et al. 2002; Sander et al. 2017), and FASTWIND (Puls et al. 2005) to fit stellar spectra and spectral energy distributions (SEDs) for a broad wavelength range to derive stellar and wind parameters. Many studies with new infrared to radio instrumentation were conducted to derive wind, and circumstellar gas and dust parameters around massive stars. Other methods include observations of masers (Habing 1996), bow-shocks (Kobulnicky et al. 2019; Henney & Arthur 2019), and spatially resolved circumstellar nebulae using a variety of techniques, including infrared interferometry (Monnier et al. 2004).

1.1. *Massive stars and exoplanets*

Exoplanets are very common around cool stars (Dressing & Charbonneau 2013), but studies failed to detect giant planets around stars with more than 3 $M_{\odot}$ (Reffert et al. 2015).† This led to the idea that giant planet formation or inward migration is suppressed around higher-mass stars. The increased radiation field from a hotter star results in short disk lifetimes, making planet formation and/or growth much less likely. In addition, any newborn planets would be susceptible to radiation stripping from the hot star and undergo atmosphere evaporation (e.g., Poppenhaeger et al. 2021).

Nonetheless, the discovery of the planet b Cen (AB)b showed that stars and stellar systems up to at least $6 - 10$ $M_{\odot}$ can host giant planets on wide orbits (Janson et al. 2021). High-contrast direct imaging observations with the Spectro-Polarimetric High-contrast Exoplanet REsearch instrument (SPHERE), mounted on the Very Large Telescope (VLT) at the European Southern Observatory (ESO), revealed a planet about ten times more massive than Jupiter in a wide circumbinary orbit around the b Cen stellar system. The planet orbits at 100 times the Jupiter-Sun distance, one of widest planet orbits discovered yet. The large distance could be key to the planet's survival in this extreme stellar environment. The more massive star in the binary has the spectral type B2.5V, corresponding to an effective temperature of $\sim 18\,000$ K. Its strong UV radiation make it doubtful that the planet could have formed in-situ by the core-accretion process. Instead, it may have been captured by the stellar system after forming in isolation, or the planet could have formed rapidly from the circumstellar gas disk through gravitational instability.

VLT SPHERE high-contrast near-infrared images and integral field spectra have enabled further discoveries, demonstrating that AO-assisted coronagraphic surveys can place constraints on the multiplicity properties of massive star companions. First results indicate that the companion mass function is populated down to the lowest stellar masses.

† The first exoplanets were discovered around millisecond radio pulsars, i.e., rapidly rotating neutron stars. However, this does not imply these planets were formed around their massive progenitor stars as they would likely be ablated or unbound during stellar death. More likely, these planets originated in the SN fallback disk around the newly-formed neutron star or from a disk of a disrupted binary companion.

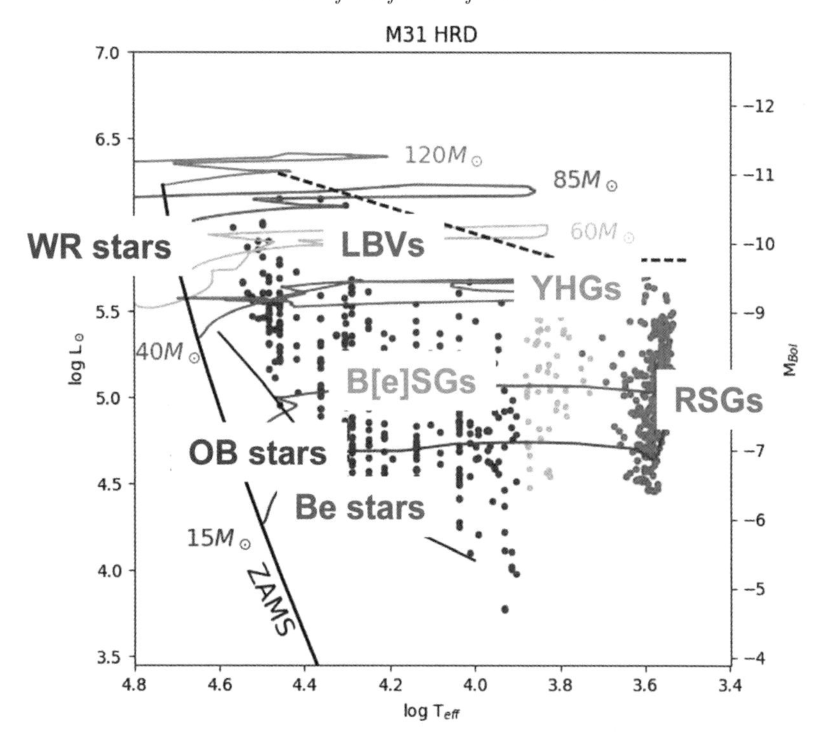

Figure 1. Representative HR diagram of the selected massive star populations (Figure from Humphreys et al. 2017, annotated).

Squicciarini et al. (2022) found that the massive star μ^2 Sco ($M \sim 9\ M_\odot$) has a substellar companion ($14.4 \pm 0.8\ M_J$) at a projected separation of 290 ± 10 au, and a probable second similar object ($18.5 \pm 1.5\ M_J$) at 21 ± 1 au. These companions are slightly more massive than the deuterium-burning limit, but they may still have formed under a planet-like scenario. Reggiani et al. (2022) probed the low-mass end of the companion mass function for 18 O-type stars and found five objects at the stellar/substellar mass boundary with estimated masses $< 0.25\ M_\odot$.

2. Science highlights of selected massive star populations

The upper Hertzsprung-Russell (HR) diagram is populated by a diverse group of luminous, and variable massive stars. Many of them show evidence for strong stellar winds and periods of enhanced or eruptive mass loss. Despite extensive observing campaigns and many theoretical studies over the past decades to investigate these massive stars, there are still many challenges to our understanding of their evolution and their final fate. This is because of low-number statistics and the variety and complexity of the involved physics, such as the impact of metallicity, mass-loss history, rotation, and binarity. In the following, recent observational science highlights for selected massive star populations (Figure 1) are presented to shine a light on the diversity of observations of outflows in massive stars and the diversity of the physical processes involved.

2.1. *Massive young stellar objects*

Massive Young Stellar Objects (MYSOs) are embedded in their dense molecular birth clouds. They are defined as infrared-bright objects that have an SED that peaks at $\sim 100\ \mu$m and a total luminosity $> 10^4\ L_\odot$. Collimated outflows occur in most astrophysical systems in which accretion, rotation, and magnetic fields interact and they are a fundamental aspect of star formation (Bally 2016). Outflows from forming stars have a profound impact on their surrounding clouds. The outflow mass directly launched by a protostar is estimated to be about $10-30\%$ of the accreted gas, i.e., $10-30\%$ of the final stellar mass. Jets and winds create cavities and inject energy and momentum on a wide-range of lengths from less than 0.01 pc to over 30 pc, which may dominate the generation of turbulence and cloud motions and determine the final stellar masses (Rosen & Krumholz 2020). As an example, Xu et al. (2022) adopted the deep learning method CASI-3D to systemically identify protostellar outflows in Orion in ^{12}CO and ^{13}CO observations obtained with the Nobeyama Radio Observatory 45m telescope (NRO 45m). They found a total outflow mass of 6332 $M_{\odot}$ and total kinetic energy of 6×10^{47} erg, sufficient to maintain the level of turbulence against dissipation in the Orion molecular cloud. The terminal shocks in outflows can dissociate molecules, sputter grains, and alter the chemical composition of the impacted media.

Protostellar outflows can be observed at radio to X-ray wavelengths in the continuum and a multitude of spectral lines (Bally 2016). Near- and mid-infrared observations and aperture synthesis with centimeter- and millimeter-wave interferometers are enabling outflow studies of highly obscured objects. The youngest outflows are best traced by molecules such as CO, SiO, H_2O, and H_2, while older outflows are best traced by shock-excited atoms and ions such as hydrogen-recombination lines, [S II], and [O II]. The outflows record the mass-accretion histories of forming stars and their symmetries provide clues about the dynamical environment of the engine. Variations in ejection velocity, mass-loss rate, and flow orientation provide information on the protostar-disk system and potential companions. For example, S- and Z-shaped symmetries indicate that the outflow axis has changed over time, maybe due to precession induced by a companion. C-shaped symmetries indicate a relative motion between the star and surrounding medium. The more luminous the source, the less collimated the outflow. Most YSOs indicating spectral type B or later (luminosities of 10^{-1} to 10^{-2} $L_{\odot}$) tend to have outflows that are highly collimated. MYSOs with luminosities of $L\sim10^5$ $L_{\odot}$ or higher become the most massive early O-type stars. Their outflows are frequently poorly collimated.

Orion-Kleinmann-Low nebula. The Orion Nebula is the nearest site of ongoing and recent massive star formation. A powerful and poorly collimated outflow emerges from the Orion-Kleinmann-Low nebula in the Orion Molecular Cloud. Many observations have confirmed a link between this outflow and an explosive event. Explanations for the explosive outflow range from a “magnetic bomb” (Matt et al. 2006) to the release of gravitational potential energy associated with the formation of a compact binary or stellar mergers (Bally & Zinnecker 2005).

AO near-infrared images obtained with the Gemini South multi-conjugate AO system (GeMS) and near-infrared imager (GSAOI), combined with previous observations, were used to study the proper motion of the high-velocity H_2 knots and three ejected stars (Bally et al. 2015, 2020) (Figure 2). The results are consistent with an origin within a few au of a massive star and a multi-body dynamic encounter that ejected the Becklin-Neugebauer object and radio source I about 500 yr ago. ALMA observations show a roughly spherically symmetric distribution of over a hundred ^{12}CO streamers with velocities extending from -150 km s^{-1} to $+145$ km s^{-1}, indicating again a dynamic age of ~500 yr and allowing the identification of the explosion center (Bally et al. 2017).

2.2. *OB dwarfs, giants, supergiants*

Some OB-type stars show evidence for strong winds and circumstellar material, indicating additional mass-loss drivers to their steady radiation-driven stellar winds. For

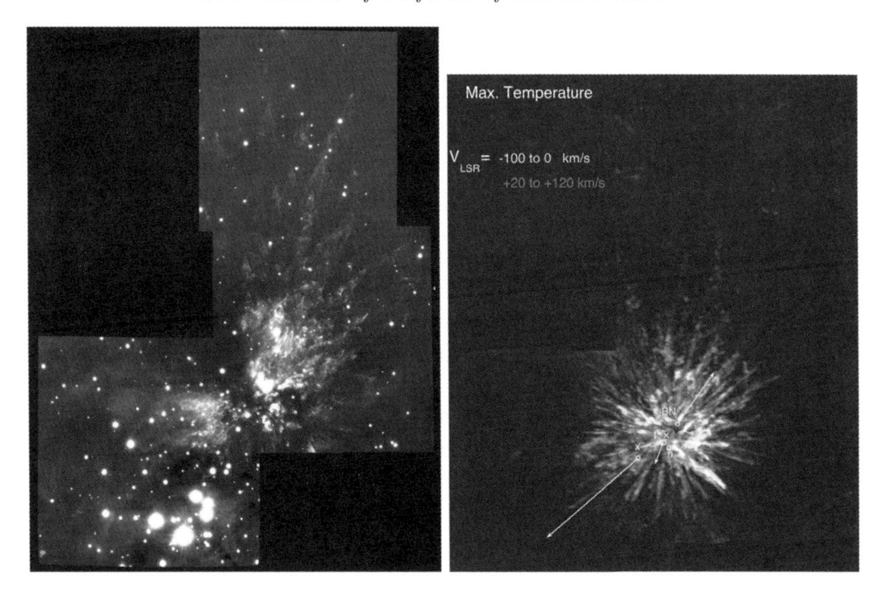

Figure 2. Explosive outflow from the Orion-Kleinmann-Low nebula in the Orion Molecular Cloud. Left: Combined H_2 (orange), [Fe II] (cyan), and K color GSAOI image (Figure from Bally et al. 2015). Right: CO $J = 2 - 1$ ALMA image (Figure from Bally et al. 2020).

example, the Be stars and B[e] supergiants (B[e]SGs) have strong and time-variable emission lines and show evidence for disk-like circumstellar material.

Be stars have ionized gaseous disks, which have been explained with the viscous decretion disk model (Rivinius et al. 2013). The mass-ejection process is likely linked to non-radial pulsation and small-scale magnetic fields, which act on top of a fast stellar rotation rate of $> 75\%$ of critical. Based on infrared observations obtained with *IRAS*, *AKARI*, and *AllWISE*, disk mass decretion rates have been estimated to be between $10^{-12}\ M_\odot\ \mathrm{yr}^{-1}$ and $10^{-9}\ M_\odot\ \mathrm{yr}^{-1}$ (Vieira et al. 2017).

B[e]SGs display outflows with a hot, fast (~ 1000 km s^{-1}) line-driven polar wind and a cool, slow (~ 100 km s^{-1}), dense equatorial wind (Zickgraf et al. 1985). Infrared excess from dust is indicative of a hot circumstellar dust disk or torus. The fast wind in the polar region is similar to the winds of normal blue supergiants. Imaging and interferometry have spatially resolved the environment of the closest and brightest Galactic B[e]SGs, providing precise measurements of the disk inclinations and disk sizes (e.g., Domiciano de Souza et al. 2007, 2008; Wheelwright et al. 2012). The kinematics of the equatorial gaseous material favor Keplerian rotation (Kraus et al. 2007, 2010; Millour et al. 2011; Wheelwright et al. 2012), likely accumulated in multiple, partial rings and possible spiral arm-like structures (Aret et al. 2012; Kraus et al. 2016; Maravelias et al. 2018). Several B[e]SGs have been discovered to be in short-period binary systems and the dynamical and spatial information indicates that the prevalent geometry is a circumbinary ring or disk (Kraus et al. 2013; Porter et al. 2022). Large dust shell structures and nebulae are detected around some objects, hinting at major mass-loss events (Liimets et al. 2022). Dust formation in the winds of hot stars is generally inhibited by the low particle densities and the harsh UV environment and thus the formation and continuous presence of the circumstellar dusty disks around the hot B[e]SGs are intriguing. Different disk formation scenarios could lead to the B[e] phenomenon, but binarity and mergers are preferred for several objects.

High-resolution, multi-wavelength, multi-epoch surveys of hundreds of Galactic and extra-galactic O- and B-type stars have been analyzed during the past decade with several stellar atmosphere codes to infer their stellar and wind parameters in a consistent manner, leading to major progress in our understanding of the mass loss in OB-type stars. Mokiem et al. (2006) and Mokiem et al. (2007b) studied optical spectra of 31 O- and early B-type stars in the Small and Large Magellanic Clouds (SMC, LMC) obtained with the Fibre Large Array Multi Element Spectrograph (FLAMES) at the VLT. Stellar parameters were determined using the stellar atmosphere code FASTWIND. One of the most important recent projects on OB-stars has been the VLT-FLAMES Tarantula Survey with multi-epoch optical spectroscopy of over 800 massive stars in the 30 Doradus region of the LMC (Evans et al. 2011). Bestenlehner et al. (2014) performed a spectral analysis of optical and near-infrared data using the non-LTE radiative transfer code CMFGEN to obtain the stellar and wind parameters of 62 O, Of, Of/WN, and WNh stars. Ramírez-Agudelo et al. (2017) determined stellar, photospheric, and wind properties of 72 presumably single O-type giants and supergiants with the non-LTE stellar atmosphere model FASTWIND. Ramachandran et al. (2019) reported a spectroscopic study of massive stars of spectral types O (23 stars) and B (297 stars) in the SMC, using observations with VLT FLAMES and archival UV data. The spectra were analyzed using PoWR stellar atmosphere models. Bouret et al. (2021) studied the evolutionary and physical properties of 13 SMC O-type giants and supergiants, using UV spectra obtained with the *Hubble Space Telescope's (HST) Cosmic Origins Spectrograph (COS)* and *Space Telescope Imaging Spectrograph (STIS)*, supplemented by optical spectra. The stellar atmosphere code CMFGEN was used to derive photospheric and wind properties. Mass-loss rates were derived from the analysis of UV P Cygni profiles of C IV and S IV resonance doublets, plus the Hα line for stars with optical spectra. The β exponent of the wind velocity law was derived from the fit of the shape of the P Cygni profile. Clumping parameters, f and v_{cl}, were derived in the UV domain. Marcolino et al. (2022) investigated the wind properties of massive stars in the Galaxy and SMC. They derived an empirical Z dependence from CMFGEN models for the winds of 96 O and B-type stars. Rickard et al. (2022) obtained UV and optical spectra of a sample of 19 O-type stars within NGC 346 in the SMC and used the non-local thermal equilibrium model atmosphere code PoWR to determine wind parameters and ionising fluxes to study stellar winds at low metallicity. Rubio-Díez et al. (2022) probed the radial clumping stratification of 25 OB stars in the intermediate and outer wind regions to derive upper limits for mass-loss rates, using optical to radio observations.

Some of the main issues identified in these studies are that the presence of inhomogeinities in the stellar winds of hot, luminous stars ("clumping") leads to severe discrepancies among different mass-loss rate diagnostics, and also between empirical estimates and theoretical predictions. This may require a reassessment of the relative roles of mass loss in steady stellar winds, in eruptive events, and through binary interaction. On the other hand, mass-loss rates inferred for low luminosity O-type stars were found to be much lower than predicted by theory ("weak wind" uncertainty). Wind line variability between observing epochs can lead to differences in the mass-loss rate estimates of up to a factor of three.

There is a strong relationship between the wind momentum and the stellar luminosity. The empirical scaling of $\dot{M} \propto L^{2.4}$, however, is steeper than theoretically expected. Objects in higher metallicity environments are observed to have stronger winds, i.e., the mass loss properties for the LMC are intermediate to massive stars in the Galaxy and SMC. For bright objects ($log\ L/L_{\odot} > 5.4$), the mass-loss rate scales with metallicity as $\dot{M} \propto Z^{0.5-0.8}$. This metallicity dependency seems to get weaker at lower luminosities.

2.3. *Red supergiants*

Red supergiants (RSGs) are evolved, helium-fusing stars originating from stars with initial masses of $8 - 40\ M_{\odot}$. They are the classical progenitors of Type II-P core-collapse SNe. The defining signature of mass loss in RSGs is the presence of circumstellar dust, observed as excess radiation in their SEDs from the silicate emission features at 10 μm and 20 μm. Several RSGs and red hypergiants show evidence for clumpy winds (e.g., Kervella et al. 2011; Ohnaka 2014; Montargès et al. 2019) and discrete, directed gaseous outflows in their optical and infrared images, spectra, and light curves (Humphreys & Jones 2022).

The driving mechanism of their strong mass loss is not understood, but leading processes are radiation pressure on dust grains or molecules, pulsations, and convective and magnetic activity (Josselin & Plez 2007; Thirumalai & Heyl 2012). The discovery of large-scale surface asymmetries or hot spots on their surfaces, which vary on timescales of months or years, supports convection as an important mechanism (e.g., Monnier et al. 2004; Haubois et al. 2009; Baron et al. 2014). Massive arcs and clumps, discrete gaseous outflows, and the presence of magnetic fields in their ejecta, suggest that enhanced convective activity together with magnetic activity may be important for the high-mass ejections (Humphreys & Jones 2022). However, studies of the atmospheres of RSGs showed that pulsation and convection alone cannot explain the elevation of material to the molecular and dust formation zones (Arroyo-Torres et al. 2015).

Beasor et al. (2020) found that quiescent mass-loss during the RSG phase is not effective at removing a significant fraction of the hydrogen-envelope prior to core-collapse. They measured the mass-loss rates of RSGs via SED modelling using mid-infrared photometry from the Stratospheric Observatory for Infrared Astronomy (SOFIA) Faint Object infraRed CAmera for the SOFIA Telescope (FORCAST) instrument and archival data from several observatories. When comparing their new mass-loss prescription to the treatment of mass-loss currently implemented in evolutionary models, they found models drastically over-predict the total mass loss, by up to a factor of 20. Thus, single stars with initial masses $< 25\ M_{\odot}$ do not lose enough mass through their quiescent winds to evolve blueward, and hence cannot create WR stars, BSGs, or LBVs as some evolutionary models have predicted.

Betelgeuse. The RSG Betelgeuse is a semi-regular variable with a period of ~ 400 d, attributed to radial pulsations, and a longer ~ 2000 d period, associated with convective and magnetic activity. Its mass-loss rate is on the order of a few $10^{-7}\ M_{\odot}\ \mathrm{yr}^{-1}$. High-spatial near- and mid-infrared imaging with the VLT Imager and Spectrometer for mid Infrared (VISIR) revealed that Betelgeuse has several clumps of dusty material within 1 arcsec of the star (Kervella et al. 2011). ALMA observations suggest that convective cells led to the production of these dusty knots and molecular plumes (Kervella et al. 2018).

Betelgeuse showed an unexpected visual dimming event by about 1 mag from November 2019 to March 2020. Spatially resolved UV spectra from *HST STIS* just prior to the event revealed an increase in the UV flux and variations in the Mg II line emission from the chromosphere over the southern hemisphere of the star, supporting an outflow of material from a convective cell that may have been enhanced by the outward motion in its 400 d radial pulsation (Dupree et al. 2020). The outflow expanded and cooled, possibly forming dust causing the deep dimming in the visual light curve. Since Betelgeuse also fainted at submillimeter wavelength, Dharmawardena et al. (2020) questioned that the dimming is due to dust formation and suggested a change in radius or in temperature. Kravchenko et al. (2021) used the tomographic method to a series of high-resolution HERMES observations at the Anglo-Australian Telescope (AAT), which

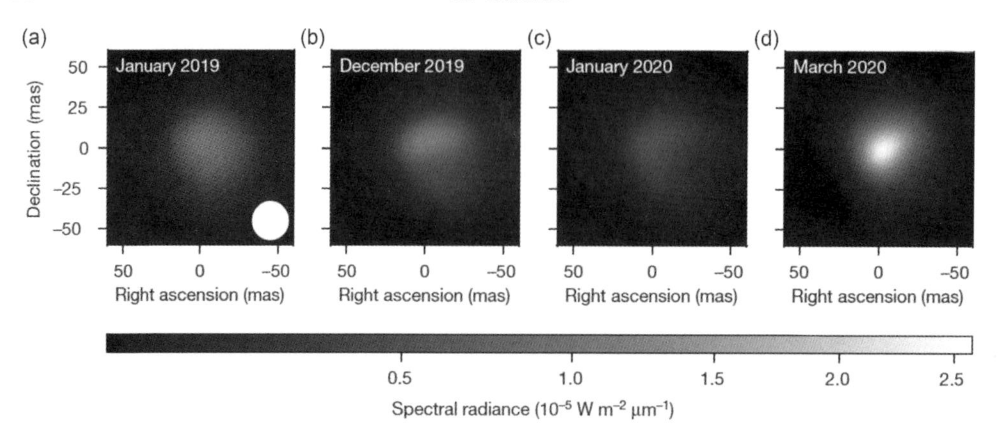

Figure 3. VLT SPHERE observations of the RSG Betelgeuse (Figure from Montargès et al. 2021).

allowed to probe different depths in the stellar atmosphere and to recover the corresponding disk-averaged velocity field. They found that the succession of two shocks along our line-of-sight in February 2018 and January 2019, combined with underlying convection and/or outward motion present at this phase of the 400 d pulsation cycle, produced a rapid expansion of a portion of the atmosphere and an outflow between October 2019 and February 2020. This resulted in a sudden increase in molecular opacity in the cooler upper atmosphere of Betelgeuse and, thus, in the decrease of the star's brightness. A visual spectro-polarimetric image obtained with VLT SPHERE showed a remarkable corresponding fading of its southern hemisphere (Montargès et al. 2021, see Figure 3). A change in diameter was ruled out via VLTI GRAVITY observations. The authors demonstrated that the fading was due to dust formation related to the gaseous outflow from a convective cell. They estimate a total mass associated with the dusty and gaseous outflow of $(0.7-3)\times 10^{-7}\ M_{\odot}$. Taniguchi et al. (2022) presented photometry of Betelgeuse with a cadence of about one observations every two days between 2017 and 2021 in the $0.45-13.5\ \mu$m wavelength range obtained by the Himawari-810 geostationary meteorological satellite. By examining also the optical and near-infrared light curves, they showed that both a decreased effective temperature and increased dust extinction may have contributed by almost equal amounts to the Great Dimming.

VY CMa. The red hypergiant VY CMa is one of the most luminous red stars in the Galaxy ($L\sim 3\times 10^5\ L_{\odot}$). The star belongs to a class of evolved massive stars that represent a short-lived evolutionary phase characterized by high mass-loss rates of up to $\sim 10^{-4}\ M_{\odot}\ \mathrm{yr}^{-1}$, extensive circumstellar ejecta, asymmetric ejections, and multiple high mass-loss events. Near-infrared spectro-interferometric observations with the Astronomical Multi-BEam combineR (AMBER) at the VLTI showed asymmetric, possibly clumpy, atmospheric layers (Wittkowski et al. 2012). Its current mass is estimated to be $\sim 17\ M_{\odot}$ with an initial mass of $25-30\ M_{\odot}$. It is one of the largest RSGs with a stellar radius of $1420\pm 120\ R_{\odot}$.

Decin et al. (2006) modelled the line profile of low excitation submillimeter CO lines, observed with the James Clerk Maxwell Telescope (JCMT) and the Swedish-ESO Submillimetre Telescope (SEST), emitted in the circumstellar envelope to study the star's mass-loss history. They demonstrated that this source underwent a phase of high mass loss about 1000 yr ago ($\sim 3.2\times 10^{-4}\ M_{\odot}\ \mathrm{yr}^{-1}$), which lasted for 100 yr. This phase was preceded by a low mass-loss phase ($\sim 1\times 10^{-6}\ M_{\odot}\ \mathrm{yr}^{-1}$), which lasted for 800 yr. The current mass-loss rate is on the order of $8\times 10^{-5}\ M_{\odot}\ \mathrm{yr}^{-1}$. Mid- and far-infrared observation with SOFIA found no extended cold dust with an age greater than about

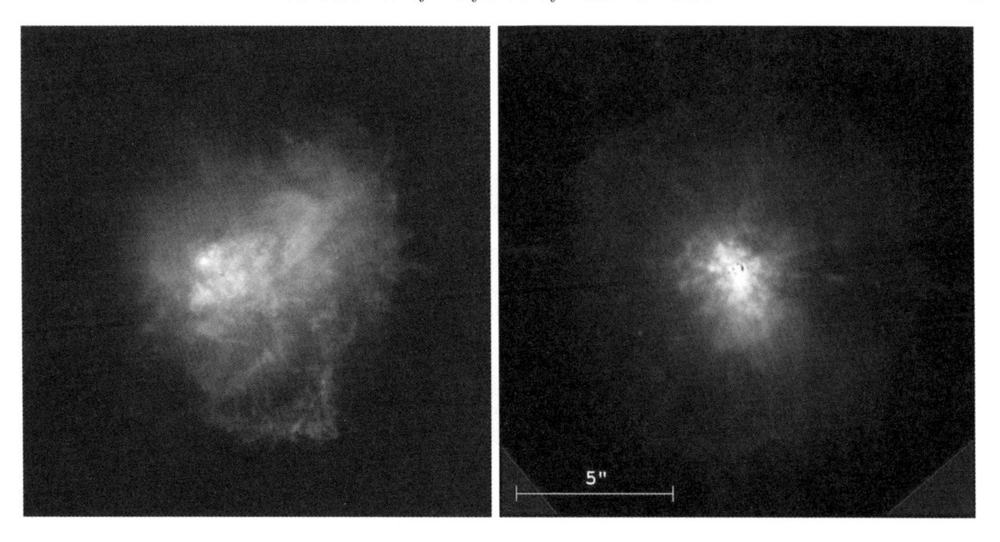

Figure 4. Left: The red hypergiant VY CMa with *HST WFPC2*. Right: The post-RSG IRC +10420 with *HST WFPC2* (Figure from Humphreys et al. 1997).

1200 yr (Shenoy et al. 2013). ALMA submillimeter observations revealed massive, dusty condensations (O'Gorman et al. 2015; Vlemmings et al. 2017; Kamiński 2019), confirming numerous high mass-loss events over the past few hundred years. ALMA gaseous sodium chloride (NaCl) emission of the inner wind shows evidence of localized mass ejections (Decin et al. 2016). Mass estimates of some of the knots and clumps, based on infrared imaging and measurements from ALMA, yield masses on the order of a few $10^{-3}\ M_{\odot}$ (Shenoy et al. 2013; O'Gorman et al. 2015; Gordon et al. 2019; Humphreys & Jones 2022).

With *HST Wide Field and Planetary Camera 2 (WFPC2)* images and *STIS* spectra, the orientation and the ejection ages of the complex circumstellar ejecta were determined (Humphreys et al. 2021; Humphreys & Jones 2022, see Figure 4). The closest knots were likely ejected within the past century and one knot as recent as in 1985–1990. The numerous knots to the south and southwest have ejection ages around 120, 200, and 250 yr ago. The random orientation of the ejecta suggests that the mass-loss events are driven by localized, large-scale ejections, maybe due to magnetic and convective regions. Several knot ejection times correspond to periods of the star's photometric variability, which suggests that discrete ejections may be common and large-scale surface or convective activity is a major source of mass loss for RSGs (Humphreys et al. 2021).

2.4. *The warm hypergiants*

Yellow hypergiants (YHGs) are rare, unstable, massive stars (de Jager 1998). They are thought to be progenitors of some SNe. As post-RSGs evolve back to the hot side of the HR diagram, they intersect a temperature domain (the "Yellow Void" or "Yellow Wall") in which their atmospheres become unstable against quasi-periodic pulsations. This can, on timescales of decades, result in episodic mass-loss. Mass-loss rates are on the order of 10^{-4} to $10^{-3}\ M_{\odot}\ \mathrm{yr}^{-1}$, resulting in large, extended circumstellar gas and dust envelopes.

IRC +10420. The YHG or intermediate post-RSG IRC +10420 is a powerful infrared source. In the past 30 yr its apparent spectral type has changed from a late F-type to a mid-A-type, maybe due to its post-RSG evolution to warmer temperatures (Oudmaijer et al. 1996). IRC +10420's infrared SED from SOFIA FORCAST imaging cannot be explained by a single mass-loss rate (Shenoy et al. 2016). During a period

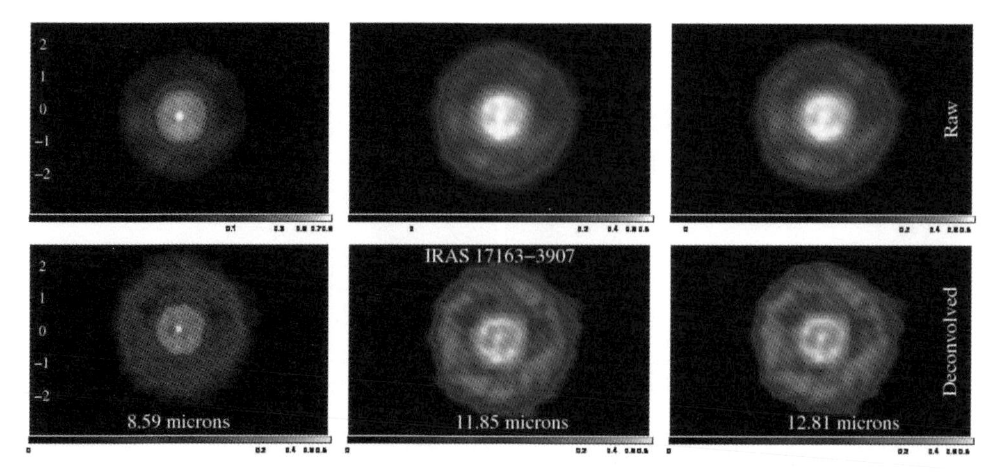

Figure 5. The YHG Fried Egg nebula with VLT VISIR (Figure from Lagadec et al. 2011).

from about 6000 yr to 2000 yr ago, the star likely had a very high mass-loss rate of $2 \times 10^{-3}\ M_{\odot}\ \mathrm{yr}^{-1}$, which then dropped to $\sim 10^{-4}\ M_{\odot}\ \mathrm{yr}^{-1}$.

HST WFPC2 optical images revealed a complex asymmetric circumstellar environment, with condensations or knots, ray-like features, and several arcs or loops within 2 arcsec of the star, and distant reflection shells. These features are evidence for high mass-loss episodes during the past few hundred years. The transverse motions of numerous knots, arcs, and condensations in its inner ejecta from second epoch *HST WFPC2* images, combined with the radial motions for several of the features, showed that they were ejected at different times, in different directions, and presumably from separate regions on the surface of the star (Tiffany et al. 2010). Vlemmings et al. (2017) used ALMA observations to study the polarization of SiO thermal/maser lines and dust continuum at 1.7 mm. Emission from magnetically aligned grains is the most likely origin of the observed continuum polarisation, which suggests a strong role of surface activity of a convective nature and magnetic fields in the star's recent mass-loss history. At milliarcsecond spatial resolution, the outflow produced by IRC +10420 has the shape of an hourglass nebula, possibly induced by high rotation and latitude dependent mass-loss (Koumpia et al. 2022).

Fried Egg nebula. One of the more publicized objects is the so-called Fried Egg nebula (IRAS 17163−3907). Huge dusty shells were observed in VLT VISIR images (Lagadec et al. 2011, see Figure 5). The first interpretation, adopting a distance of 4 kpc, was that these shells were caused by successive ejections ∼ 400 yr apart and that the total circumstellar mass exceeded 4 $M_{\odot}$. Using a large set of optical to infrared spectroscopic, photometric, spectropolarimetric, and interferometric data, and adopting the Gaia DR2 distance of 1.2 kpc, Koumpia et al. (2020) identified three distinct mass-loss episodes 30, 100, 125 yr ago, which are characterized by different mass-loss rates and resulting in a total mass lost of $6.5 \times 10^{-3}\ M_{\odot}$. The authors argued that the second bi-stability jump at $T_{\mathrm{eff}} \sim 8800$ K, due to the recombination of Fe III to Fe II (Petrov et al. 2016), is a promising mechanism for the mass-loss ejections. However, the observations do not allow yet to discriminate between the photospheric pulsations and wind bi-stability mechanisms as the prevalent mass-loss mechanism.

2.5. *Classical and giant eruption LBVs*

Classical LBVs experience "outbursts" and episodes of enhanced mass loss on timescales of years to decades. During an outburst, they transit in the HR diagram

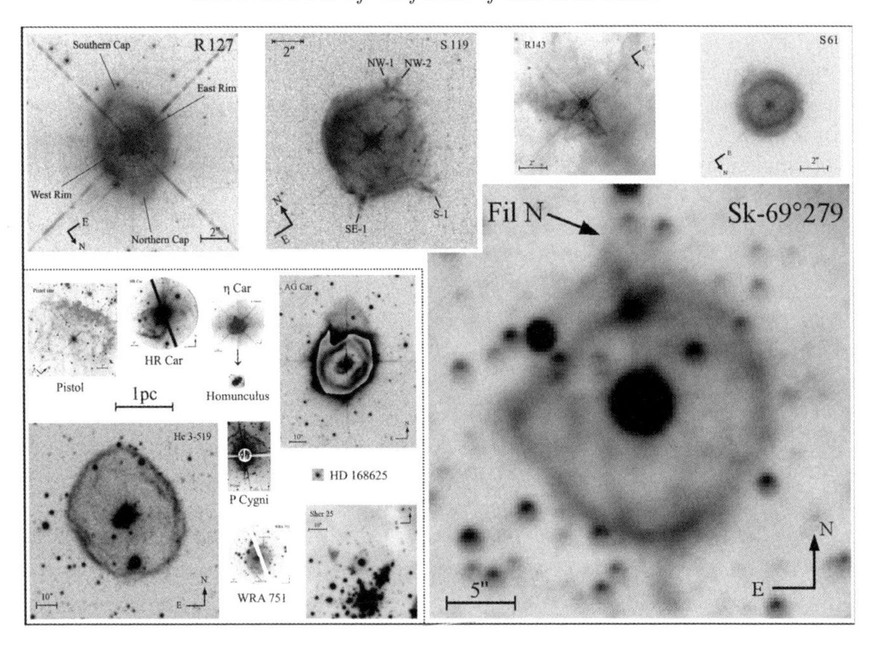

Figure 6. Galactic and LMC LBV nebulae in *HST* images (Figure from Weis 2011).

from a hot quiescent state ($T_{\rm eff} \sim 12\,000 - 30\,000$ K) to lower temperatures ($T_{\rm eff} \sim 7\,000 - 8\,000$ K), i.e., their spectral appearance changes from a hot supergiant to a much cooler A or F-type supergiant. Their visual magnitudes change by $1-2$ mag, at almost constant bolometric luminosity. They are generally divided into high-luminous LBVs ($log\ L/L_\odot > 5.8$), which evolve from the most massive stars with $M > 50\ M_\odot$, and less-luminous LBVs, which have initial masses of the order $M \sim 25-40\ M_\odot$ and which have presumably lost more than half of their initial mass during a previous RSG phase. Giant eruption LBVs, represented by the famous Galactic stars η Car and P Cygni, exhibit eruptions with visual magnitude changes of $2-3$ mag or more during which they can expel large amounts of material.

The terminal velocities of LBV winds are in the range of $100-250$ km s^{-1}, significantly lower than those of normal OB supergiants. LBV mass-loss rates are up to $10^{-5} - 10^{-4}\ M_\odot$ yr^{-1}, a factor of $10-100$ larger than those of normal supergiants. During their quiescent phases, LBVs lose mass likely via ordinary line-driving and the bistability mechanism is a good candidate to account for their transitional mass-loss variations (Vink 2021). The large mass-loss rates and giant eruptions observed in some LBVs suggest that also continuum-driven winds and/or explosions could play a role (Shaviv 1998, 2001; Owocki et al. 2004; Smith & Owocki 2006; Rest et al. 2012; Davidson & Humphreys 2012; Owocki & Shaviv 2016; González et al. 2022).

The morphology, kinematics, and chemical composition of their circumstellar nebulae trace the star's mass-loss history and evolution. Typically, LBV nebulae have sizes of $0.2-5$ pc, ionized masses of $1-4\ M_\odot$, and expansion velocities of several tenths to a few hundreds km s^{-1} (Weis 2003, see Figure 6). Bordiu et al. (2021) found that molecular gas may account for $>30\%$ of the total mass lost around LBVs. At least 50% of the LBV nebulae are bipolar and several nebulae show in addition equatorial ring-like structures. To explain the observed nebular morphologies, a range of models have been proposed (see, e.g., Nota et al. 1995; Frank 1999; Frank et al. 1998; Owocki & Gayley 1997; Dwarkadas & Owocki 2002; Maeder & Desjacques 2001; Soker 2004). LBV nebulae are also unique laboratories to study the dust formation and dust survival in harsh conditions. The dust grains are unusually large and have an unusual composition compared to grains in the interstellar medium and the asymptotic giant branch stars. They may trace the cool outbursts or giant eruptions (Kochanek 2011) or they could have formed in a prior RSG phase (Waters et al. 1997).

It is yet unclear what causes the transitions of LBVs across the HR diagram and the giant LBV eruptions. Instabilities in the outer layers and core instabilities are proposed. The most prominent still discussed instability mechanisms are 1) radiation and turbulent pressure instabilities (Humphreys & Davidson 1984; Lamers 1986; Stothers 2003; Lamers & Fitzpatrick 1988; Ulmer & Fitzpatrick 1998), 2) ionization-induced dynamical instabilities (Maeder 1992; Stothers & Chin 1993, 1994, 1997; Glatzel & Kiriakidis 1993; Kiriakidis et al. 1993; Glatzel & Kiriakidis 1998; Lovekin & Guzik 2014; Jiang et al. 2018; Grassitelli et al. 2021), 3) deep rooted energy release or addition (Smith & Arnett 2014; Dessart et al. 2010; Smith & Arnett 2014; Meakin & Arnett 2007; Quataert & Shiode 2012; Shiode & Quataert 2014; Podsiadlowski et al. 2010; Smith & Arnett 2014; Dessart et al. 2010; Owocki et al. 2019). A single mechanism may be operating over a wide range of energy and mass, different mechanisms may produce the same observables, or several mechanisms may cooperate. Rotation (Langer 1998; Zhao & Fuller 2020; Shi & Fuller 2022), binary interaction (Kenyon & Gallagher 1985; Gallagher 1989; Smith 2011; Owocki et al. 2019; Quataert et al. 2016; Soker 2001, 2004, 2007; Kashi & Soker 2009; Smith et al. 2003; Koenigsberger 2004), and stellar mergers (Iben 1999; Portegies Zwart & van den Heuvel 2016; Hirai et al. 2021) may also play a crucial role in shaping the LBV phenomenon.

Eta Carinae. Eta Car is the prototype of the giant eruption LBVs. In the 1840/50s, it experienced an eruption during which it expelled up to 40 $M_{\odot}$ in 20 yr, which resulted in its Homunculus nebula. It is an eccentric, massive binary system with strong winds (the primary has a current mass-loss rate of $\sim 2.5 \times 10^{-4}$ $M_{\odot}$ yr^{-1}) and a phase-dependent wind-wind-collision (WWC) zone. The primary star is often classified as an LBV and the secondary is a WR or O-type star with a faster, less dense wind. Due to its dramatic nature and brightness, η Car has been extensively observed at all wavelength ranges and with a multitude of observing techniques (see Figure 7). Some selected recent studies are mentioned below.

Morphology and proper motions studies of the ejecta have been conducted at different wavelengths and spatial scales. Artigau et al. (2011) found with near-infrared adaptive optics imaging with the Near-Infrared Coronagraphic Imager (NICI) and the Nasmyth Adaptive Optics System (NAOS) Near-Infrared Imager and Spectrograph (CONICA), NaCO, a lobed pattern, the "butterfly nebula," outlined by bright Brγ and H_2 emission and light scattered by dust. Proper motions measured from the combined NICI and NaCO images together with radial velocities show that the knots and filaments that define the bright rims of the butterfly nebula were ejected at two different epochs corresponding approximately to the giant eruption and a second eruption around 1890. Abraham et al. (2020) presented images of η Car in the recombination lines H30α and He30α and the underlying continuum with 50 mas resolution (110 au), obtained with ALMA. They estimated the proper motions of nearby ejecta and derived ejection dates of 1952.6, 1957.1, and 1967.6, which are all close to periastron passages. Mehner et al. (2016) used large-scale integral field unit observations in the optical with the Multi Unit Spectroscopic Explorer (MUSE) at the VLT to reveal the detailed three-dimensional structure of its outer ejecta. Morpho-kinematic modeling of these ejecta were conducted with the code SHAPE (Steffen et al. 2010), which revealed a spatially coherent structure. Infrared long-baseline interferometry and synthesis imaging provides intensity and velocity distributions with milliarcsecond spatial resolution and offers opportunities to study the inner environment around η Car with a spatial resolution of a few au. Using

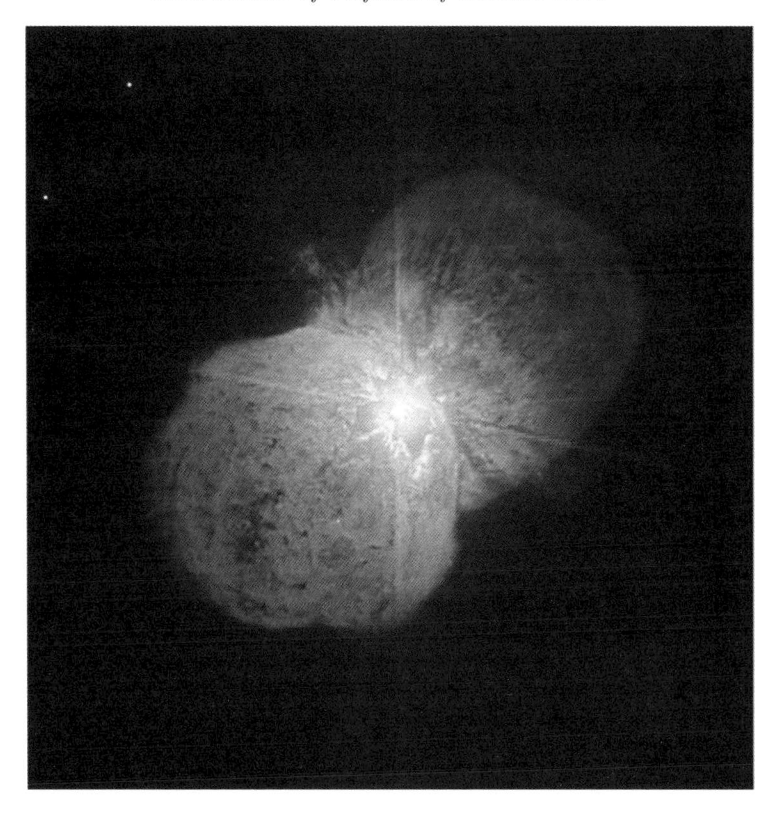

Figure 7. Eta Car with VLT NACO (Credit: ESO).

the Multi AperTure mid-Infrared SpectroScopic Experiment (MATISSE) at the VLTI, Weigelt et al. (2021) performed Brα imaging of η Car's distorted wind. The asymmetries found are likely caused by the WWC. A fit of the observed continuum visibilities with the visibilities of a stellar wind CMFGEN model provides a diameter of the primary stellar wind of 2.84 ± 0.06 mas (6.54 ± 0.14 au). Analysis of the orbital phase-locked X-ray variability provides our most direct constraints on the momentum ratio of the two winds, the orbital eccentricity, and the shape of the colliding wind bow shock formed around the secondary star (Corcoran et al. 2017; Espinoza-Galeas et al. 2022). Observed orbit-to-orbit X-ray emission variations must arise from changes in the stellar wind properties of one or both of the stars. Explanations include significant changes in the primary star's wind mass-loss rate, variations in wind velocity, or wind clumping properties. At even higher energies, using γ-ray *FERMI-LAT* observations, Martí-Devesa & Reimer (2021) found that the wind collision region of this system is perturbed from orbit to orbit, affecting particle transport within the shock.

2.6. *Wolf-Rayet stars.*

Classical WR stars are generally assumed to be evolved, helium-burning massive stars that have lost most of their hydrogen envelope through prior mass loss, exposing the helium core. They are the progenitor candidates of SN Ib/Ic and long-duration gamma ray bursts. Classification as a WR star is performed based on spectroscopic terminology, i.e., based on the presence of strong and broad emission lines which are produced by strong winds. Such spectra can also originate in objects with very large initial stellar mass and luminosity or in binary-stripped objects (Stanway et al. 2016; Götberg et al.

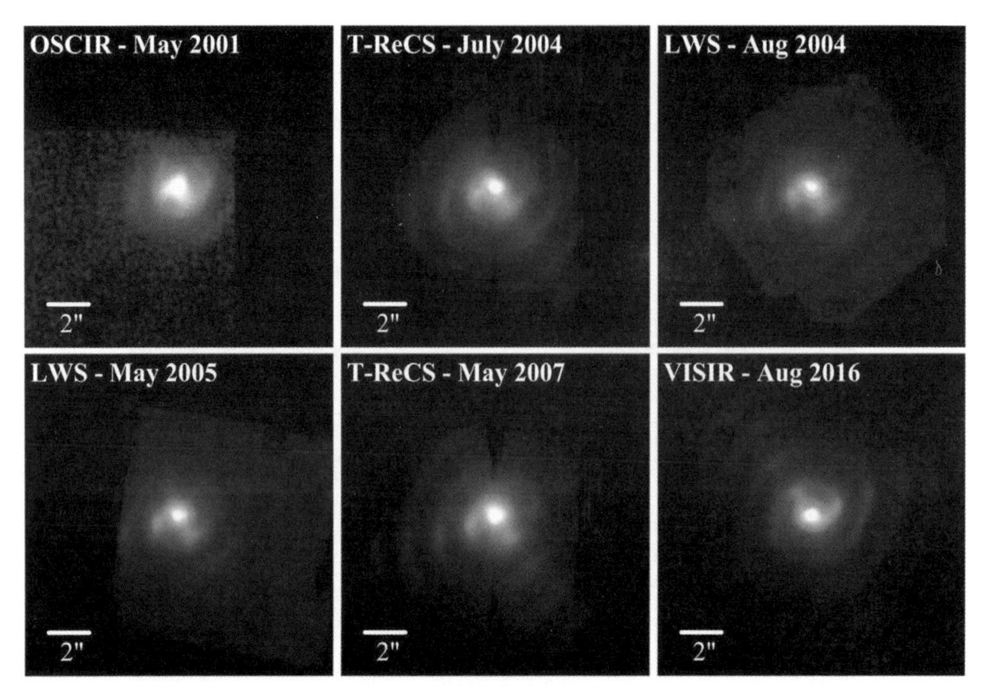

Figure 8. N-band images of WR 112 (Figure from Lau et al. 2020).

2020). Thus, this stellar group may include objects still in their core hydrogen-burning phase of evolution. The mechanism of the hydrogen-envelope stripping and how it varies with metallicity is a long-standing unsolved issue. It is not clear if all single O-type stars can evolve to become a WR star, or only those in interacting binaries.

Classical WR stars have strong winds with large mass-loss rates (Crowther 2007), driven by radiation pressure and typically a factor of 10 larger than O-star winds with the same luminosity. The wind strength is found to be dependent on the Fe abundance, but with a metallicity dependence that is less steep than for OB stars (Vink & de Koter 2005). Empirical mass-loss rates have been derived from radio emission, optical emission line equivalent width relationships, and spectral analysis of optical and near-infrared data using non-LTE radiative transfer codes like CMFGEN (Nugis & Lamers 2000; Bestenlehner et al. 2014). Sander et al. (2019) and Hamann et al. (2019) revised the correlations between the mass-loss rate and the luminosity for WC, WO, and WN stars based on Gaia DR2 data. Sander et al. (2019) found for WC and WO stars luminosities ranging from $log\ L/L_{\odot} = 4.9 - 6.0$, which indicates that the WC stars are likely formed from a broader initial mass range than previously assumed. They also obtained mass-loss rates ranging between $log\ \dot{M} = -5.1$ and -4.1, with $\dot{M} \propto L^{0.68}$. Hamann et al. (2019) found that the correlations between mass-loss rate and luminosity show a large scatter for the WN stars, with shallower slopes than found previously.

In colliding wind systems, the enhanced density in the WWC produces favorable conditions for dust formation. Combined with the Coriolis force due to the orbital motion, this gives rise to so-called pinwheel nebulae that are characterized by a dust plume following an Archimedian spiral pattern.

WR 112. WR 112 is a dust-forming late-type carbon-rich Wolf–Rayet (WC)–OB binary system with a dusty circumstellar nebula that exhibits a complex asymmetric morphology, which traces the orbital motion and dust formation in the colliding winds of the central binary (Figure 8). Lau et al. (2017) presented high spatial resolution mid-infrared images of its nebula obtained with VLT VISIR over a period of 9 yr. The observations revealed a morphology resembling a series of arc-like filaments and broken shells, that the authors interpreted as a consecutive series of stagnant dust shells exhibiting no proper motion, which may have formed in the outflow during a previous red or yellow supergiant phase. Lau et al. (2020) presented a multi-epoch proper motion and morphological analysis of the circumstellar dust using seven high spatial resolution N-band imaging observations spanning almost 20 yr, leading to a revision of their previous morphological interpretation of the geometry. They found evidence of proper motion of the circumstellar dust consistent with a nearly edge-on spiral with a period of $P \sim 20$ yr. They spectroscopically derived a terminal stellar wind velocity of $v = 1230 \pm 260$ km s^{-1}. Fitting optically thin dust SED models, they determined a dust production rate of $\dot{M}_{\rm dust} = 2.7^{1.0}_{1.3} \times 10^{-6}$ $M_\odot$ yr^{-1}, which demonstrates that WR 112 is one of the most prolific dust-making WC systems known.

3. Summary

There has been much observational progress in the past decade in our understanding of the mass loss in the most massive stars. This can be attributed to ground-based 10-m class observatories equipped with multi-object spectrographs, AO, and interferometric instrumentation, and space-based observatories with capabilities in the UV, infrared, X-ray.

There are still many unsolved questions related to mass loss and outflows in the most massive stars. Observational measurements have large uncertainties. It is also not yet established how rotation and binarity affect the mass loss. The majority of massive stars are binary systems that will at one point in their evolution interact and the finding of relatively low steady wind mass-loss rates has shifted the attention to binary interaction that can create and shape outflows.

Due to their brightness, massive stars are excellent targets for almost any observing technique and wavelength range, enabling a multi-wavelength approach that provides different diagnostics and consistency checks. Upcoming large survey missions (e.g., Euclid, Roman, Rubin) will provide an improved census of evolved massive stars, massive star transients, and SN precursors. High-multiplex spectroscopic surveys (e.g., 4MOST, MOONS) will allow their follow-up and detailed analysis of their stellar winds and outflows.

References

Abraham, Z., Beaklini, P. P. B., Cox, P., Falceta-Gonçalves, D., & Nyman, L.-Å. 2020, MNRAS, 499, 2493
Aret, A., Kraus, M., Muratore, M. F., & Borges Fernandes, M. 2012, MNRAS, 423, 284
Arroyo-Torres, B., Wittkowski, M., Chiavassa, A., et al. 2015, A&A, 575, A50
Artigau, É., Martin, J. C., Humphreys, R. M., et al. 2011, AJ, 141, 202
Bally, J., Ginsburg, A., Silvia, D., & Youngblood, A. 2015, A&A, 579, 130
Bally, J. 2016, ARA&A, 54, 491
Bally, J., Ginsburg, A., Arce, H., et al. 2017, ApJ, 837, 60
Bally, J., Ginsburg, A., Forbrich, J., & Vargas-González, J. 2020, ApJ, 889, 178
Bally, J., & Zinnecker, H. 2005, AJ, 129, 2281
Baron, F., Monnier, J. D., Kiss, L. L., et al. 2014, ApJ, 785, 46
Beasor, E. R., Davies, B., Smith, N., et al. 2020, MNRAS, 492, 5994
Bestenlehner, J. M., Gräfener, G., Vink, J. S., et al. 2014, A&A, 570, A38
Björklund, R., Sundqvist, J. O., Puls, J., & Najarro, F. 2021, A&A, 648, A36
Bordiu, C., Bufano, F., Cerrigone, L., et al. 2021, MNRAS, 500, 5500
Bouret, J. C., Martins, F., Hillier, D. J., et al. 2021, A&A, 647, A134
Brennan, S. J., Fraser, M., Johansson, J., et al. 2022, MNRAS, 513, 5666

Carrasco-González, C., Sanna, A., Rodríguez-Kamenetzky, A., et al. 2021, ApJL, 914, L1
Castor, J. I., Abbott, D. C., & Klein, R. I. 1975, ApJ, 195, 157
Corcoran, M. F., Liburd, J., Morris, D., et al. 2017, ApJ, 838, 45
Cortes, P. C., Le Gouellec, V. J. M., Hull, C. L. H., et al. 2021, ApJ, 907, 94
Crowther, P. A. 2007, ARA&A, 45, 177
Cunningham, N. J., Moeckel, N., & Bally, J. 2009, ApJ, 692, 943
Davidson, K., & Humphreys, R. M. 2012, Nature, 486, E1
de Jager, C. 1998, A&A Rev., 8, 145
Decin, L., Hony, S., de Koter, A., et al. 2006, A&A, 456, 549
Decin, L., Richards, A. M. S., Millar, T. J., et al. 2016, A&A, 592, A76
Dessart, L., Livne, E., & Waldman, R. 2010, MNRAS, 405, 2113
Dharmawardena, T. E., Mairs, S., Scicluna, P., et al. 2020, ApJL, 897, L9
Domiciano de Souza, A., Kervella, P., Bendjoya, P., & Niccolini, G. 2008, A&A, 480, L29
Domiciano de Souza, A., Driebe, T., Chesneau, O., et al. 2007, A&A, 464, 81
Dressing, C. D., & Charbonneau, D. 2013, ApJ, 767, 95
Dupree, A. K., Strassmeier, K. G., Matthews, L. D., et al. 2020, ApJ, 899, 68
Dwarkadas, V. V., & Owocki, S. P. 2002, ApJ, 581, 1337
Espinoza-Galeas, D., Corcoran, M. F., Hamaguchi, K., et al. 2022, ApJ, 933, 136
Evans, C. J., Taylor, W. D., Hénault-Brunet, V., et al. 2011, A&A, 530, A108
Foley, R. J., Berger, E., Fox, O., et al. 2011, ApJ, 732, 32
Frank, A. 1999, NewA, 43, 31
Frank, A., Ryu, D., & Davidson, K. 1998, ApJ, 500, 291
Fullerton, A. W., Massa, D. L., & Prinja, R. K. 2006, ApJ, 637, 1025
Gallagher, J. S. 1989, in ASSL, Vol. 157, IAU Colloq. 113: Physics of Luminous Blue Variables, ed. K. Davidson, A. F. J. Moffat, & H. J. G. L. M. Lamers, 185
Glatzel, W., & Kiriakidis, M. 1993, MNRAS, 263, 375
Glatzel, W., & Kiriakidis, M. 1998, MNRAS, 295, 251
González, R. F., Zapata, L. A., Raga, A. C., et al. 2022, A&A, 659, A168
Gordon, M. S., Jones, T. J., Humphreys, R. M., et al. 2019, AJ, 157, 57
Götberg, Y., de Mink, S. E., McQuinn, M., et al. 2020, A&A, 634, A134
Gräfener, G., Koesterke, L., & Hamann, W. R. 2002, A&A, 387, 244
Grassitelli, L., Langer, N., Mackey, J., et al. 2021, A&A, 647, A99
Groh, J. H. 2017, Philos. Trans. Royal Soc. A, 375, 20170219
Groh, J. H., Meynet, G., & Ekström, S. 2013, A&A, 550, L7
Habing, H. J. 1996, A&A Rev., 7, 97
Hamann, W. R., Gräfener, G., Liermann, A., et al. 2019, A&A, 625, A57
Haubois, X., Perrin, G., Lacour, S., et al. 2009, A&A, 508, 923
Henney, W. J., & Arthur, S. J. 2019, MNRAS, 489, 2142
Hillier, D. J., & Miller, D. L. 1998, ApJ, 496, 407
Hirai, R., Podsiadlowski, P., Owocki, S. P., Schneider, F. R. N., & Smith, N. 2021, MNRAS, 503, 4276
Humphreys, R. M., & Davidson, K. 1984, Science, 223, 243
Humphreys, R. M., Smith, N., Davidson, K., Jones, T. J., et al. 1997, AJ, 114, 2778
Humphreys, R. M., Davidson, K., Hahn, D., et al. 2017, ApJ, 844, 40
Humphreys, R. M., Davidson, K., Richards, A. M. S., et al. 2021, AJ, 161, 98
Humphreys, R. M., & Jones, T. J. 2022, AJ, 163, 103
Iben, Jr., I. 1999, in ASP Conf. Ser., Vol. 179, Eta Carinae at The Millennium, ed. J. A. Morse, R. M. Humphreys, & A. Damineli, 367
Janson, M., Gratton, R., Rodet, L., et al. 2021, Nature, 600, 231
Jiang, Y.-F., Cantiello, M., Bildsten, L., et al. 2018, Nature, 561, 498
Josselin, E., & Plez, B. 2007, A&A, 469, 671
Justham, S., Podsiadlowski, P., & Vink, J. S. 2014, ApJ, 796, 121
Kamiński, T. 2019, A&A, 627, A114
Kashi, A., & Soker, N. 2009, NewA, 14, 11

Kenyon, S. J., & Gallagher, J. S., I. 1985, ApJ, 290, 542
Kervella, P., Perrin, G., Chiavassa, A., et al. 2011, A&A, 531, A117
Kervella, P., Decin, L., Richards, A. M. S., et al. 2018, A&A, 609, A67
Kiriakidis, M., Fricke, K. J., & Glatzel, W. 1993, MNRAS, 264, 50
Kobulnicky, H. A., Chick, W. T., & Povich, M. S. 2019, AJ, 158, 73
Kochanek, C. S. 2011, ApJ, 743, 73
Koenigsberger, G. 2004, RMxAA, 40, 107
Koumpia, E., Oudmaijer, R. D., Graham, V., et al. 2020, A&A, 635, A183
Koumpia, E., Oudmaijer, R. D., de Wit, W. -J., et al. 2022, MNRAS, 515, 2766
Kraus, M., Borges Fernandes, M., & de Araújo, F. X. 2007, A&A, 463, 627
Kraus, M., Borges Fernandes, M., & de Araújo, F. X. 2010, A&A, 517, A30
Kraus, M., Oksala, M. E., Nickeler, D. H., et al. 2013, A&A, 549, A28
Kraus, M., Cidale, L. S., Arias, M. L., et al. 2016, A&A, 593, A112
Kravchenko, K., Jorissen, A., Van Eck, S., et al. 2021, A&A, 650, L17
Kudritzki, R.-P., & Puls, J. 2000, ARA&A, 38, 613
Lagadec, E., Zijlstra, A. A., Oudmaijer, R. D., et al. 2011, A&A, 534, L10
Lamers, H. J. G. L. M. 1986, A&A, 159, 90
Lamers, H. J. G. L. M., & Fitzpatrick, E. L. 1988, ApJ, 324, 279
Lamers, H. J. G. L. M., Nota, A., Panagia, N., Smith, L. J., & Langer, N. 2001, ApJ, 551, 764
Lamers, H. J. G. L. M., Snow, T. P., & Lindholm, D. M. 1995, ApJ, 455, 269
Langer, N. 1998, A&A, 329, 551
Lau, R. M., Hankins, M. J., Schödel, R., et al. 2017, ApJL, 835, L31
Lau, R. M., Hankins, M. J., Han, Y., et al. 2020, ApJ, 900, 190
Liimets, T., Kraus, M., Moiseev, A., et al. 2022, Galaxies, 10, 41
Lovekin, C. C., & Guzik, J. A. 2014, MNRAS, 445, 1766
Lucy, L. B., & Solomon, P. M. 1970, ApJ, 159, 879
Maeder, A. 1992, in Instabilities in Evolved Super- and Hypergiants, ed. C. de Jager & H. Nieuwenhuijzen, 138
Maeder, A., & Desjacques, V. 2001, A&A, 372, L9
Maravelias, G., Kraus, M., Cidale, L. S., et al. 2018, MNRAS, 480, 320
Marcolino, W. L. F., Bouret, J. C., Rocha-Pinto, H. J., Bernini-Peron, M., & Vink, J. S. 2022, MNRAS, 511, 5104
Martí-Devesa, G., & Reimer, O. 2021, A&A, 654, A44
Matt, S., Frank, A., & Blackman, E. G. 2006, ApJL, 647, L45
Meakin, C. A., & Arnett, D. 2007, ApJ, 667, 448
Mehner, A., Davidson, K., Humphreys, R. M., et al. 2015, A&A, 578, A122
Mehner, A., Steffen, W., Groh, J. H., et al. 2016, A&A, 595, A120
Millour, F., Meilland, A., Chesneau, O., et al. 2011, A&A, 526, A107
Mokiem, M. R., de Koter, A., Evans, C. J., et al. 2006, A&A, 456, 1131
Mokiem, M. R., de Koter, A., Vink, J. S., et al. 2007a, A&A, 473, 603
Mokiem, M. R., de Koter, A., Evans, C. J., et al. 2007b, A&A, 465, 1003
Monnier, J. D., Millan-Gabet, R., Tuthill, P. G., et al. 2004, ApJ, 605, 436
Montargès, M., Homan, W., Keller, D., et al. 2019, MNRAS, 485, 2417
Montargès, M., Cannon, E., Lagadec, E., et al. 2021, Nature, 594, 365
Nota, A., Livio, M., Clampin, M., & Schulte-Ladbeck, R. 1995, ApJ, 448, 788
Nugis, T., & Lamers, H. J. G. L. M. 2000, A&A, 360, 227
O'Gorman, E., Vlemmings, W., Richards, A. M. S., et al. 2015, A&A, 573, L1
Ohnaka, K. 2014, A&A, 568, A17
Oudmaijer, R. D., Groenewegen, M. A. T., Matthews, H. E., et al. 1996, MNRAS, 280,1062
Owocki, S. P., & Gayley, K. G. 1997, in Astron. Soc. Pac. Conf. Ser., Vol. 120, Luminous Blue Variables: Massive Stars in Transition, ed. A. Nota & H. Lamers, 121
Owocki, S. P., Gayley, K. G., & Shaviv, N. J. 2004, ApJ, 616, 525
Owocki, S. P., Hirai, R., Podsiadlowski, P., & Schneider, F. R. N. 2019, MNRAS, 485, 988
Owocki, S. P., & Shaviv, N. J. 2016, MNRAS, 462, 345

Petrov, B., Vink, J. S., & Gräfener, G. 2016, MNRAS, 458, 1999
Podsiadlowski, P., Ivanova, N., Justham, S., & Rappaport, S. 2010, MNRAS, 406, 840
Poppenhaeger, K., Ketzer, L., & Mallonn, M. 2021, MNRAS, 500, 4560
Portegies Zwart, S. F., & van den Heuvel, E. P. J. 2016, MNRAS, 456, 3401
Porter, A., Blundell, K., & Lee, S. 2022, MNRAS, 509, 1720
Puls, J., Urbaneja, M. A., Venero, R., et al. 2005, A&A, 435, 669
Puls, J., Vink, J. S., & Najarro, F. 2008, A&A Rev., 16, 209
Puls, J., Kudritzki, R. P., Herrero, A., et al. 1996, A&A, 305, 171
Quataert, E., Fernández, R., Kasen, D., Klion, H., & Paxton, B. 2016, MNRAS, 458, 1214
Quataert, E., & Shiode, J. 2012, MNRAS, 423, L92
Ramachandran, V., Hamann, W. R., Oskinova, L. M., et al. 2019, A&A, 625, A104
Ramírez-Agudelo, O. H., Sana, H., de Koter, A., et al. 2017, A&A, 600, A81
Reffert, S., Bergmann, C., Quirrenbach, A., Trifonov, T., & Künstler, A. 2015, A&A, 574, A116
Reggiani, M., Rainot, A., Sana, H., et al. 2022, A&A, 660, A122
Rest, A., Prieto, J. L., Walborn, N. R., et al. 2012, Nature, 482, 375
Rickard, M. J., Hainich, R., Hamann, W. R., et al. 2022, arXiv e-prints, arXiv:2207.09333
Rivinius, T., Carciofi, A. C., & Martayan, C. 2013, A&A Rev., 21, 69
Rosen, A. L., & Krumholz, M. R. 2020, AJ, 160, 78
Rubio-Díez, M. M., Sundqvist, J. O., Najarro, F., et al. 2022, A&A, 658, A61
Sander, A. A. C., Hamann, W. R., Todt, H., Hainich, R., & Shenar, T. 2017, A&A, 603, A86
Sander, A. A. C., Hamann, W. R., Todt, H., et al. 2019, A&A, 621, A92
Shaviv, N. J. 1998, ApJL, 494, L193
Shaviv, N. J. 2001, MNRAS, 326, 126
Shenar, T., Sana, H., Marchant, P., et al. 2021, A&A, 650, A147
Shenoy, D., Humphreys, R. M., Jones, T. J., et al. 2016, AJ, 151, 51
Shenoy, D. P., Jones, T. J., Humphreys, R. M., et al. 2013, AJ, 146, 90
Shi, Y., & Fuller, J. 2022, MNRAS, 513, 1115
Shiode, J. H., & Quataert, E. 2014, ApJ, 780, 96
Smith, L. J., Nota, A., Pasquali, A., et al. 1998, ApJ, 503, 278
Smith, L. J., Stroud, M. P., Esteban, C., & Vilchez, J. M. 1997, MNRAS, 290, 265
Smith, N. 2011, MNRAS, 415, 2020
Smith, N. 2014, ARA&A, 52, 487
Smith, N. 2017, Philos. Trans. Royal Soc. A, 375, 20160268
Smith, N., Andrews, J. E., Filippenko, A. V., et al. 2022, MNRAS, 515, 71
Smith, N., & Arnett, W. D. 2014, ApJ, 785, 82
Smith, N., Davidson, K., Gull, T. R., Ishibashi, K., & Hillier, D. J. 2003, ApJ, 586, 432
Smith, N., & Owocki, S. P. 2006, ApJL, 645, L45
Smith, N., Miller, A., Li, W., et al. 2010, AJ, 139, 1451
Soker, N. 2001, MNRAS, 325, 584
Soker, N. 2004, ApJ, 612, 1060
Soker, N. 2007, ApJ, 661, 482
Squicciarini, V., Gratton, R., Janson, M., et al. 2022, A&A, 664, 9
Stanway, E. R., Eldridge, J. J., & Becker, G. D. 2016, MNRAS, 456, 485
Steffen, W., Koning, N., Wenger, S., Morisset, C. & Magnor, M. 2010, IEEE Transactions on Visualization and Computer Graphics, Vol. 17, Issue 4, 454
Stothers, R. B. 2003, ApJ, 589, 960
Stothers, R. B., & Chin, C.-W. 1993, ApJL, 408, L85
Stothers, R. B., & Chin, C.-W. 1994, ApJL, 426, L43
Stothers, R. B., & Chin, C.-W. 1997, ApJ, 489, 319
Taniguchi, D., Yamazaki, K., & Uno, S. 2022, NatAs, 6, 930
Thirumalai, A., & Heyl, J. S. 2012, MNRAS, 422, 1272
Tiffany, C., Humphreys, R. M., Jones, T. J., & Davidson, K. 2010, AJ, 140, 339
Ulmer, A., & Fitzpatrick, E. L. 1998, ApJ, 504, 200
Ustamujic, S., Orlando, S., Miceli, M., et al. 2021, A&A, 654, A167

Vieira, R. G., Carciofi, A. C., Bjorkman, J. E., et al. 2017, MNRAS, 464, 3071
Vink, J. S. 2021, arXiv e-prints, arXiv:2109.08164
Vink, J. S., & de Koter, A. 2005, A&A, 442, 587
Vink, J. S., de Koter, A., & Lamers, H. J. G. L. M. 2001, A&A, 369, 574
Vlemmings, W. H. T., Khouri, T., Martí-Vidal, I., et al. 2017, A&A, 603, A92
Wallström, S. H. J., Muller, S., Lagadec, E., et al. 2015, A&A, 574, A139
Wallström, S. H. J., Lagadec, E., Muller, S., et al. 2017, A&A, 597, A99
Waters, L. B. F. M., Morris, P. W., Voors, R. H. M., & Lamers, H. J. G. L. M. 1997, in Astron. Soc. Pac. Conf. Ser., Vol. 120, Luminous Blue Variables: Massive Stars in Transition, ed. A. Nota & H. Lamers, 326
Weigelt, G., Hofmann, K. H., Schertl, D., et al. 2021, A&A, 652, A140
Weis, K. 2003, A&A, 408, 205
Weis, K. 2011, Proc. Int. Astron. Union, IAUS 272, Active OB stars: structure, evolution, mass loss, and critical limits, ed. C. Neiner, G. Waade, G. Meynet & G. Peters, 372
Wheelwright, H. E., de Wit, W. J., Weigelt, G., Oudmaijer, R. D., & Ilee, J. D. 2012, A&A, 543, A77
Wittkowski, M., Hauschildt, P. H., Arroyo-Torres, B., & Marcaide, J. M. 2012, A&A, 540, L12
Xu, D., Offner, S. S. R., Gutermuth, R., Kong, S., & Arce, H. G. 2022, ApJ, 926, 19
Zhao, X., & Fuller, J. 2020, MNRAS, 495, 249
Zickgraf, F. J., Wolf, B., Stahl, O., Leitherer, C., & Klare, G. 1985, A&A, 143, 421

Winds of Stars and Exoplanets
Proceedings IAU Symposium No. 370, 2023
A. A. Vidotto, L. Fossati & J. S. Vink, eds.
doi:10.1017/S1743921322004239

Observations of planetary winds and outflows

Leonardo A. Dos Santos

Space Telescope Science Institute, 3700 San Martin Drive, Baltimore, MD 21218, USA

Abstract. We have recently hit the milestone of 5,000 exoplanets discovered. In stark contrast with the Solar System, most of the exoplanets we know to date orbit extremely close to their host stars, causing them to lose copious amounts of gas through atmospheric escape at some stage in their lives. In some planets, this process can be so dramatic that they shrink in timescales of a few million to billions of years, imprinting features in the demographics of transiting exoplanets. Depending on the transit geometry, ionizing conditions, and atmospheric properties, a planetary outflow can be observed using transmission spectroscopy in the ultraviolet, optical or near-infrared. In this review, we will discuss the main techniques to observe evaporating exoplanets and their results. To date, we have evidence that at least 28 exoplanets are currently losing their atmospheres, and the literature has reported at least 42 non-detections.

Keywords. (stars:) planetary systems, planets and satellites: general, techniques: spectroscopic

1. Introduction

The discovery of 51 Peg b, a Jupiter-mass planet orbiting a Sun-like star with a period of only 4.3 days (Mayor & Queloz 1995), was initially received by the astronomical community with skepticism. But less than one year after this unexpected discovery, several other short-period gas giants were announced by competing teams (Schilling 1996), leading us to come to terms with these so-called "hot Jupiters" likely being a natural outcome of planet formation and evolution. One of the first questions that were posed during these early years of exoplanet science was whether hot Jupiters could survive the high mass loss rates driven by the extreme stellar irradiation at short periods (Guillot et al. 1996). The current consensus is that hot Jupiters are massive enough to retain their atmospheres for billions of years, but the same cannot be said about other hot exoplanets (e.g., Lecavelier Des Etangs 2007; Koskinen et al. 2007).

Hydrodynamic atmospheric escape was originally formulated by Watson et al. (1981) to explain the early evolution of the Earth and Venus. The idea behind this formulation stemmed from the insight of Gross (1972), who argued that, for planets with exospheres hotter than $\sim$10 000 K, a selective escape of gases (as in Öpik 1963) would be impossible. Instead, what follows is a bulk motion of gas in the upper atmosphere, or a so-called planetary outflow. More than two decades later, however, this process would be invoked to explain, at least partially, the observation of extended atmospheres in transiting exoplanets (e.g., Vidal-Madjar et al. 2003; Lecavelier des Etangs et al. 2010) and later some demographic features in the exoplanet population (e.g., Szabó & Kiss 2011; Owen & Wu 2013).

The main technique used to observe outflows in exoplanets is called transmission spectroscopy. This is the same method that yielded the first detection of sodium in an exoplanet (Charbonneau et al. 2002), and remains one of the most prolific techniques to study atmospheres in extrasolar worlds. When a planet transits, part of the host star's light is filtered through the thin layer of gas at the limbs of the planet, imprinting wavelength-dependent signatures in the in-transit spectrum. This dependency emerges mainly due to a combination of the density, velocity, altitude and chemical composition of the absorbing material.

Tipically, the ratio of the area covered by the lower-atmosphere and the disk of a star is in the order of 10^{-3} to 10^{-4} in the optical and near-infrared (Seager & Sasselov 2000). This level of precision requires strong spectroscopic features to be detectable (e.g., Wyttenbach et al. 2015; Sing et al. 2016). At higher altitudes, where the atmosphere is gravitationally unbound from the planet, the diffuse gas can extend to several planetary radii and produce deep in-transit absorption signatures (e.g., Ehrenreich et al. 2015). It is precisely at high altitudes that we can observe signatures of planetary outflows.

This review has the following structure: In Sect. 2, we will discuss the basic formulation that serves as the backbone of the transit spectroscopy technique; in Sect. 3, we shall go over the main results of searches for escape of hydrogen using the Lyman-α and Balmer-series lines; in Sect. 4, we discuss the observations of exospheric metals, the smoking-gun signal of hydrodynamic escape in exoplanets; in Sect. 5, we will discuss metastable He transmission spectroscopy, currently the most productive technique to observe atmospheric escape; finally, in Sect. 6, we draw some of the main conclusions stemming from these observations and propose some new perspectives for future research in this sub-field of exoplanet science.

2. The basics of transit spectroscopy

In this manuscript, we shall adopt that the transmission spectrum ϕ of an exoplanet in function of transit phase θ and wavelength λ is given by:

$$\phi(\theta, \lambda) = 1 - \frac{f_{\rm in}(\theta, \lambda)}{F_{\rm out}(\lambda)}. \tag{2.1}$$

where $F_{\rm out}$ is the out-of-transit spectrum of the host star and $f_{\rm in}$ is the observed in-transit spectrum†. It is also convenient to define the transmission spectrum in the rest frame of the planet by Doppler shifting the spectra according to:

$$\lambda_{\rm p}(\theta) = \lambda\left(\frac{c}{\Delta v(\theta)} + 1\right), \tag{2.2}$$

where Δv is the difference between radial velocity of the planet at a particular phase θ and the reference velocity of the observer, and $\lambda_{\rm p}$ will be the resulting wavelength in the rest frame of the planet. Finally, we can define the transmission spectrum Φ independent from the planetary phase by taking the mean of $\phi(\theta, \lambda_{\rm p})$ over the range of θ observed in transit:

$$\Phi(\lambda_{\rm p}) = \frac{1}{\Delta\theta}\int \phi(\theta, \lambda_{\rm p})d\theta. \tag{2.3}$$

At low spectral resolution we cannot resolve the variation of the in-transit absorption with respect to the planetary Doppler velocity, and the transmission spectrum can be simplified to:

$$\Phi(\lambda) = 1 - \frac{F_{\rm in}(\lambda)}{F_{\rm out}(\lambda)}. \tag{2.4}$$

In the formulation described above, we averaged the in-transit signature over the phase space and study the signature in function of wavelength. As we will see in Sections 3 and 4, light curves are also routinely used to study transit spectra and search for in-transit

† We use lower-case f to denote the dependence of the in-transit spectra to the orbital phase θ. For the out-of-transit flux, we adopt upper-case F to indicate that it does not depend on the planetary phase.

excess absorption that could indicate the presence of an atmosphere. In this method, we instead average signals over the wavelength space, and analyze its dependence in fuction of transit phase. As a recommendation for the reader, a more detailed treatise on transit spectroscopy can be found in Deming et al. (2021).

3. Escape of H: Lyman-α and Balmer-series spectroscopy

Classically, observations of atmospheric escape in exoplanets have been performed in ultraviolet (UV), which probes escape of hydrogen (H) and metallic species (see Sect. 4). The spectral feature of strongest interest is the Lyman-α (hereafter Lyα) line at 1215.67 Å, which traces atomic H. The *Hubble Space Telescope* (*HST*) is currently the only instrument capable of observing the Lyα, and it is possibly going to remain in this position until the launch of the next flagship NASA space telescope (National Academies of Sciences, Engineering, and Medicine 2021).

Since the interstellar medium (ISM) is rich in neutral H, the stellar Lyα line is partially or completely absorbed when observed from the Solar System. For stars with low radial velocities, the ISM absorption takes place near the core of the line; those with large radial velocities in relation to the Solar System and the ISM manage to dodge the absorption, and their Lyα cores are observable; see, e.g., the cases of Kepler-444 (Bourrier et al. 2017b) and Barnard's Star (France et al. 2020). Save a few exceptions, it is likely that the Lyα line is completely absorbed by the ISM for F, G and K-type stars beyond 60 pc; for M dwarfs, this limiting distance is much shorter.

Some other H features can be observed at optical wavelengths, such as the Balmer series (which include Hα, Hβ, Hγ), and provide another window to observe H escape in exoplanets. Observing in the optical has its advantages: there is no strict need to use a space telescope, ISM absorption is not a limiting issue, and it can be performed at high resolution. The disadvantages are that only highly-irradiated hot Jupiters display an in-transit absorption in the Balmer series, and no detection has so far been obtained for smaller or less irradiated planets (see Sect. 3.3).

3.1. *Hot Jupiters*

The first exoplanet to have a definitive detection of escaping H was the hot Jupiter HD 209458 b, as originally reported by Vidal-Madjar et al. (2003) and later confirmed by Ehrenreich et al. (2008). Using *HST* and the Space Telescope Imaging Spectrograph (STIS), Vidal-Madjar et al. (2003) detected a flux decrease of $15\% \pm 4\%$ in the blue wing of the Lyα line of the host star during the transit of the planet; since the transit depth at optical wavelengths is only $\sim 1.5\%$, the authors argued that the excess absorption seen in Lyα is due to a large cloud of H surrounding HD 209458 b, which in turn is fed by atmospheric escape.

One particular point of contention in the literature related to Lyα detections is regarding the Doppler velocities at which the signatures are measured; in the case of HD 209458 b, the in-transit planetary absorption takes place at velocities as high as -130 km s^{-1} in the stellar rest frame, indicating that the detected escaping material is accelerated away from the star. One-dimensional hydrodynamic escape models are unable to explain such high velocities (Murray-Clay et al. 2009), thus requiring other processes to explain them. The exact mechanism behind this effect has been the subject of an intense debate in the literature, and the most discussed contenders are radiation pressure and charge exchange in the interface between the stellar and planetary winds (e.g., Holmström et al. 2008; Lecavelier Des Etangs et al. 2008; Bourrier & Lecavelier des Etangs 2013; Vidotto & Bourrier 2017; Wang & Dai 2018; Debrecht et al. 2020).

Using *HST*, but this time with the 1-st order CCD/G430M setup at optical wavelengths, Ballester et al. (2007) reported on an excess absorption of $0.03\% \pm 0.006\%$ during the transit of HD 209458 b. The authors argue that this feature is caused by a large population of hot H atoms in the planet's upper atmosphere, which absorb the stellar light in the Balmer jump and continuum.

Another early discovery of evaporation was that of the extensively studied hot Jupiter HD 189733 b (Lecavelier des Etangs et al. 2010), for which the authors detect an in-transit Lyα absorption of $14.4\% \pm 3.6\%$. Similar to HD 209458 b, this absorption takes place at highly blueshifted Doppler velocities. The main point of discussion for this planet is that there is strong evidence that its escape signals and its high-energy environment are variable (Lecavelier des Etangs et al. 2012; Bourrier et al. 2013, 2020; Pillitteri et al. 2022); as we shall see in the next sections, this variability has not only been observed in Lyα, but other wavelengths as well.

The search for escape of atomic H is complicated by the fact that they rely predominantly on *HST*, which is oversubscribed. However, excited H has been detected in the archetypal hot Jupiters HD 209458 b and HD 189733 b (Jensen et al. 2012), and in the ultra-hot Jupiter KELT-9 b (Yan & Henning 2018; Cauley et al. 2019; Wyttenbach et al. 2020; Sánchez-López et al. 2022) using the Balmer series H lines. Similar to the Lyα observations, HD 189733 b also displays signals of variability in the Hα line (Cauley et al. 2017a). Other hot Jupiters with reported Hα detections are WASP-12 b (Jensen et al. 2018), KELT-20 b (Casasayas-Barris et al. 2018), WASP-52 b (Chen et al. 2020), WASP-33 b (Yan et al. 2021b), and WASP-121 b (Yan et al. 2021a).

One of the main differences between the Lyα and Hα detections is that the latter tend to display excess in-transit absorption in the order of 1%, which is shallower than the former. Another key difference is that ISM absorption is not a limitation for these observations, and we have access to the core of the absorption. In fact, observations at high spectral resolution show that the excess in-transit signals in the Balmer series do not show a net blueshift, and are thus confined to relatively low Doppler velocities when compared to Lyα. From the modeling perspective, these low-velocity signatures are advantageous because they do not require expensive three-dimensional simulations. This, in turn, means that we can use simplified formulations to extract mass loss rates for the observed exoplanet, such as the Parker-wind (Parker 1958) approximation, as seen in Wyttenbach et al. (2020) and Yan et al. (2021b).

3.2. *Neptunes and sub-Neptunes*

Although the first observations of atmospheric escape in exoplanets were obtained for hot Jupiters, Ehrenreich et al. (2011) predicted that evaporating Neptune-sized worlds could not only be observed as well, but would show excess in-transit absorption just as deep as their larger counterparts. What they did not predict is that this signal could, in fact, be even larger than that. Upon observing the warm Neptune Gl 436 b (also known as GJ 436 b) with *HST*/STIS, Ehrenreich et al. (2015) found that the Lyα blue wing of the host star is obscured by a factor of $56.3\% \pm 3.5\%$ when the planet transits (see Fig. 1; see also Kulow et al. 2014). Such a signal can only be explained by the presence of a large cloud of atomic H around the planet, fed by an atmospheric escape rate in the order of 10^9 g s^{-1} and accelerated away from the star. Further observations would later show that Lyα transit of Gl 436 b is not only deep, but also extremely asymmetric and long (Lavie et al. 2017), stable across several years and observable with *HST*/COS (Dos Santos et al. 2019). Since then, Gl 436 b has become the archetypal evaporating Neptune, and its observations have been extensively used to test modeling frameworks for atmospheric escape (e.g., Bourrier et al. 2015, 2016; Kislyakova et al. 2019;

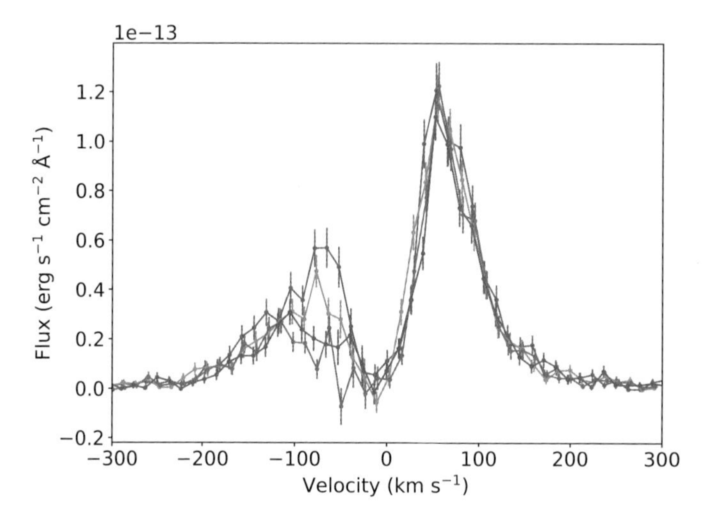

Figure 1. Lyman-α flux time-series during the transit of Gl 436 b, where blue and orange correspond to the spectra before the transit, green during the transit, and red after the transit (Ehrenreich et al. 2015). The deep absorption in the blue wing, between Doppler velocities [-120, -50] km s^{-1}, is explained by a large cloud of H around Gl 436 b fed by atmospheric escape.

Khodachenko et al. 2019; Villarreal D'Angelo et al. 2021; Attia et al. 2021; Carolan et al. 2021). Perhaps another warm Neptune that has become almost as iconic as Gl 436 b in the last few years is HAT-P-11 b, which displays signatures of atmospheric escape not only in Lyα (Ben-Jaffel et al. 2022), but also in ionized carbon and metastable helium (see Sections 4 and 5).

Similar escape signatures in other Neptunes have been observed in Lyα, and each of them stand out for a particular reason. GJ 3470 b was observed with *HST* in the Panchromatic Comparative Exoplanetology Treasury (PanCET) program with both the STIS and COS spectrographs (Bourrier et al. 2018b, 2021), yielding a signal of 35% $\pm$ 7% in the blue wing, which is also explained by a large H exosphere similar to Gl 436 b. A key difference with GJ 3470 b is that it displays an in-transit excess absorption in the Lyα red wing as well, indicating the presence of material inflowing into the star. To this day, the exact physical mechanism behind this inflow remains a mystery, but Bourrier et al. (2018b) tentatively suggests that it could be caused by an elongated layer of dense atomic H extending beyond the Roche lobe. Using COS observations, Dos Santos et al. (2019) detected a similar, but episodic absorption in the red wing of Gl 436 during one of the observed transits of Gl 436 b in the PanCET program.

In this context, the tentative detection of exopsheric H in K2-18 b by Dos Santos et al. (2020b) stands out because this mini-Neptune is not, by any means, a hot exoplanet. Since it orbits an M dwarf with a period of approximately 30 days, K2-18 b is in fact a temperate world. The authors conclude that, due to how faint the host star is in the far-UV, more observations are necessary to confirm the detection. That notwithstanding, a primordial atmosphere of only a few percent mixing ratio of H can lead to temperatures in the upper atmosphere as high as 10 000 K, even in a temperate planet (Gross 1972). At these conditions, the kinetic energy of particles in the upper atmosphere exceed the gravitational potential of the planet, leading to a rapid atmospheric expansion and consequent escape. Furthermore, a recent study provided further support that planets at amenable levels of irradiation can sustain a large cloud of atomic H detectable during transits (Owen et al. 2021), but this hypothesis still requires further observations to be put under test. Another planet with a tentative Lyα detection is 55 Cnc b (Ehrenreich et al. 2012).

Table 1. List of non-detections of H escape reported in the literature.

Planet name	Obs. method	Reference
HD 147506 b	Hα	Jensen et al. (2012)
HD 149026 b	Hα	Jensen et al. (2012)
HAT-P-32 b	Hα	Mallonn & Strassmeier (2016)
KELT-3 b	Hα	Cauley et al. (2017b)
Gl 436 b	Hα	Cauley et al. (2017b)
TRAPPIST-1 system	Lyα	Bourrier et al. (2017a)
Kepler-444 system	Lyα	Bourrier et al. (2017b)
HD 97658 b	Lyα	Bourrier et al. (2017c)
55 Cnc e	Lyα	Bourrier et al. (2018a)
GJ 1132 b	Lyα	Waalkes et al. (2019)
π Men c	Lyα	García Muñoz et al. (2020)
WASP-29 b	Lyα	Dos Santos et al. (2021)
K2-25 b	Lyα	Rockcliffe et al. (2021)
GJ 9827 b & d	Lyα and Hα	Carleo et al. (2021)
HD 63433 b	Lyα	Zhang et al. (2022d)

The case of the mini-Neptune HD 63433 c stands out for being the youngest transiting exoplanet with atmospheric escape detected in Lyα (Zhang et al. 2022d). It orbits a G5-type star with an orbital period of 20.5 d. Interestingly, the inner planet in the system, with a period of 7 d, does not display a Lyα signal, again providing support to the hypothesis that exospheric H in highly-irradiated Neptunes ionizes too quickly to be detectable in our observations. Similarly young exoplanets with signatures of evaporation are expected to be important to disentangle the roles of different escape mechanisms, such as photoevaporation and core-powered mass loss (e.g., Gupta & Schlichting 2020; King & Wheatley 2021). However, their observations are challenging due stellar activity modulation (Rackham et al. 2019, 2022), and even when detections have been observed, the interpretation can be complicated since their masses are usually not known (see, however, the case of K2-100 b in Barragán et al. 2019).

3.3. *Non-detections*

There are several reasons why atmospheric escape of H can remain undetected, even for planets that are expected to be evaporating. In Lyα, these reasons boil down to: (i) ISM absorption, which absorbs the flux at Doppler velocities where the absorption was supposed to take place; and ii) The host star luminosity yields a low signal-to-noise ratio, which is the usual suspect for M dwarfs. For Hα, the most likely limitation is the amount of ionized H in the atmosphere, which may not be high enough to produce a detectable signal.

Non-detections of atmospheric escape are severely under-reported, even though they can be just as informative as secure detections. In Table 1 we compile a list of non-detections of H escape that have been reported in refereed publications.

4. Hydrodynamic escape of metals observed in the UV

Other signatures of escape can be observed in UV wavelengths, among them the metal lines of carbon (C), nitrogen (N), oxygen (O), silicon (Si), magnesium (Mg), sulfur (S) and iron (Fe). Since these species are much heavier than H and He, they can only be lifted to the upper atmosphere when the escape is not selective; in other words, metals can only escape when the outflow is in a hydrodynamic regime. Similarly to Lyα, these metal lines are present in emission in stars of types between F and M. The advantage of observing these lines is that they do not have ISM absorption, or it is not as dramatic as in Lyα. The disadvantage is that metal lines are intrinsically weaker than Lyα, which means the detector will register lower count rates, yielding lower signal-to-noise ratios.

For this reason, most of the detections of escaping metals have been obtained for hot Jupiters, where the signatures are stronger.

4.1. *Hot Jupiters*

Vidal-Madjar et al. (2004) first reported on the detection of O I and C II in the upper atmosphere of HD 209458 b using *HST*/STIS. According to the authors, the high velocity disperson and depth of the in-transit absorption suggests that the escaping metals are outflowing at supersonic velocities above the Roche lobe, an effect also known as geometric blow-off (Lecavelier des Etangs et al. 2004). In this Roche-lobe filling regime, the mass loss rates of hot exoplanets can be enhanced significantly; in the case of HD 209458 b, Erkaev et al. (2007) found this factor to be in the order of 50%. Using observations with the COS spectrograph, Linsky et al. (2010) reported on detections of C II and Si III in HD 209458 b, which is in conflict with the non-detection of Si III in Vidal-Madjar et al. (2004); however, these COS detections were later contested (Ballester & Ben-Jaffel 2015). Observations in the near-UV have yielded additional evidence for hydrodynamic escape in this planet associated with the presence of Mg I (Vidal-Madjar et al. 2013). According to the authors, the Mg feature probes the thermosphere and the exobase, precisely where the escape takes place; however, they also detect a tentative signal of a Mg comet-like tail in the exosphere of the planet. Finally, Schlawin et al. (2010) discussed a tentative detection of Si IV in the limb-brightened transit of HD 209458 b.

The archetypal hot Jupiter HD 189733 b was also among the early discoveries of escaping metals (Ben-Jaffel & Ballester 2013). Despite a significant stellar variability, the transit observations obtained with COS indicated the presence of O I and a possible early ingress associated with C II. HD 189733 b has since been observed again in the PanCET program, and the analysis of that dataset is currently under way.

Another category of planets that have become a testbed for atmospheric escape is that of the ultrahot Jupiters (UHJ), namely those that orbit closely to stars of type F or earlier. Because they orbit more massive stars, their escaping signatures are frequently detected in a regime of geometric blow-off. WASP-12 b was the first UHJ to have a detection of escaping metals. Using the COS spectrograph, Fossati et al. (2010) reported on excess in-transit absorption signatures in the core of the Mg II resonant lines at moderate significance, and on significantly enhanced transit depths measure in wide-band NUV light curves. The wide-band excess absorption are attributed to a collection of different absorbing metals in the exosphere of WASP-12 b.

The latest UHJ in which escaping metals have been detected is WASP-121 b (Sing et al. 2019). Based on STIS observations, the authors find evidence of Mg II and Fe II ions filling the Roche-lobe of the planet (see Fig. 2), and deeper broadband NUV light curves compared to optical wavelengths.

4.2. *Neptunes and sub-Neptunes*

To date, only two sub-Jovian worlds have been shown to display signatures of escaping metals, both of them obtained with *HST*/COS. The warm Neptune HAT-P-11 b has an excess in-transit absorption of $15\% \pm 4\%$ in the blue wing of the ground-state C II line at 133.45 nm, as well as a post-transit tail absorption of $12.5\% \pm 4\%$ (Ben-Jaffel et al. 2022). The authors argue that this signal is consistent with the planet's atmosphere having a sub-solar metallicity and an extended magnetotail.

The second, and perhaps most intriguing detection is that of the super-Earth π Men c (García Muñoz et al. 2021). Benefitting from the brightness of the host star, the authors reported on an in-transit absorption of $3.9\% \pm 1.1\%$ in the blue wing of the excited-state C II line at 1335 Å. Based on this single-transit observation, the authors concluded

Table 2. List of non-detections of escaping metals reported in the literature.

Planet name	Instrument	Reference
WASP-13 b	*HST*/COS	Fossati et al. (2015)
55 Cnc e	Ground-based spectrographs	Ridden-Harper et al. (2016)
Gl 436 b	*HST*/STIS & COS	Loyd et al. (2017); Dos Santos et al. (2019)
WASP-18 b	*HST*/COS	Fossati et al. (2018)
WASP-29 b	*HST*/COS	Dos Santos et al. (2021)
HD 189733 b	XMM-Newton optical monitor	King et al. (2021)
GJ 3470 b	*HST*/COS	Bourrier et al. (2021)

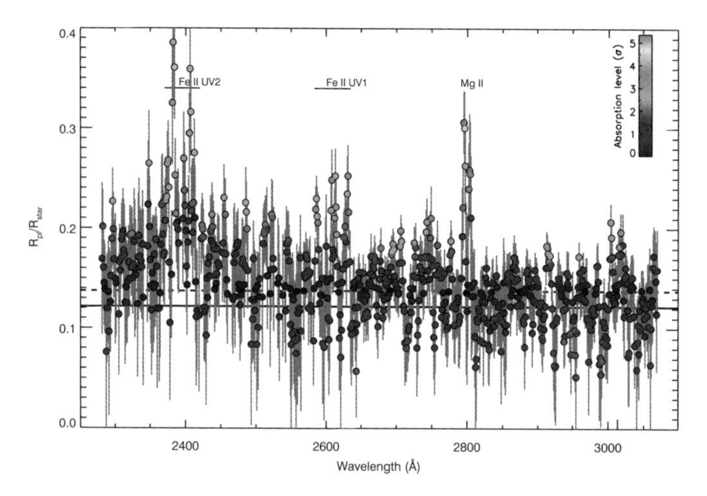

Figure 2. Transmission spectrum of WASP-121 b in the near-UV with detections of Mg II and Fe II (Sing et al. 2019). Reproduced with the permission of AAS Journals.

that π Men c possesses a thick atmosphere with more than 50% heavy volatiles in mass fraction, and that the escaping C fills the Roche lobe of the planet.

Although STIS observations hinted at a tentative detection of Si III in Gl 436 b (Lavie et al. 2017), an ensemble of COS data was later used to show that the observed signal was not present (Loyd et al. 2017), and that the STIS data was likely contaminated by stellar activity modulation (Dos Santos et al. 2019).

4.3. *Non-detections*

Similarly to Lyα, observations of metals in the UV suffer from the low signal-to-noise ratios, and this is probably the main limitation for transmission spectroscopy in these wavelengths. As seen in the case of π Men c (García Muñoz et al. 2021), some of the in-transit signals we are looking for are in the order of only a few percent, which requires high levels of contrast in order to be detected. Additionally, as shown by Dos Santos et al. (2019), stellar activity can also pose as a false positive. We list the non-detections of escaping metals reported in the literature in Table 2.

5. Metastable He spectroscopy in the near-infrared

Classically, the near-infrared helium (He) triplet located at 1.083 μm has been used to probe the chromosphere and transition region of cool stars (e.g., Andretta & Jones 1997). The presence of He in the upper atmospheres of exoplanets was originally predicted by the theoretical models of Seager & Sasselov (2000), but early observations of HD 209458 b were unable to detect a signal (Moutou et al. 2003). Several years later, Oklopčić & Hirata

(2018) predicted that escaping He could produce signals as deep as 6% in the core of the triplet of HD 209458 b, which could be detectable at high spectral resolution.

Neutral He atoms can exist in two states: singlet (1^1S, electrons with anti-parallel spin) or triplet (2^3S, electrons with parallel spin). Since the radiative decay of triplet He into singlet state is relatively long, the former is also known as a metastable state. The formation of this line depends on the balance of rates that either populate or depopulate the triplet state: recombination, collisional excitation and de-excitation, charge exchange, and photoionization. According to Oklopčić (2019), planets orbiting late-type and active stars tend to display prominent in-transit He absorption due to their favorably high levels of extreme UV flux. Poppenhaeger (2022) further proposed that metastable He absorption also has a dependence on the iron abundance in the corona of stellar hosts, since most of the extreme-UV flux comes from coronal iron emission lines in cool stars. As we shall see shortly, this trend has mostly been held in our observations.

The most important advantage of observing metastable He is that this technique does not necessarilly require a space telescope, and can be observed from the ground. In fact, ground-based facilities can perform experiments at much higher spectral resolutions that those achieved from space. In this regime, the Doppler anomaly of the planet can be resolved during the transit (e.g., Wyttenbach et al. 2015), which helps in discerning if the signal is of planetary nature or stellar. As opposed to Lyα observations, the in-transit absorption is seen in the core and wings of the He triplet, which means we are not only probing the accelerated particles well above the exobase, but also the outflowing gas near the thermosphere. This allows us to use simpler, one-dimensional models to interpret the observations (e.g., Oklopčić & Hirata 2018; Lampón et al. 2020; Dos Santos et al. 2022; Linssen et al. 2022) and extract more precise mass loss rates than those determined from Lyα data (e.g., Bourrier & Lecavelier des Etangs 2013). The disadvantage of ground-based He spectroscopy is that, due to spectral normalization, information about the planetary continuum absorption is lost, but since the signals are relatively deep, the impact of this limitation is not of great importance. Other disadvantages include telluric contamination and lower sensitivities than space telescopes.

5.1. *Hot Jupiters*

For a change, the first discovery of metastable He in exoplanet was not in HD 209458 b, but rather the hot Jupiter WASP-107 b (Spake et al. 2018). In this study, the authors observed a single transit with *HST* and the Wide-Field Camera 3 (WFC3) instrument and measured a transit depth of $0.049\% \pm 0.011\%$ in a low-resolution bandpass of 98 Å. Later, this feature would be observed again from the ground and at high spectral resolution with the CARMENES spectrograph installed on the 3.5 m telescope at the Calar Alto Observatory (Allart et al. 2019) and with the Keck II/NIRSPEC spectrograph (Kirk et al. 2020). Recently, Spake et al. (2021) reported on the observation of the He tail that trails WASP-107 b, also detected with the NIRSPEC instrument.

Several other hot Jupiters have since been observed to be evaporating and exhibit in-transit He absorption. The CARMENES spectrograph has been particularly productive, yielding detections for HD 189733 b (with variability; Salz et al. 2018), WASP-69 b (Nortmann et al. 2018), HD 209458 b (Alonso-Floriano et al. 2019), HAT-P-32 b (Czesla et al. 2022) and a tentative detection for the UHJ WASP-76 b (Casasayas-Barris et al. 2021). Another productive instrument for He spectroscopy in hot Jupiters has been the NIRSPEC spectrograph, which was responsible for detections in HD 198733 b (Zhang et al. 2022a), WASP-52 b and a tentative signal for WASP-177 b (Kirk et al. 2022). Using the GIANO spectrograph installed on the Telescopio Nazionale Galileo (TNG), Guilluy et al. (2020) reproduced the He signature of HD 189733 b.

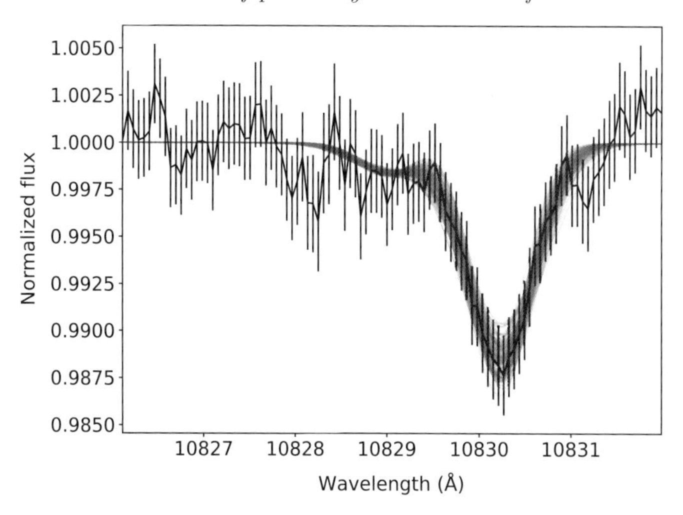

Figure 3. Transmission spectrum of HAT-P-11 b near the metastable He triplet observed with the CARMENES spectrograph (black symbols; Allart et al. 2018) and a family of transmission spectra simulations that were fit to the data based on an isothermal Parker-wind model (red curves; Dos Santos et al. 2022).

Following up on the increasing interest in atmospheric escape in exoplanets, Vissapragada et al. (2020) presented the first results detections of He using a new method: ultra-narrowband photometry of the He triplet with the Wide-field Infrared Camera (WIRC) installed on the 200-inch Hale Telescope at Palomar Observatory. They use a custom-made filter centered on 1083.3 nm in vacuum, with an FWHM of 0.635 nm, and a maximum transmission of 95.6%. Naturally, since measurements are performed in photometry, the in-transit absorption is not spectrally resolved, so the results do not encode information about velocities. However, the instrument has demonstrated a significant productivity, yielding detections for the hot Jupiters WASP-69 b (Vissapragada et al. 2020), HAT-P-18 b (Paragas et al. 2021), and tentative detections for WASP-52 b and NGTS-5 b (Vissapragada et al. 2022). This tentative observation for WASP-52 b, with an in-transit depth of $0.29\% \pm 0.13\%$ in the filter's bandpass, is in slight tension with the firm detection reported in Kirk et al. (2022), for which an in-transit depth of $0.66\% \pm 0.14\%$ was measured.

5.2. *Neptunes and sub-Neptunes*

The first reports of He observations in transiting hot Jupiters were concomitant with the discoveries in sub-Jovian worlds. Similar to Lyα results we listed in Sect. 3, Neptunes can also display deep in-transit signals that sometimes rival those of their larger counterparts. Using an *HST*/WFC3 archival dataset, Mansfield et al. (2018) demonstrated that the warm Neptune HAT-P-11 b has a transit depth of $\sim 0.355\%$ in a 49 Å-wide channel centered in the He triplet. This study was simultaneous to that of Allart et al. (2018), who reported a detection of He in HAT-P-11 b obtained with the CARMENES spectrograph. At high spectral resolution, this signature is resolved with an average depth of $1.08\% \pm 0.05\%$ (see Fig. 3).

Along with HAT-P-11 b, another warm Neptune to have both Lyα and He detections is GJ 3470 b (Ninan et al. 2020), the latter obtained with the Habitable Zone Planet Finder (HPF) spectrograph installed on the Hobby-Eberly Telescope (HET). This signal was also measured with CARMENES (Palle et al. 2020), and a large mass loss rate of $\sim 10^{11}$ g s^{-1} was inferred. Based on this observation, Lampón et al. (2021) concludes that

Table 3. List of non-detections of metastable He reported in the literature.

Planet name	Instrument	Reference
HD 209458 b†	VLT/ISAAC	Moutou et al. (2003)
Gl 436 b	CAO 3.5m/CARMENES	Nortmann et al. (2018)
KELT-9 b	CAO 3.5m/CARMENES	Nortmann et al. (2018)
WASP-12 b	*HST*/WFC3	Kreidberg & Oklopčić (2018)
WASP-52 b†	Palomar/WIRC	Vissapragada et al. (2020)
K2-100 b	Subaru/IRD	Gaidos et al. (2020)
WASP-127 b	Gemini/Phoenix	Dos Santos et al. (2020a)
AU Mic b	Subaru/IRD & Keck II/NIRSPEC	Hirano et al. (2020)
GJ 1214 b†	Keck II/NIRSPEC	Kasper et al. (2020); Spake et al. (2022)
HD 97658 b	Keck II/NIRSPEC	Kasper et al. (2020)
55 Cnc e	Keck II/NIRSPEC	Zhang et al. (2021)
TRAPPIST-1 system	Subaru/IRD & HET/HPF	Krishnamurthy et al. (2021)
K2-136 c	Subaru/IRD	Gaidos et al. (2021)
V1298 Tau b & c	Palomar/WIRC	Vissapragada et al. (2021)
WASP-80 b	TNG/GIANO & Palomar/WIRC	Fossati et al. (2022), Vissapragada et al. (2022)
GJ 9827 d	Keck II/NIRSPEC	Kasper et al. (2020)
τ Boo b (in emission)	CAO 3.5m/CARMENES	Zhang et al. (2020)
GJ 9827 b & d	CAO 3.5m/CARMENES	Carleo et al. (2021)
HD 63433 system	Keck II/NIRSPEC	Zhang et al. (2022d)
WASP-177 b†	Palomar/WIRC	Vissapragada et al. (2022)

Notes: The † symbol denotes planets with alternate results that yielded a detection.

GJ 3470 b is in a photon-limited escape regime, where the mass loss rate is limited by the incident flux of ionizing photons (Owen & Alvarez 2016).

Using the NIRSPEC spectrograph, Kasper et al. (2020) concluded that the sub-Neptune GJ 1214 b does not have a detectable He signature. However, the observation of one transit with the CARMENES spectrograph reported by Orell-Miquel et al. (2022) yielded a tentative detection. According to Orell-Miquel et al., stellar activity alone cannot have caused a false-positive, and argue that telluric contamination is the probable culprit of the non-detection observed by Kasper et al.. This argument was further contested by Spake et al. (2022), who observed an additional NIRSPEC transit in an epoch of minimal telluric contamination and still obtained a non-detection.

The Palomar/WIRC narrowband photometry has proven to be precise enough to detect He outflows in Neptunes as well, with a firm detection for HAT-P-26 b (Vissapragada et al. 2022) and a tentative detection for the young sub-Neptune V1298 Tau d (Vissapragada et al. 2021). Young sub-Neptunes are target of strong importance for atmospheric escape observations because we think it is in their youth that most of photoevaporation takes place (e.g., Owen & Wu 2013). Using the NIRSPEC spectrograph, Zhang et al. (2022c,b) reported on the first discoveries of atmospheric escape in the young mini-Neptunes HD 73583 b, TOI-1430 b, TOI-2076 b and TOI-1683 b.

5.3. *Non-detections*

Many more planets than those listed in this review have been observed in He spectroscopy (priv. comm.), and the data analyses are currently under way. We list all the reported He non-detections in Table 3, where we also include those that have been positively detected using different instruments or analyses (these cases are marked with the symbol †).

6. Conclusions and future perspectives

Atmospheric escape has been studied in the Solar System since the beginning of the 20th Century. However, observations of upper atmospheres in hot exoplanets in the last

Table 4. List of planets with detections of atmospheric escape.

Planet name	Signature(s)
HD 209458 b	Lyα, Balmer-series, metals, He
HD 189733 b	Lyα, Balmer-series, metals, He
KELT-9 b	Balmer-series
WASP-12 b	Balmer-series, metals
KELT-20 b	Balmer-series
WASP-52 b	Balmer-series, He
WASP-33 b	Balmer-series
WASP-121 b	Balmer-series, metals
Gl 436 b	Lyα
GJ 3470 b	Lyα, He
K2-18 b	Lyα (tentative)
HD 63433 c	Lyα
HAT-P-11 b	Lyα, C II, He
π Men c	C II
WASP-107 b	He
WASP-69 b	He
HAT-P-32 b	He
WASP-76 b	He (tentative)
WASP-177 b	He (tentative)
GJ 1214 b	He (tentative)
HAT-P-18 b	He
NGTS-5 b	He (tentative)
HAT-P-26 b	He
V1298 Tau d	He (tentative)
HD 73583 b	He
TOI-1430 b	He
TOI-2076 b	He
TOI-1683 b	He

two decades have advanced our understanding about the physics of evaporation by leaps and bounds. To date, we have observed escape in 28 exoplanets, including tentative detections (see a complete list in Table 4). These worlds have sizes varying from Jupiter-size to mini-Neptunes, and irradiation levels ranging from the most extremely-irradiated planet known (KELT-9 b) to Earth-like bolometric fluxes (K2-18 b).

Our observational efforts have shown that, so far, metastable He spectroscopy is the most productive avenue to observe escape in hot exoplanets with a H-dominated atmosphere orbiting active stars. Lyα observations, on the other hand, seem to yield detections for planets in relatively milder irradiation conditions. According to Owen et al. (2021), the reason for that is due to lack of observable flux in the core of the Lyα line, which means that we have access only to signatures that occur at high Doppler velocities. In order for H atoms to achieve these high velocities, they need to stay neutral for a long time and produce a detectable exospheric tail. For the cases where H ionizes too quickly in the exosphere, it is thus recommended to observe in the Balmer series lines. Some hot Jupiters, like HD 209458 b and HD 189733 b, have an optimal set of parameters that allows the detection of Lyα, He, Hα, and metals. Some Neptunes, like GJ 3470 b and HAT-P-11 b, also possess an optimal set of parameters that enables the observation of escape in more than one spectral channel. These cases seem to be, however, rare.

Despite these observational efforts, many questions related to the atmospheric evolution of exoplanets remain open. For instance, what are the mechanisms that carve the hot Neptune desert (e.g., Davis & Wheatley 2009; Szabó & Kiss 2011; Mazeh et al. 2016)? Based on a survey of escaping He in Saturn-sized ho gas giants with Palomar/WIRC, Vissapragada et al. (2022) concluded that the upper edge of the Neptune desert is stable against evaporation, with measured escape rates that remove less than 10% of these planet's masses. This suggests that other, additional mechanisms are necessary to carve the desert, such as a history of migration (e.g., Owen & Lai 2018; Attia et al. 2021).

More observations and modeling are required to test these hypotheses. Another persistent open question in this field is whether close-in gas giant exoplanets have hydrodynamically unstable thermospheres (Salz et al. 2016), which was originally proposed by Watson et al. (1981) and confirmed for only a handful of exoplanets to date. More observations of exospheric metals will help elucidate this puzzle, since they trace hydrodynamic escape directly.

For the future, as we mentioned in Sect. 5, observing escape in young sub-Neptunes will also be important because it may give us clues about the respective roles of photoevaporation (driven by X-rays and extreme-UV irradiation; e.g., Lammer et al. 2003; Volkov et al. 2011; Tripathi et al. 2015; Erkaev et al. 2016) and core-powered mass loss (e.g., Ginzburg et al. 2018; Gupta & Schlichting 2019). The main challenge in this endeavor is that young stars are active, and the activity poses a problem to measure planetary masses through the radial velocity method and, in addition, can produce false-positive detections of escape (e.g., Dos Santos et al. 2019).

With a sample of 28 exoplanets with signals of atmospheric escape, we have by now gathered a sample with which we can begin interpreting at a comparative level. In order to answer some of the open questions described above, we will benefit from carrying out a uniform analysis of this sample with a common theoretical framework (see, e.g., Sing et al. 2016). To cite an example, this approach will enable us to find correlations between measured properties of evaporating exoplanets, such as their bulk density, incoming high-energy flux, and mass-loss rates (as predicted by the energy-limited formulation). Studies that have already begun performing this comparative exoplanetology approach for evaporating exoplanets are Lampón et al. (2021) and Vissapragada et al. (2022).

Finally, with the successful launch and commissioning of *JWST*, we will have yet another instrument capable of observing the metastable He line, and with space-based precision. Although its capabilities for He transmission spectroscopy remain to be tested, it has three instrument configurations that can measure spectra at 1.083 μm: NIRISS/SOSS (2nd order only), NIRSpec/G140M and NIRSpec/G140H. The downside of *JWST* is that its lower resolution may not be able to spectrally resolve the in-transit absorption, but a more precise instrument could enable us to observe fainter signatures than those accessible from the ground.

References

Allart, R., Bourrier, V., Lovis, C., et al. 2019, A&A, 623, A58
Allart, R., Bourrier, V., Lovis, C., et al. 2018, Science, 362, 1384
Alonso-Floriano, F. J., Snellen, I. A. G., Czesla, S., et al. 2019, A&A, 629, A110
Andretta, V. & Jones, H. P. 1997, ApJ, 489, 375
Attia, O., Bourrier, V., Eggenberger, P., et al. 2021, A&A, 647, A40
Ballester, G. E. & Ben-Jaffel, L. 2015, ApJ, 804, 116
Ballester, G. E., Sing, D. K., & Herbert, F. 2007, Nature, 445, 511
Barragán, O., Aigrain, S., Kubyshkina, D., et al. 2019, MNRAS, 490, 698
Ben-Jaffel, L. & Ballester, G. E. 2013, A&A, 553, A52
Ben-Jaffel, L., Ballester, G. E., García Muñoz, A., et al. 2022, Nature Astronomy, 6, 141
Bourrier, V., de Wit, J., Bolmont, E., et al. 2017a, AJ, 154, 121
Bourrier, V., Dos Santos, L. A., Sanz-Forcada, J., et al. 2021, A&A, 650, A73
Bourrier, V., Ehrenreich, D., Allart, R., et al. 2017b, A&A, 602, A106
Bourrier, V., Ehrenreich, D., King, G., et al. 2017c, A&A, 597, A26
Bourrier, V., Ehrenreich, D., & Lecavelier des Etangs, A. 2015, A&A, 582, A65
Bourrier, V., Ehrenreich, D., Lecavelier des Etangs, A., et al. 2018a, A&A, 615, A117
Bourrier, V. & Lecavelier des Etangs, A. 2013, A&A, 557, A124
Bourrier, V., Lecavelier des Etangs, A., Dupuy, H., et al. 2013, A&A, 551, A63
Bourrier, V., Lecavelier des Etangs, A., Ehrenreich, D., et al. 2018b, A&A, 620, A147

Bourrier, V., Lecavelier des Etangs, A., Ehrenreich, D., Tanaka, Y. A., & Vidotto, A. A. 2016, A&A, 591, A121
Bourrier, V., Wheatley, P. J., Lecavelier des Etangs, A., et al. 2020, MNRAS, 493, 559
Carleo, I., Youngblood, A., Redfield, S., et al. 2021, AJ, 161, 136
Carolan, S., Vidotto, A. A., Villarreal D'Angelo, C., & Hazra, G. 2021, MNRAS, 500, 3382
Casasayas-Barris, N., Orell-Miquel, J., Stangret, M., et al. 2021, A&A, 654, A163
Casasayas-Barris, N., Pallé, E., Yan, F., et al. 2018, A&A, 616, A151
Cauley, P. W., Redfield, S., & Jensen, A. G. 2017a, AJ, 153, 217
Cauley, P. W., Redfield, S., & Jensen, A. G. 2017b, AJ, 153, 81
Cauley, P. W., Shkolnik, E. L., Ilyin, I., et al. 2019, AJ, 157, 69
Charbonneau, D., Brown, T. M., Noyes, R. W., & Gilliland, R. L. 2002, ApJ, 568, 377
Chen, G., Casasayas-Barris, N., Pallé, E., et al. 2020, A&A, 635, A171
Czesla, S., Lampón, M., Sanz-Forcada, J., et al. 2022, A&A, 657, A6
Davis, T. A. & Wheatley, P. J. 2009, MNRAS, 396, 1012
Debrecht, A., Carroll-Nellenback, J., Frank, A., et al. 2020, MNRAS, 493, 1292
Deming, D., Stevenson, K. B., & Ehrenreich, D. 2021, in ExoFrontiers; Big Questions in Exoplanetary Science, ed. N. Madhusudhan, 7–1
Dos Santos, L. A., Bourrier, V., Ehrenreich, D., et al. 2021, A&A, 649, A40
Dos Santos, L. A., Ehrenreich, D., Bourrier, V., et al. 2020a, A&A, 640, A29
Dos Santos, L. A., Ehrenreich, D., Bourrier, V., et al. 2020b, A&A, 634, L4
Dos Santos, L. A., Ehrenreich, D., Bourrier, V., et al. 2019, A&A, 629, A47
Dos Santos, L. A., Vidotto, A. A., Vissapragada, S., et al. 2022, A&A, 659, A62
Ehrenreich, D., Bourrier, V., Bonfils, X., et al. 2012, A&A, 547, A18
Ehrenreich, D., Bourrier, V., Wheatley, P. J., et al. 2015, Nature, 522, 459
Ehrenreich, D., Lecavelier Des Etangs, A., & Delfosse, X. 2011, A&A, 529, A80
Ehrenreich, D., Lecavelier Des Etangs, A., Hébrard, G., et al. 2008, A&A, 483, 933
Erkaev, N. V., Kulikov, Y. N., Lammer, H., et al. 2007, A&A, 472, 329
Erkaev, N. V., Lammer, H., Odert, P., et al. 2016, MNRAS, 460, 1300
Fossati, L., France, K., Koskinen, T., et al. 2015, ApJ, 815, 118
Fossati, L., Guilluy, G., Shaikhislamov, I. F., et al. 2022, A&A, 658, A136
Fossati, L., Haswell, C. A., Froning, C. S., et al. 2010, ApJ Letters, 714, L222
Fossati, L., Koskinen, T., France, K., et al. 2018, AJ, 155, 113
France, K., Duvvuri, G., Egan, H., et al. 2020, AJ, 160, 237
Gaidos, E., Hirano, T., Mann, A. W., et al. 2020, MNRAS, 495, 650
Gaidos, E., Hirano, T., Omiya, M., et al. 2021, Research Notes of the American Astronomical Society, 5, 238
García Muñoz, A., Fossati, L., Youngblood, A., et al. 2021, ApJ Letters, 907, L36
García Muñoz, A., Youngblood, A., Fossati, L., et al. 2020, ApJ Letters, 888, L21
Ginzburg, S., Schlichting, H. E., & Sari, R. 2018, MNRAS, 476, 759
Gross, S. H. 1972, Journal of Atmospheric Sciences, 29, 214
Guillot, T., Burrows, A., Hubbard, W. B., Lunine, J. I., & Saumon, D. 1996, ApJ Letters, 459, L35
Guilluy, G., Andretta, V., Borsa, F., et al. 2020, A&A, 639, A49
Gupta, A. & Schlichting, H. E. 2019, MNRAS, 487, 24
Gupta, A. & Schlichting, H. E. 2020, MNRAS, 493, 792
Hirano, T., Krishnamurthy, V., Gaidos, E., et al. 2020, ApJ Letters, 899, L13
Holmström, M., Ekenbäck, A., Selsis, F., et al. 2008, Nature, 451, 970
Jensen, A. G., Cauley, P. W., Redfield, S., Cochran, W. D., & Endl, M. 2018, AJ, 156, 154
Jensen, A. G., Redfield, S., Endl, M., et al. 2012, ApJ, 751, 86
Kasper, D., Bean, J. L., Oklopčić, A., et al. 2020, AJ, 160, 258
Khodachenko, M. L., Shaikhislamov, I. F., Lammer, H., et al. 2019, ApJ, 885, 67
King, G. W., Corrales, L., Wheatley, P. J., et al. 2021, MNRAS, 506, 2453
King, G. W. & Wheatley, P. J. 2021, MNRAS, 501, L28
Kirk, J., Alam, M. K., López-Morales, M., & Zeng, L. 2020, AJ, 159, 115

Kirk, J., Dos Santos, L. A., López-Morales, M., et al. 2022, AJ, 164, 24
Kislyakova, K. G., Holmström, M., Odert, P., et al. 2019, A&A, 623, A131
Koskinen, T. T., Aylward, A. D., & Miller, S. 2007, Nature, 450, 845
Kreidberg, L. & Oklopčić, A. 2018, Research Notes of the American Astronomical Society, 2, 44
Krishnamurthy, V., Hirano, T., Stefánsson, G., et al. 2021, AJ, 162, 82
Kulow, J. R., France, K., Linsky, J., & Loyd, R. O. P. 2014, ApJ, 786, 132
Lammer, H., Selsis, F., Ribas, I., et al. 2003, ApJ Letters, 598, L121
Lampón, M., López-Puertas, M., Czesla, S., et al. 2021, A&A, 648, L7
Lampón, M., López-Puertas, M., Lara, L. M., et al. 2020, A&A, 636, A13
Lavie, B., Ehrenreich, D., Bourrier, V., et al. 2017, A&A, 605, L7
Lecavelier Des Etangs, A. 2007, A&A, 461, 1185
Lecavelier des Etangs, A., Bourrier, V., Wheatley, P. J., et al. 2012, A&A, 543, L4
Lecavelier des Etangs, A., Ehrenreich, D., Vidal-Madjar, A., et al. 2010, A&A, 514, A72
Lecavelier Des Etangs, A., Vidal-Madjar, A., & Desert, J. M. 2008, Nature, 456, E1
Lecavelier des Etangs, A., Vidal-Madjar, A., McConnell, J. C., & Hébrard, G. 2004, A&A, 418, L1
Linsky, J. L., Yang, H., France, K., et al. 2010, ApJ, 717, 1291
Linssen, D., Oklopčić, A., & MacLeod, M. 2022, arXiv e-prints, arXiv:2209.03677
Loyd, R. O. P., Koskinen, T. T., France, K., Schneider, C., & Redfield, S. 2017, ApJ Letters, 834, L17
Mallonn, M. & Strassmeier, K. G. 2016, A&A, 590, A100
Mansfield, M., Bean, J. L., Oklopčić, A., et al. 2018, ApJ Letters, 868, L34
Mayor, M. & Queloz, D. 1995, Nature, 378, 355
Mazeh, T., Holczer, T., & Faigler, S. 2016, A&A, 589, A75
Moutou, C., Coustenis, A., Schneider, J., Queloz, D., & Mayor, M. 2003, A&A, 405, 341
Murray-Clay, R. A., Chiang, E. I., & Murray, N. 2009, ApJ, 693, 23
National Academies of Sciences, Engineering, and Medicine. 2021, Pathways to Discovery in Astronomy and Astrophysics for the 2020s (Washington, DC: The National Academies Press)
Ninan, J. P., Stefansson, G., Mahadevan, S., et al. 2020, ApJ, 894, 97
Nortmann, L., Pallé, E., Salz, M., et al. 2018, Science, 362, 1388
Oklopčić, A. 2019, ApJ, 881, 133
Oklopčić, A. & Hirata, C. M. 2018, ApJ Letters, 855, L11
Öpik, E. J. 1963, Geophysical Journal, 7, 490
Orell-Miquel, J., Murgas, F., Pallé, E., et al. 2022, A&A, 659, A55
Owen, J. E. & Alvarez, M. A. 2016, ApJ, 816, 34
Owen, J. E. & Lai, D. 2018, MNRAS, 479, 5012
Owen, J. E., Murray-Clay, R. A., Schreyer, E., et al. 2021, arXiv e-prints, arXiv:2111.06094
Owen, J. E. & Wu, Y. 2013, ApJ, 775, 105
Palle, E., Nortmann, L., Casasayas-Barris, N., et al. 2020, A&A, 638, A61
Paragas, K., Vissapragada, S., Knutson, H. A., et al. 2021, ApJ Letters, 909, L10
Parker, E. N. 1958, ApJ, 128, 664
Pillitteri, I., Micela, G., Maggio, A., Sciortino, S., & Lopez-Santiago, J. 2022, A&A, 660, A75
Poppenhaeger, K. 2022, MNRAS, 512, 1751
Rackham, B. V., Apai, D., & Giampapa, M. S. 2019, AJ, 157, 96
Rackham, B. V., Espinoza, N., Berdyugina, S. V., et al. 2022, arXiv e-prints, arXiv:2201.09905
Ridden-Harper, A. R., Snellen, I. A. G., Keller, C. U., et al. 2016, A&A, 593, A129
Rockcliffe, K. E., Newton, E. R., Youngblood, A., et al. 2021, AJ, 162, 116
Salz, M., Czesla, S., Schneider, P. C., et al. 2018, A&A, 620, A97
Salz, M., Schneider, P. C., Czesla, S., & Schmitt, J. H. M. M. 2016, A&A, 585, L2
Sánchez-López, A., Lin, L., Snellen, I. A. G., et al. 2022, A&A, 666, L1
Schilling, G. 1996, Science, 273, 429
Schlawin, E., Agol, E., Walkowicz, L. M., Covey, K., & Lloyd, J. P. 2010, ApJ Letters, 722, L75
Seager, S. & Sasselov, D. D. 2000, ApJ, 537, 916

Sing, D. K., Fortney, J. J., Nikolov, N., et al. 2016, Nature, 529, 59
Sing, D. K., Lavvas, P., Ballester, G. E., et al. 2019, AJ, 158, 91
Spake, J. J., Oklopčić, A., & Hillenbrand, L. A. 2021, AJ, 162, 284
Spake, J. J., Oklopčić, A., Hillenbrand, L. A., et al. 2022, arXiv e-prints, arXiv:2209.03502
Spake, J. J., Sing, D. K., Evans, T. M., et al. 2018, Nature, 557, 68
Szabó, G. M. & Kiss, L. L. 2011, ApJ, 727, L44
Tripathi, A., Kratter, K. M., Murray-Clay, R. A., & Krumholz, M. R. 2015, ApJ, 808, 173
Vidal-Madjar, A., Désert, J.-M., Lecavelier des Etangs, A., et al. 2004, ApJ Letters, 604, L69
Vidal-Madjar, A., Huitson, C. M., Bourrier, V., et al. 2013, A&A, 560, A54
Vidal-Madjar, A., Lecavelier des Etangs, A., Désert, J.-M., et al. 2003, Nature, 422, 143
Vidotto, A. A. & Bourrier, V. 2017, MNRAS, 470, 4026
Villarreal D'Angelo, C., Vidotto, A. A., Esquivel, A., Hazra, G., & Youngblood, A. 2021, MNRAS, 501, 4383
Vissapragada, S., Knutson, H. A., Greklek-McKeon, M., et al. 2022, arXiv e-prints, arXiv:2204.11865
Vissapragada, S., Knutson, H. A., Jovanovic, N., et al. 2020, AJ, 159, 278
Vissapragada, S., Stefánsson, G., Greklek-McKeon, M., et al. 2021, AJ, 162, 222
Volkov, A. N., Johnson, R. E., Tucker, O. J., & Erwin, J. T. 2011, ApJ Letters, 729, L24
Waalkes, W. C., Berta-Thompson, Z., Bourrier, V., et al. 2019, AJ, 158, 50
Wang, L. & Dai, F. 2018, ApJ, 860, 175
Watson, A. J., Donahue, T. M., & Walker, J. C. G. 1981, Icarus, 48, 150
Wyttenbach, A., Ehrenreich, D., Lovis, C., Udry, S., & Pepe, F. 2015, A&A, 577, A62
Wyttenbach, A., Mollière, P., Ehrenreich, D., et al. 2020, A&A, 638, A87
Yan, D., Guo, J., Huang, C., & Xing, L. 2021a, ApJ Letters, 907, L47
Yan, F. & Henning, T. 2018, Nature Astronomy, 2, 714
Yan, F., Wyttenbach, A., Casasayas-Barris, N., et al. 2021b, A&A, 645, A22
Zhang, M., Cauley, P. W., Knutson, H. A., et al. 2022a, arXiv e-prints, arXiv:2204.02985
Zhang, M., Knutson, H. A., Dai, F., et al. 2022b, arXiv e-prints, arXiv:2207.13099
Zhang, M., Knutson, H. A., Wang, L., Dai, F., & Barragán, O. 2022c, AJ, 163, 67
Zhang, M., Knutson, H. A., Wang, L., et al. 2022d, AJ, 163, 68
Zhang, M., Knutson, H. A., Wang, L., et al. 2021, AJ, 161, 181
Zhang, Y., Snellen, I. A. G., Mollière, P., et al. 2020, A&A, 641, A161

Winds of Stars and Exoplanets
Proceedings IAU Symposium No. 370, 2023
A. A. Vidotto, L. Fossati & J. S. Vink, eds.
doi:10.1017/S1743921322004458

The effect of winds in red supergiants: modeling for interferometry

Gemma González-Torà[1,2], Markus Wittkowski[1], Ben Davies[2] and Bertrand Plez[3]

[1]European Southern Observatory (ESO), Karl Schwarzschildstrasse 2, 85748 Garching bei München, Germany

[2]Astrophysics Research Institute, Liverpool John Moores University, 146 Brownlow Hill, Liverpool L3 5RF, United Kingdom

[3]LUPM, Université de Montpellier, CNRS, 34095 Montpellier, France

Abstract. Red supergiants (RSGs) are evolved massive stars in a stage preceding core-collapse supernova. Understanding evolved-phases of these cool stars is key to understanding the cosmic matter cycle of our Universe, since they enrich the cosmos with newly formed elements. However, the physical processes that trigger mass loss in their atmospheres are still not fully understood, and remain one of the key questions in stellar astrophysics. We use a new method to study the extended atmospheres of these cold stars, exploring the effect of a stellar wind for both a simple radiative equilibrium model and a semi-empirical model that accounts for a chromospheric temperature structure. We then can compute the intensities, fluxes and visibilities matching the observations for the different instruments at the Very Large Telescope Interferometer (VLTI). Specifically, when comparing with the atmospheric structure of HD 95687 based on published VLTI/AMBER data, we find that our model can accurately match these observations in the K-band, showing the enormous potential of this methodology to reproduce extended atmospheres of RSGs.

Keywords. stars: atmospheres, stars: massive, stars: evolution, stars: fundamental parameters, stars: mass-loss, supergiants

---

1. Introduction

When evolved massive stars leave the main sequence and start the red supergiant (RSG) phase, stellar winds can impact the final fate in their evolutionary path (Chiosi & Maeder 1986). These mass-loss events are initiated in the extended atmospheres of RSGs: part of their material is ejected and transported up to several radii. Beyond that, the temperature is cold enough to start condensing the ejected material into dust grains.

However, the mechanism that triggers these mass-loss events in the extended atmospheres of RSGs is still poorly understood. There have been some attempts to explain the mechanism of stellar winds in RSGs (e.g., Kudritzki & Puls 2000; Josselin & Plez 2007; Kee et al. 2021), but there is still no consensus.

The current models use the following atmospheric structure: first a stellar modelization up to the photosphere (defined where $\tau_{\rm Ross} = 2/3$) with model atmosphere grids such as MARCS (Gustafsson et al. 2008), and then adding the contribution of dust modeling, such as DUSTY (Ivezic & Elitzur 1997). The atmospheric extension from the photosphere to where the dust is formed is usually left empty, since we do not know the physical processes that trigger these mass-loss events.

In this work, we aim to "fill" the atmospheric extension from the photosphere to the dusty shell. We use the model developed by Davies & Plez (2021), where they explored the extension of the atmospheres close to the stellar surface at radii smaller than the inner dust shells, in the optical and near-IR, by adding the influence of a stellar wind in the MARCS model atmospheres. This work is based on preliminary results from González-Torà et al., submitted to A&A.

For the purpose to study the extension of the atmosphere, we use interferometric data. Interferometry uses an array of telescopes to increase the angular resolution of the observations. By using interferometry, we have high-spatial resolution data of the stellar atmospheres of RSGs. This is an additional information that spectroscopy does not fully provide. Therefore, it is a very powerful tool to study the structure of extended atmospheres in detail.

2. Model

Our model is based on Davies & Plez (2021), where they start with a MARCS model atmosphere and plug-in the effect of a stellar wind. This model assumes local thermodynamic equilibrium (LTE), hydrostatic equilibrium, and spherical symmetry. We extend the model with a radius stratification up to $\sim 8.5\,R_\star$, where $R_\star$ is defined as the radius where the Rosseland opacity $\tau_{\mathrm{Ross}} = 2/3$. For a detailed discussion about the limitations and assumptions, the reader is referred to Davies & Plez (2021).

To determine the wind density, we use the mass continuity expression,

$$\dot{M} = 4\pi r^2 \rho(r) v(r) \tag{2.1}$$

where ρ and v are the density and velocity as a function of the stellar radial coordinate r respectively. The wind density $\rho_{\mathrm{wind}}(r)$ has the shape proposed by Harper et al. (2001):

$$\rho_{\mathrm{wind}} = \frac{\rho_{\mathrm{phot.}}}{(R_{\mathrm{max}}/R\star)^2}\left(1-\left(\frac{0.998}{(R_{\mathrm{max}}/R_\star)}\right)^\gamma\right)^\beta \tag{2.2}$$

where R_{max} is the arbitrary outer-most radius of the model, in our case $8.5\,R_\star$. The β and γ parameters define the smoothness of the extended wind region and were initially set in the semi-empirical 1D model of α Ori by Harper et al. (2001): $\beta_{\mathrm{Harp}} = -1.10$ and $\gamma_{\mathrm{Harp}} = 0.45$.

The velocity profile is found assuming a fiducial wind limit of $v_\infty = 25 \pm 5$ km/s, that is the value matched to Richards & Yates (1998); van Loon et al. (2005), and Equation 2.1.

For the temperature profile we first used simple radiative transfer equilibrium (R.E.), that will result in a smoothly decreasing temperature profile for the extended atmosphere. We also defined a different temperature profile based on spatially-resolved radio continuum data of α Ori by Harper et al. (2001). The main characteristic of this profile is a temperature inversion in the chromosphere of the star, that peaks at $\sim 1.4\,R_\star$, and decreases again.

Figure 1 shows the the density (upper panel) and the temperature structures (lower panel) of our model, the latest for simple R.E. (blue squares) and a temperature chromospheric inversion (red circles).

3. Results

3.1. *For base model*

We compute the spectra, intensities and squared visibility amplitudes ($|V|^2$) for a base model of $T_{\mathrm{eff}} = 3500$ K, $\log g = 0.0$, $[Z] = 0$, $\xi = 5$ km/s, $M = 15\,M_\odot$, $R_\star = 690\,R_\odot$ and $R_{\mathrm{max}} = 8.5\,R_\star$, corresponding to a RSG similar to HD 95687 (Arroyo-Torres et al. 2015). The density parameters in Equation 2.2 are $\beta_{\mathrm{Harp}} = -1.10$ and $\gamma_{\mathrm{Harp}} = 0.45$ as in

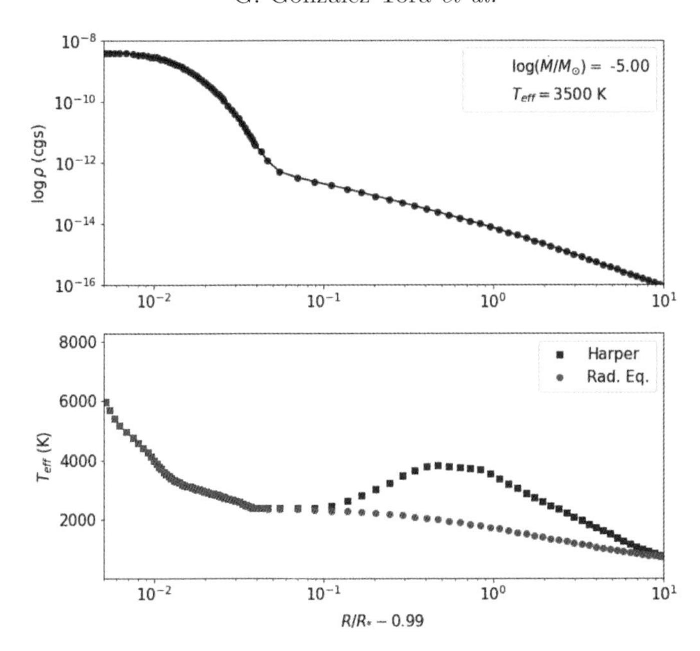

Figure 1. Density and temperature profiles (Harper temperature inversion in blue, simple R.E. in red) for the extended model.

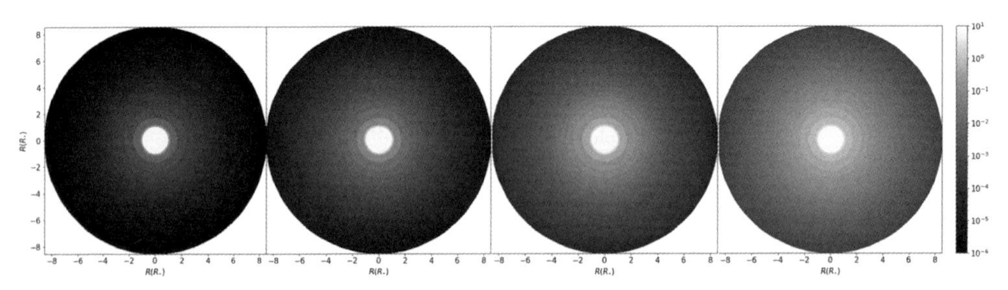

Figure 2. 2D profile of our atmospheric model, from left to right: $\dot{M} = 10^{-7}$, 10^{-6}, 10^{-5} and 10^{-4} $M_\odot$/yr. This is the case of simple R.E.

Harper et al. (2001) and the wind limit $v_\infty = 25$ km/s. The temperature profile is initially set to simple R.E. We use mass-loss rates of $\dot{M} = 10^{-4}$, 10^{-5}, 10^{-6} and 10^{-7} $M_\odot$/yr, and a simple MARCS model without any wind. As an example, we simulate a star with an angular diameter at the photosphere of $\theta_{\rm Ross} = 3$ mas, a baseline of $B = 60$ m and without any additional over-resolved component.

Figure 2 shows the 2D intensities for mass-loss rates of $\dot{M} = 10^{-7}$, 10^{-6}, 10^{-5} and 10^{-4} $M_\odot$/yr (panels from left to right), for a star with an extended radius of $R_{\rm max} = 8.5\, R_\star$ for a cut in the transition CO (2-0) ($\lambda = 2.29\,\mu$m). We see that, unlike MARCS, all models show extension at $R > 1\, R_\star$. As expected, as we increase the mass-loss rate we also have more extension.

Figure 3 shows the flux for a wavelength range of $1.8\,\mu\text{m} < \lambda < 5.0\,\mu\text{m}$ corresponding to the K, L and $M-$bands, for simple MARCS model (orange), $\dot{M} = 10^{-7}$ (purple), 10^{-6} (blue), 10^{-5} (dark green) and 10^{-4} $M_\odot$/yr (red). Compared to a MARCS model, as we increase the mass-loss rate we start seeing more features in the spectra (e.g., the water absorption in $1.8\,\mu\text{m} < \lambda < 2.0\,\mu\text{m}$ and $3.0\,\mu\text{m} < \lambda < 3.5\,\mu\text{m}$, the SiO and CO emissions in $\lambda > 4.0\,\mu\text{m}$).

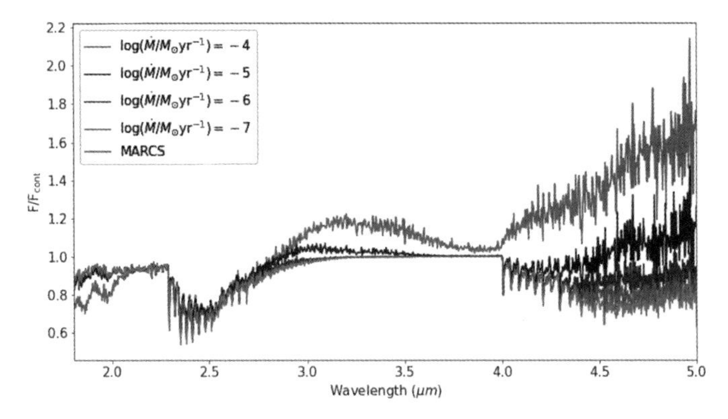

Figure 3. Normalized flux for the different $\dot{M}$. This is the case of simple R.E.

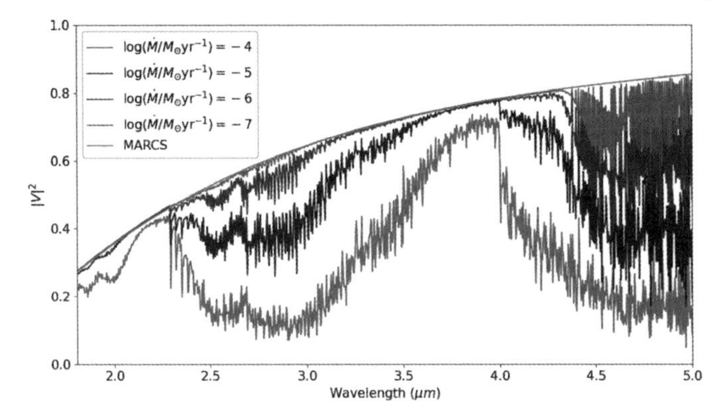

Figure 4. Same as Figure 3 but for the $|V|^2$.

To calculate the squared $|V|^2$ we used the Hankel transform as in Davis et al. (2000). Figure 4 shows the same as Figure 3 but for $|V|^2$. We the same features as Figure 3 (e.g., the water absorption in $1.8\,\mu\text{m} < \lambda < 2.0\,\mu\text{m}$ and $3.0\,\mu\text{m} < \lambda < 3.5\,\mu\text{m}$, the SiO and CO emissions in $\lambda > 4.0\,\mu\text{m}$), and the extra CO extension in $2.3\,\mu\text{m} < \lambda < 3.1\,\mu\text{m}$. These features show that our model is able to extend the atmosphere, unlike simple MARCS which has no features in the $|V|^2$.

3.2. *Deviations from base model*

We can change the β and γ parameters of Equation 2.2 that define the density profile. This will affect the wind density profile close to the photosphere, as we decrease the value of the parameters the density becomes steeper in that region. This will have an effect in the lines formed close to the stellar surface (e.g., water layers Kravchenko et al. 2020), and therefore are sensitive to variations of the density profile in this region.

As for the temperature profile, the main difference between R.E. and the chromospheric temperature inversion is seen in the CO for the K-band. Even for lower mass-loss rates $\log \dot{M}/M_\odot < -5$ we observe the CO lines very depleted in the flux when using the temperature inversion profile, and already in emission for $\log \dot{M}/M_\odot \sim -4$. This is not the case of R.E., since the CO is always in absorption in that region (Figure 3). When compared to the observations, we always see the CO lines in absorption, confirming that R.E. may be more appropriate to fit the observational data. Depending on the wavelength range of observations, the temperature peak at the chromosphere varies: CO MOLsphere data derived a temperature of ~ 2000 K at $1.2 - 1.4\,R_\star$ (Lim et al. 1998), while for the optical and ultraviolet the peak temperature is ~ 5000 K at similar radii (Ohnaka et al. 2013). O'Gorman et al. (2020) suggested these components co-exist in different structures at similar radii in an inhomogeneous atmosphere. Observations at different wavelengths may then be sensitive to different such structures.

3.3. *Case study*

We compare our model to published VLTI/AMBER data of the RSG HD 95687 available by Arroyo-Torres et al. (2015).The data were taken using the AMBER medium-resolution mode ($R \sim 1500$) in the $K - 2.1\,\mu$m and $K - 2.3\,\mu$m bands.

For our model fit, we check both the spectra and $|V|^2$. We use a range of mass-loss rates of $-7 < \log \dot{M}/M_{\odot} < -4$ with a grid spacing of $\Delta \dot{M}/M_{\odot} = 0.25$, and density parameters $-1.1 < \beta < -1.60$ in steps of $\Delta\beta = 0.25$ and $0.05 < \gamma < 0.45$ in steps of $\Delta\gamma = 0.2$. For the temperature profile we use R.E., since the temperature inversion would either show depleted CO lines ($\lambda = 2.5 - 3\,\mu$m) for the spectra, which do not match with the observations, or not enough extension for the $|V|^2$. Therefore it is not possible to find a model with the temperature inversion profile that fits both spectra and $|V|^2$ simultaneously (see Section 3.2).

Figure 5 shows the MARCS model fit to the data of Arroyo-Torres et al. (2015) (blue), our best MARCS+wind model fit with $\log \dot{M}/M_{\odot} = -5.50$, $\beta_{\rm Harp} = -1.60$ and $\gamma_{\rm Harp} = 0.05$ (red), compared to the data of HD 95687 (gray). We see that our model can fit well both flux and $|V|^2$. There is a big improvement with respect to MARCS for the $|V|^2$, especially in the region where the CO is present: $2.3\,\mu\mathrm{m} < \lambda < 3.1\,\mu$m. In addition, the density profile is steeper than expected, but the $\dot{M}$ is reasonable when compared with typical mass-loss prescriptions (e.g., de Jager et al. 1988; Schröder & Cuntz 2005; Beasor et al. 2020).

4. Conclusions

This is the first extended atmosphere model to our knowledge that can reproduce in great detail both the spectra and $|V|^2$ simultaneously. Therefore, we have shown the immense potential of this semi-empirical model of MARCS+wind, not only to match the spectral features without the need of dust, but also the visibilities obtained by interferometric means.

To fit both the water and CO extensions, the density shape should be steeper close to the surface of the star than previously expected by Harper et al. (2001). Regarding the temperature profile, we find that the R.E. reproduces the spectra better than the chromospheric temperature inversion, since we do not observe any emission in the CO bands. The possible reason that R.E. fits better than the temperature inversion could be the presence of different spatial cells with different temperatures in the hot luke-warm chromospheres of RSGs (O'Gorman et al. 2020).

In the future, we want to compare this model with more wavelength ranges, to see the effects in different wavelengths such as the L or M-bands.

5. Q&A

Question: Why are there 4 panels in Figure 5 instead of 2? It seems like the wavelength regions overlap.

Answer: This is because the visibilities were taken with different array configurations and therefore different baselines. We decided to separate both the fluxes and visibilities following Arroyo-Torres et al. (2015) to avoid confusion.

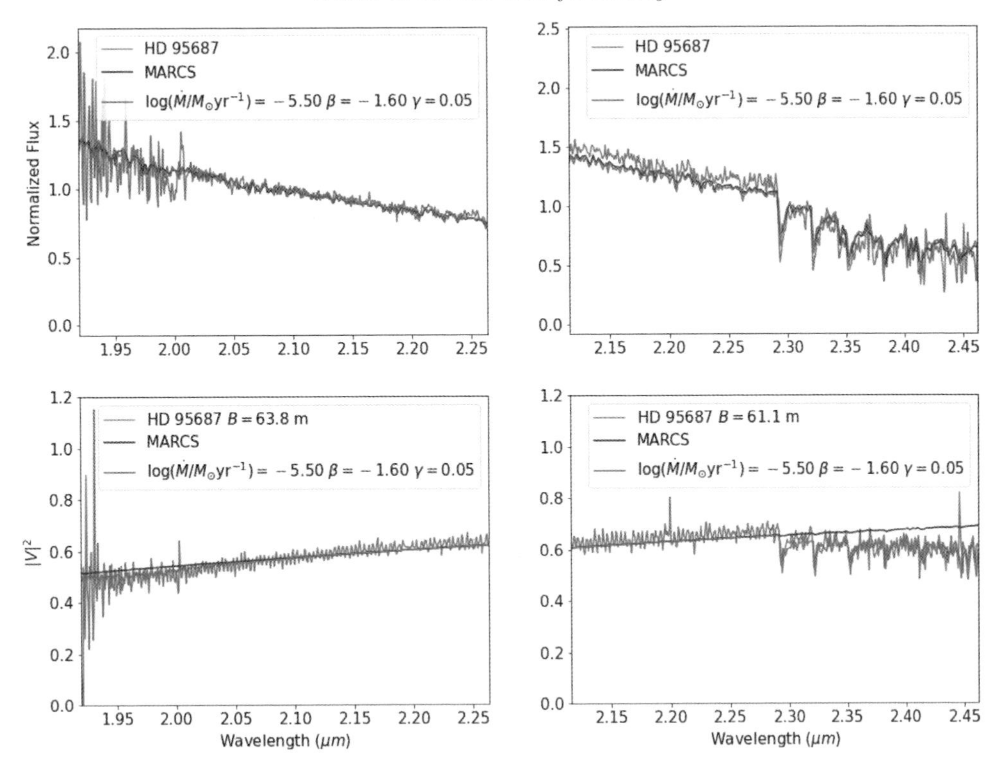

Figure 5. Normalized flux and $|V|^2$ for HD 95687 as observed with VLTI/AMBER (in gray), the best fitting results for this work in red, and the best MARCS fitting in blue.

References

Arroyo-Torres, B., Wittkowski, M., Chiavassa, A., et al. 2015, A&A, 575, A50

Beasor, E. R., Davies, B., Smith, N., et al. 2020, MNRAS, 492, 5994

Chiosi, C. & Maeder, A. 1986, ARA&A, 24, 329

Davies, B. & Plez, B. 2021, MNRAS, 508, 5757

Davis, J., Tango, W. J., & Booth, A. J. 2000, MNRAS, 318, 387

de Jager, C., Nieuwenhuijzen, H., & van der Hucht, K. A. 1988, AApS, 72, 259

Gustafsson, B., Edvardsson, B., Eriksson, K., et al. 2008, A&A, 486, 951

Harper, G. M., Brown, A., & Lim, J. 2001, ApJ, 551, 1073

Ivezic, Z. & Elitzur, M. 1997, MNRAS, 287, 799

Josselin, E. & Plez, B. 2007, A&A, 469, 671

Kee, N. D., Sundqvist, J. O., Decin, L., de Koter, A., & Sana, H. 2021, A&A, 646, A180

Kravchenko, K., Wittkowski, M., Jorissen, A., et al. 2020, A&A, 642, A235

Kudritzki, R.-P. & Puls, J. 2000, ARA&A, 38, 613

Lim, J., Carilli, C. L., White, S. M., Beasley, A. J., & Marson, R. G. 1998, Nature, 392, 575

O'Gorman, E., Harper, G. M., Ohnaka, K., et al. 2020, A&A, 638, A65

Ohnaka, K., Hofmann, K. H., Schertl, D., et al. 2013, A&A, 555, A24

Richards, A. M. S. & Yates, J. A. 1998, Irish Astronomical Journal, 25, 7

Schröder, K. P. & Cuntz, M. 2005, ApJ Letters, 630, L73

van Loon, J. T., Cioni, M. R. L., Zijlstra, A. A., & Loup, C. 2005, A&A, 438, 273

Winds of Stars and Exoplanets
Proceedings IAU Symposium No. 370, 2023
A. A. Vidotto, L. Fossati & J. S. Vink, eds.
doi:10.1017/S1743921322004604

The porous envelope and circumstellar wind matter of the closest carbon star, CW Leonis

Hyosun Kim[1], Ho-Gyu Lee[1], Youichi Ohyama[2], Ji Hoon Kim[3], Peter Scicluna[4], You-Hua Chu[2], Nicolas Mauron[5] and Toshiya Ueta[6]

[1]Korea Astronomy and Space Science Institute, 776, Daedeokdae-ro, Yuseong-gu, Daejeon 34055, Republic of Korea

[2]Institute of Astronomy and Astrophysics, Academia Sinica, 11F of Astronomy-Mathematics Building, AS/NTU, No.1, Sec. 4, Roosevelt Rd, Taipei 10617, Taiwan, R.O.C.

[3]SNU Astronomy Research Center, Department of Physics and Astronomy, Seoul National University, 1 Gwanak-ro, Gwanak-gu, Seoul 08826, Korea

[4]European Southern Observatory, Alonso de Cordova 3107, Santiago RM, Chile

[5]Laboratoire Univers et Particules, Universite de Montpellier and CNRS, Batiment 13, CC072, Place Bataillon, F-34095 Montpellier, France

[6]Department of Physics and Astronomy, University of Denver, 2112 E Wesley Ave., Denver, CO 80208, USA

Abstract. Recent abrupt changes of CW Leonis may indicate that we are witnessing the moment that the central carbon star is evolving off the Asymptotic Giant Branch (AGB) and entering into the pre-planetary nebula (PPN) phase. The recent appearance of a red compact peak at the predicted stellar position is possibly an unveiling event of the star, and the radial beams emerging from the stellar position resemble the feature of the PPN Egg Nebula. The increase of light curve over two decades is also extraordinary, and it is possibly related to the phase transition. Decadal-period variations are further found in the residuals of light curves, in the relative brightness of radial beams, and in the extended halo brightness distribution. Further monitoring of the recent dramatic and decadal-scale changes of this most well-known carbon star CW Leonis at the tip of AGB is still highly essential, and will help us gain a more concrete understanding on the conditions for transition between the late stellar evolutionary phases.

Keywords. stars: AGB and post-AGB, (stars:) binaries: general, stars: carbon, (stars:) circumstellar matter, stars: evolution, stars: individual (CW Leonis), stars: late-type, stars: mass loss, stars: winds, outflows

1. Introduction

Many pre-planetary nebulae (PPN) consist of newly-formed inner bipolar/multipolar lobes and outer spirals/rings/arcs that are the fossil records of stellar wind matter accumulated during the asymptotic giant branch (AGB) phase. The coexistence of two such morphologically distinct circumstellar structures is a mystery; however, it is widely believed that binaries play a key role. The most direct clue to resolving the mystery of the shape transition along stellar phase evolution may be offered by catching the moment when an AGB star is evolving off the current phase toward the PPN phase. Recent dramatic changes of CW Leonis likely indicate that we are witnessing the moment of transition between these late stellar evolutionary phases.

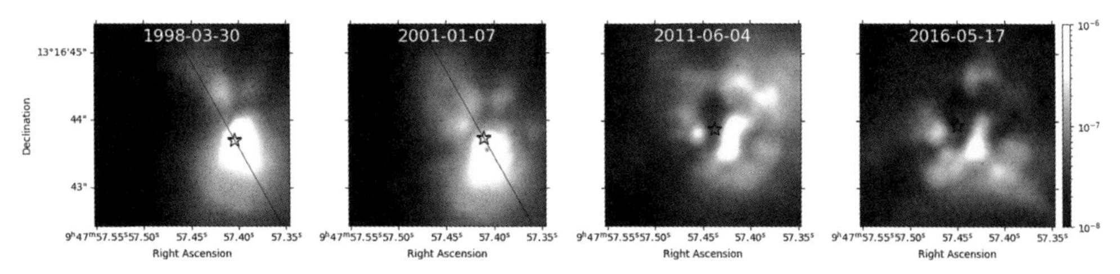

Figure 1. Temporal change of brightness in the central 1.5 arcsec region of CW Leonis. The star symbol indicates the proper-motion-corrected position of the star at each epoch, denoted at the top of each panel. The bipolar-like structure (black line) before 2011, which had likely given a misleading impression for the origin and evolutionary phase of CW Leonis, disappeared in the 2011 and 2016 epochs. From left to right, the Hubble Space Telescope images are taken with the F606W filter ($\sim 0.6\,\mu$m) at the epochs of 1998-03-30 (Prop. ID: 6856, PI: J. Trauger), 2001-01-07 (Prop. ID: 8601, PI: P. Seitzer), 2011-06-04 (Prop. ID: 12205, PI: T. Ueta), and 2016-05-17 (Prop. ID: 14501, PI: H. Kim).

2. Previous Views on CW Leonis

CW Leonis is the closest (distance of about 123 pc; Groenewegen et al. 2012) and the most well-studied carbon-rich AGB star (or carbon star). Multi-wavelength observations suggest that CW Leonis is likely a binary system (e.g., Jeffers et al. 2014; Decin et al. 2015). Its non-concentric ring-like pattern over 200 arcsec is remarkable (Mauron & Huggins 2000), which can be modeled by a spiral-shell structure introduced by an eccentric orbit binary at the center (e.g., Cernicharo et al. 2015). However, neither the carbon star nor the companion has been identified because of obscuration by the dense circumstellar matter ejected from this extreme carbon star at the tip of AGB.

Several near-infrared observations were executed with adaptive optics and speckle interferometry in 1995–2003 (Tuthill et al. 2000; Osterbart et al. 2000; Weigelt et al. 2002; Murakawa et al. 2005) achieving high-resolutions (< 0.1 arcsec) but losing stellar positional information at the cost of field sizes; several clumps were revealed, but their relationship with the central star was unclear. The vigorous debate about which clump corresponds to the carbon star ended in vain through a monitoring study over 2000–2008 showing that the clumps faded out around 2005 (Stewart et al. 2016).

In the optical, before 2011, the core region exhibited an extended bipolar nebula without any distinct point source (Haniff & Buscher 1998; Skinner et al. 1998; Leão et al. 2006), from which this object was thought to have an invisible star residing in a dusty disk lying perpendicular to the bipolar structure. The bipolar-like structure, however, disappeared in the latest Hubble Space Telescope images taken in 2011 and 2016 (Kim et al. 2015, 2021), suggesting a completely different view for CW Leonis.

3. Recent Dramatic Changes and Porous Envelope Scenario

Surprisingly, the latest optical images of CW Leonis, taken using Hubble Space Telescope, in 2011 and 2016 revealed dramatic changes in the core region of the circumstellar envelope from those taken about 10-year earlier in 1998 and 2001 (Figure 1). Besides the disappearance of the long-believed bipolar-like nebula, several important features are identified and become evidences for a porous envelope of the central star (Kim et al. 2021).

In contrast to its absence at the previous epochs, a local brightness peak appears exactly at the expected stellar position (Figure 2, top middle) and it is identified as the reddest spot in the color map (Figure 2, top left). It is compact; its full width at half maximum above the adjacent diffuse emission is slightly larger than the standard point-spread function. This red compact spot at a local peak is interpreted as the direct starlight,

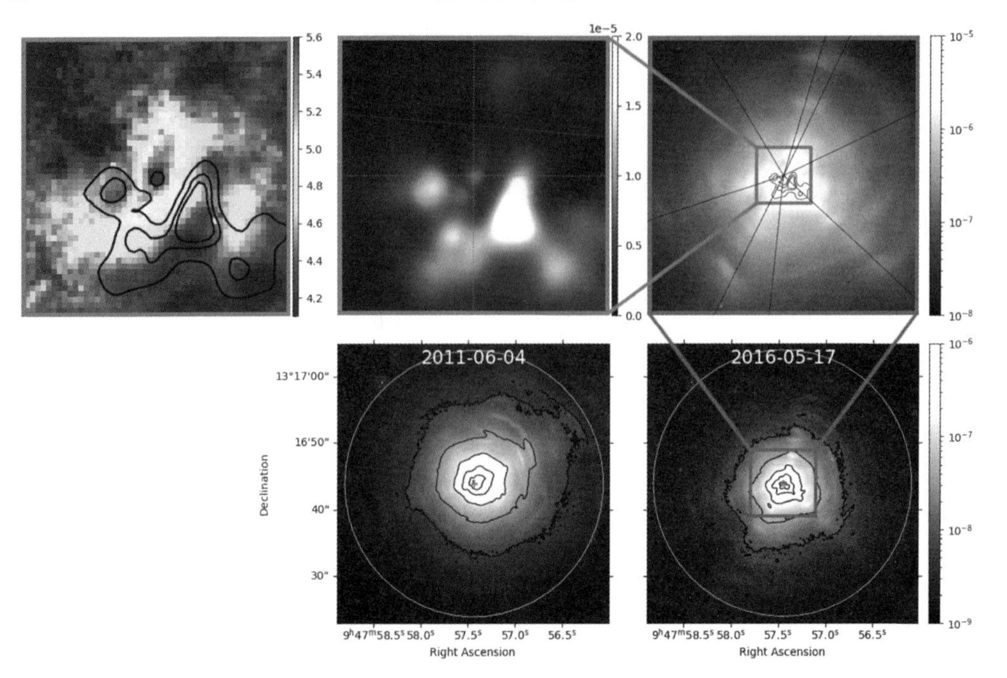

Figure 2. The Hubble Space Telescope image of CW Leonis taken with the F814W filter ($\sim 0.8\,\mu$m) in 2016 (bottom right), compared to the 2011 epoch image (bottom left). The radial beams and multiple rings appearing prominent in the central 5 arcsec region are the intriguing features (top right). The color map for central 1 arcsec region shows the reddest spot coincides with the stellar position, marked by black contours of the F814W brightness (top left) same as the image in the top middle panel. The color bars range the F814W brightness in logarithmic scales, but for the color map in the top left panel being in linear scales for the magnitude difference between the F606W and F814W images.

escaping through one of the gaps in the clumpy envelopes shrouding the star. This radial beam may pulsate around the line of sight with a small angle and be coincidentally aligned with the line of sight at the observed epoch in 2016.

The outer part of the observed image exhibits eight straight lines of brightness that are radially stretched from the central star (Figure 2, top right). In the context of the porous envelope scenario, these searchlight beams indicate the trajectories of starlight penetrating the holes in the inner envelopes, along which adjacent dust particles in the circumstellar envelopes are illuminated. Any other interpretation for their origin is precluded because of the straightness of these beams regardless of the considerably fast stellar proper motion.

The extended halo brightness distribution becomes fairly symmetric about the central star in the 2016 image, compared to the elongated distribution to the northwest in the 2011 image (Figure 2, bottom panels). This change is again beyond the scope of dynamics of matter that would require a very long time to move the whole halo with an extremely large extent. It is speculated that one of the radial beams, that is related to the new emergence of central red compact brightness peak, alters its angle toward the line of sight in 2016 from a slightly misaligned angle pointing toward the northwestern direction in 2011, explaining the redistribution of the halo brightness.

The radial beams appearing in the plane of the sky do not seem to shift their positions much with time (Figure 3). Their position angles with respect to the predicted stellar positions at the individual epochs are almost fixed. Therefore, it is natural that we assume the precession angle for the radial beam relevant to the extended brightness

Figure 3. Searchlight beams of CW Leonis are fixed in the relative positions with respect to the proper-motion-corrected stellar position and their relative brightnesses are varying with time. The snapshots are taken from an animation in the Research Gallery of http://hubblesite.org. Image credit: ESA/Hubble, NASA, Toshiya Ueta (Univ. of Denver), Hyosun Kim (KASI).

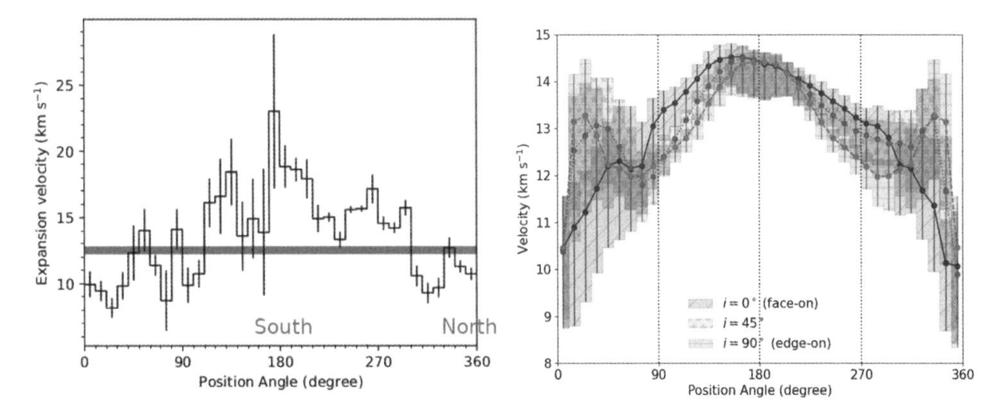

Figure 4. (Left) the expansion velocities in 36 sectors derived from differential proper motion of recurrent ring-like pattern of CW Leonis (Kim et al. 2021); (right) the correspondance in an eccentric-orbit binary model with the pericenter of the mass-losing star to the north (at the position angle of 0°), viewed at three different inclination angles 0° (blue, solid line, hatched area), 45° (green, dotted line, star-filled area), and 90° (red, dashed line, horizontally lined area), respectively (Kim 2022).

distribution is small. In order to verify this scenario, another epoch imaging observations with the same setup is anticipated.

In contrast, the relative brightnesses of the radial beams significantly change and the period seems to be about 10 years (Figure 3). The brightest beams are toward north in the 2001 and 2011 epochs (downward in the figure) while toward south in the 2016 epoch (upward). Indeed, besides the stellar pulsation of 640-day period, a decadal variation has been suggested based on near-infrared and optical photometric data (Dyck et al. 1991; Kim et al. 2021). In particular, the increases of K-band flux and the point source contribution in it during 1980–1990 are quite similar to the event found in 2016. To assess whether the variations are indeed periodic, more frequent longer-term monitoring observations are desired.

4. Anisotropic Wind Expansion induced by an Eccentric Binary

The multiple shell pattern wrapping around the star is one of the most intriguing features of CW Leonis. Our analysis using differential proper motion of the pattern indicates expansion of shells of ejected material from the star. The derived speeds of the expanding shells depend on the direction (Figure 4, left). The expansion speeds vary not only across different position angles within the Hubble Space Telescope image (about 7 $\mathrm{km\,s^{-1}}$ faster to the south), but their average speed in the plane of the sky was also about 2 $\mathrm{km\,s^{-1}}$ slower than the wind speed along the line of sight that is derived from molecular line observations in radio wavelengths. This variation of measured speeds indicates an overall nonspherical geometry of the wind matter. We further find that these observations (Figure 4, left panel) are compatible with a binary model having an eccentric orbit (see the right panel of Figure 4, from Kim 2022).

We regard, however, the velocity measurement was somewhat uncertain. The 2016 image was not as deep as the 2011 image, reducing the number of rings used in the analysis. Another obstacle was the relatively small expansion length (the average of positional difference of individual rings between the 2011 and 2016 epochs) due to the short 5-year interval, which was only slightly larger than the size of point-spread function. With these reasons, high-resolution high-sensitivity imaging monitoring is further needed.

5. Conclusion

The recent drastic changes in optical images suggest that the previously-seen bipolar-like structure could not be a concrete structure but could possibly be parts of searchlight beams with varying relative brightnesses along time. These radial beams reveal the pathways of starlight illuminating dusty material after escaping through the gaps in the clumpy envelope enshrouding the star. The appearance of a distinct brightness peak exactly at the predicted stellar position and the abnormal shift of large halo distribution both can be explained by a hypothesized radial beam toward us that is slowly precessing with a small angle and aligned with the line of sight at the latest observation epoch.

Although the complexity of CW Leonis has been well known since its discovery, the three dimensional morphology of its central core and evolving beams is uniquely revealed with recent Hubble Space Telescope optical monitoring of the core images. It also allows to trace the expansion velocity of shells and clumps as seen in the evolving light of the central star. Furthermore, on-going efforts on three dimensional hydrodynamic models fitting the data likely suggests a small inclination of the orbit (close to face-on) with the pericenter of the mass-losing star at North. Further systematic monitoring of this canonical high mass-loss carbon star is mandatory to strengthen our interpretation coupled with our modelling. Hubble Space Telescope allows us to be very near of establishing a robust understanding of the mass loss of this strategic but mysterious AGB (or soon post-AGB) star.

Acknowledgments

HK acknowledges support by the National Research Foundation of Korea (NRF) grant (No. 2021R1A2C1008928) and Korea Astronomy and Space Science Institute (KASI) grant (Project No. 2022-1-840-05), both funded by the Korea Government (MSIT).

References

Cernicharo, J., Marcelino, N., Agúndez, M., & Guélin, M. 2015, A&A, 575, A91
Decin, L., Richards, A. M. S., Neufeld, D., et al. 2015, A&A, 574, AA5
Dyck, H. M., Benson, J. A., Howell, R. R., et al. 1991, AJ, 102, 200
Groenewegen, M. A. T., Barlow, M. J., Blommaert, J. A. D. L., et al. 2012, A&A, 543, L8
Haniff, C. A., & Buscher, D. F. 1998, A&A, 334, L5
Jeffers, S. V., Min, M., Waters, L. B. F. M., et al. 2014, A&A, 572, AA3
Kim, H. 2022, in prep.
Kim, H., Lee, H.-G., Mauron, N., & Chu, Y.-H. 2015, ApJ Letters, 804, L10
Kim, H., Lee, H.-G., Ohyama, Y., et al. 2021, ApJ, 914, 35

Leão, I. C., de Laverny, P., Mékarnia, D., de Medeiros, J. R., & Vandame, B. 2006, A&A, 455, 187
Mauron, N., & Huggins, P. J. 2000, A&A, 359, 707
Menten, K. M., Reid, M. J., Kamiński, T., & Claussen, M. J. 2012, A&A, 543, AA73
Murakawa, K., Suto, H., Oya, S., et al. 2005, A&A, 436, 601
Osterbart, R., Balega, Y. Y., Blöcker, T., Men'shchikov, A. B., & Weigelt, G. 2000, A&A, 357, 169
Skinner, C. J., Meixner, M., & Bobrowsky, M. 1998, MNRAS, 300, L29
Stewart, P. N., Tuthill, P. G., Monnier, J. D., et al. 2016, MNRAS, 455, 3102
Tuthill, P. G., Monnier, J. D., Danchi, W. C., & Lopez, B. 2000, ApJ, 543, 284
Weigelt, G., Balega, Y. Y., Blöcker, T., et al. 2002, A&A, 392, 131

Winds of Stars and Exoplanets
Proceedings IAU Symposium No. 370, 2023
A. A. Vidotto, L. Fossati & J. S. Vink, eds.
doi:10.1017/S1743921322003581

Is the magnetospheric accretion active in the Herbig Ae/Be stars?

Giovanni Pinzón[1], **Jesús Hernández**[2] **and Javier Serna**[2]

[1]Observatorio Astronómico Nacional, Universidad Nacional de Colombia, Bogotá, Colombia

[2]Instituto de Astronomía, Universidad Nacional Autónoma de México, Ensenada, B.C, México

Abstract. This contribution is based on the work published by (Pinzón et al. 2021) in which we computed rotation rates for a sample of 79 young stars (~3 Myr) in a wide range of stellar masses (from T Tauri Stars to Herbig Ae/Be stars) in in the Orion Star Formation Complex (OSFC). We study whether the magnetospheric accretion scenario (MA), valid for young low mass stars, may be applied over a wide range of stellar masses of not. Under the assumption that stellar winds powered by stellar accretion are the main source for the stellar spin down, the hypothesis of an extension of MA toward higher masses seems plausible. A comparison with Ap/Bp stars suggest that HAeBes should suffer a loss of angular momentum by a factor between 12 and 80 during the first 10 Myr in order to match the magnetic Ap/Bp zone in HR diagram.

Keywords. stars, formation, rotation.

---

1. Overview

Herbig Ae/Be stars (HAeBes) are young stellar objects with similar properties to their lower mass counterparts, the T Tauri stars (TTs). They both have comparable ages, infrared and ultraviolet excess levels associated with the presence of an accreting circumstellar disk, p-cygni line profiles in the most prominent emission lines, and forbidden line emission that confirms the existence of stellar winds in such systems (Hubrig et al. 2014). Although, the internal structure of HAeBes differs from that of TTs; while the former is mainly radiative, TTs are fully-convective stars. Consequently, there are a low number of magnetic HAeBes, less than 10% in the majority of studied star formation regions. The origin of magnetism in the few magnetic HAeBes is uncertain. In the merging scenario of Ferrario (2009), the magnetic field in HAeBes is the consequence of merging two magnetic TTs when at least one of them has reached the end of its radiative track in the HRD.

In the MA framework, TTs generate their own magnetic field that is able to truncate the disk at a few stellar radii from the stellar surface. From such location, gas in the disk with higher angular momentum falls onto the stellar surface along the magnetic field lines. The energy released during accretion process excites large fluxes of Alfvén waves along open field lines in the magnetosphere, that is, stellar winds, which play an important role in the outward transfer of angular momentum (Boehm et al. 1994; Matt et al. 2008). While this scenario describes angular momentum in TTs, its extrapolation toward magnetic HAeBes is not straightforward. High resolution spectral analysis of the region between the star and the inner disk, suggest that accretion process in HAeBes differs from that in TTs. Futhermore, the simultaneous presence of redshifted and blushifted

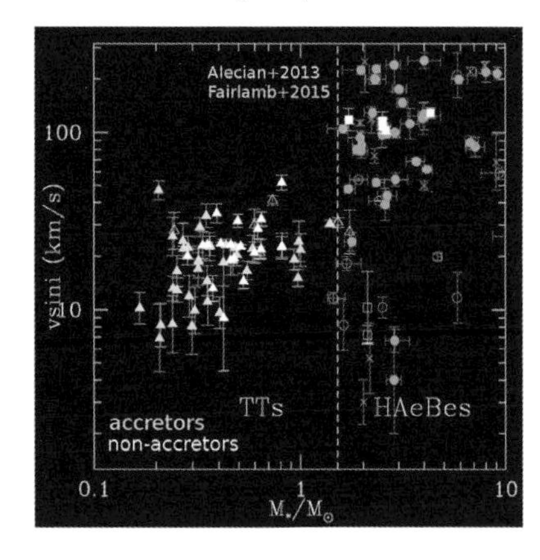

Figure 1. Left. Sample of TTs and HAeBes in OSFC. **Right.** Snapshot of the projected rotational velocities for the same objects. TTs (triangles) and HAeBes (rectangles). TTs with active accretion are indicated with bigger triangles in blue. Complementary data of HAeBes taken from (Alecian 2013) and (Fairlamb 2015) have been included in gray. Binaries are indicated with open circles and upper limit values with crosses.

absorption features in Balmer lines is a characteristic present in TTs spectra. In contrast, most promiment emission lines HAeBes generally lack such simultaneous absorption features(Cauley 2015).

2. Rotation of TTs and HAeBes

Using Hectochelle ($R\sim32000$) and FIES ($R\sim68000$) high resolution spectra we computed the projected rotational velocities (vsin$i$) for the sample of 79 TTs and HAeBes with confirmed membership (Hernández 2014) indicated in Figure 1. We obtained vsin$i$ applying two methods: (1) analysis of the cross correlation function of the object spectrum against a rotational template (Tonry & Davis 1979) and (2) Fourier Transform (FT) of selected line profiles in the object spectrum (Serna 2021). Results shown in Figure 1 confirm similar vsin$i$ values for accretors and non-accretors as expected for TTs. While HAeBes rotate on average, to one third of their break-up limit (~300 km/s for $3M_{\odot}$), TTs rotation remains below of 10% of this limit as expected from MA scenario.

3. Stellar winds in HAeBes

Stellar wind indicators such as p-cygni line profiles are observed in the FIES spectra of HAeBes. In Figure 2 we can a see a dip in the flux at -200 km/s in the residual emission of $H\beta$ 4861Å. In addition to this, the presence of the [OI]λ6300, which is originated in outermost regions of magnetosphere is a clear indicator of stellar winds. On the other hand, the significant Balmer excess observed in the most of the HAeBes spectra when compared with the expected flux of reference main-sequence stars, suggest that accretion and winds could be related in someway. The interplay among accretion and mass loss in the winds, regulates the angular momentum evolution during the first Myr of stellar evolution.

We used a stellar spin evolution model that includes accretion powered stellar winds in the context of MA framework, as the main source of angular momentum loss from the stellar surfaces in both, TTs and HAeBes. Briefly, the model assumes that stars are contracting and that posses a dipolar magnetic field with strenghts in the range 0.5 to 3 kG. Star-disk interaction is included through the net torque onto the star, which has two main contributions: (1) magnetic braking with the disk and (2) stellar wind

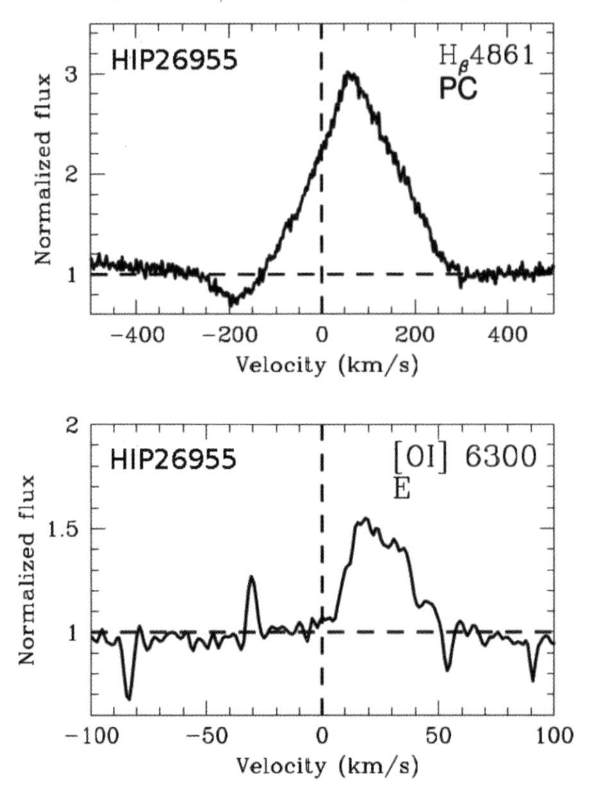

Figure 2. Residual emission for the $H\beta$ 4861Å and [OI] 6300.31Å lines in FIES spectrum of the Herbig Ae/Be star HIP26955. The continuum, normalized to 1.0 is marked with a horizontal dashed line while the stellar rest velocity is labeled with a vertical dashed line.

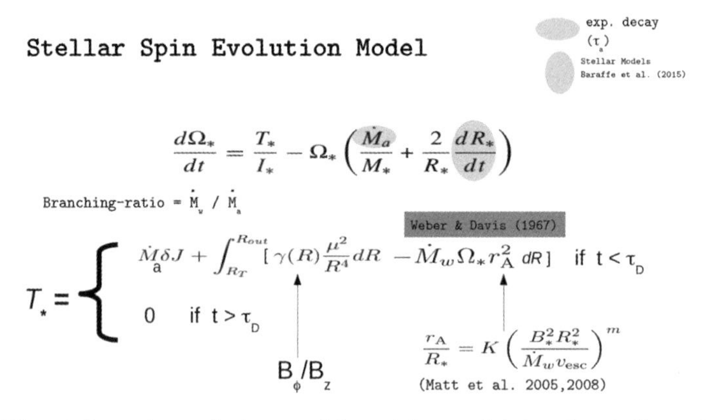

Figure 3. The stellar spin evolution model used for explaining the stellar rotation in our sample of TTs and HAeBes.

torque. At each instant the net torque onto star is computed and used to obtain the stellar angular frequency ($\Omega_* = vR_*$). Stellar radii and moment of inertia are obtained from evolutionary models of Baraffe et al. (2015). We suppose that cirmumstellar disk surrounding young star has a finite timelife $\tau_{disk} = 3$ Myr and that accretion and mass loss rate follow an exponential time decay as indicated in Figure 3.

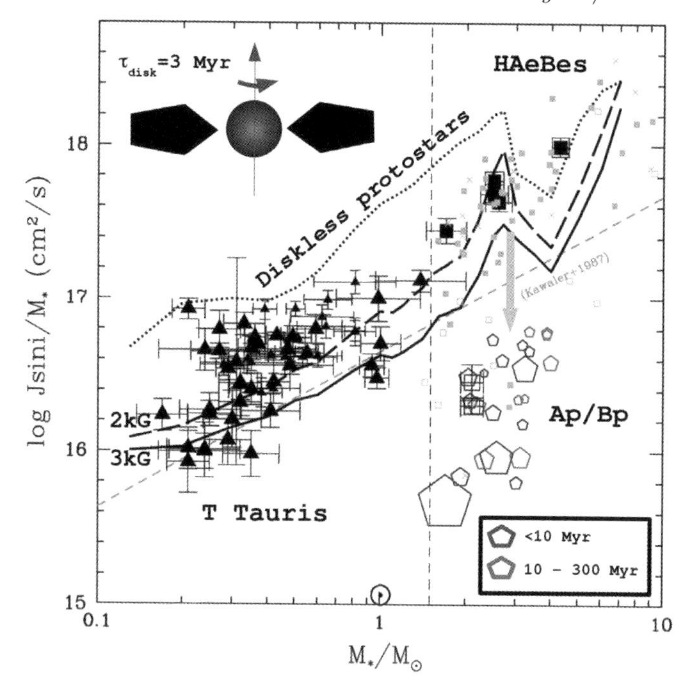

Figure 4. Specific angular momentum for TTs in the σ-Orionis cluster (triangles) and HAeBes (rectangles in black). The dashed grey line represents the main-sequence trend. Pentagons correspond to the sample of Ap/Bp stars of Aurière (2007) with size symbols are proportional to the dipolar field strength. Solid lines represent a snapshot of angular momentum at 3 Myr for two $B_* = 2$ and 3 kG.

4. Main Results

• Using stellar radii for our well characterized sample of TTs and HAeBes, we computed the specific angular momenta (per unit of mass J/M_*). For comparison purposes and under assumption that disks survive along the Hayashi track in the HR diagram, we calculated the synthetic $< J/M_* >$ values. Solid black lines in Figure 4 correspond for magnetic field strenghts of 2 and 3 kG under the following assumptions : (1) MA operates along the whole spectral types of our sample and (2) the mass loss rate behaves as the accretion rate but with one tenth of its intensity.

• The best fit to the data is obtained for $B_* = 2.0$ kG. For this particular case and through the use of the Kawaler law, we estimate that specific angular momentum is transferred outward during the contraction towards the main sequence by a factor of $\sim$3.2, equivalent to $\sim 3.2 \times 10^{17}\ cm^2\ s^{-1}$.

• We have included $< J/M_* >$ values for the sample of Ap/Bp stars of Aurière (2007) which are expected to be evolved HAeBes. We find that specific angular momentum must be lost by a factor between 12 and 80 from HAeBes to Ap/Bp stars, depending on the intensity of the dipolar field.

Although detailed phenomena of TTs and HAeBes are not considered, our approach result useful for testing the validiy of MA using stellar rotation as a proxy. However, the results obtained for the angular momentum in HAeBes do not explain the low rotation of Ap/Bp stars.

References

Alecian, E., Wade, G., et al. 2013, MNRAS, 429, 1001

Aurière, M., Wade, G., 2007, et al. A&A, 475, 1053

Baraffe, I. and Homeier, D., 2015, A&A, 577, 42
Boehm, T. and Catalá, C., 1994, A&A, 290, 167
Cauley, P., Johns-Krull, C., et al. 2015, ApJ, 810, 5
Fairlamb, R., Oudmaijer, R., et al. 2015, MNRAS, 453, 976
Ferrario, L., Pringle, J., et al. 2009, MNRAS, 400, L71
Hernández, J., Calvet, N., et al. 2014, ApJ, 794, 36
Hubrig, A. A., Ilyin, I., et al. 2014, in European Phys. Journal Web of Conf., Vol. 64, 08006
Matt, S. and Pudritz, R., 2008, ApJ, 678, 1109
Pinzón, G., Hernández, J., Serna, J., et al. 2021, AJ, 162, 90
Serna, J., Hernández, J., et al. 2021, ApJ, 923, 177
Tonry, J. and Davis, M., 1979, AJ, 84, 1511

Winds of Stars and Exoplanets
Proceedings IAU Symposium No. 370, 2023
A. A. Vidotto, L. Fossati & J. S. Vink, eds.
doi:10.1017/S174392132300011X

Short-term variations of surface magnetism and prominences of the young sun-like star V530 Per

Cang Tianqi[1,2], **Pascal Petit**[2], **Jean-François Donati**[2] **and Colin Folsom**[2,3]

[1]Department of Astronomy, Beijing Normal University, Xinjiekouwaidajie 19, 100875, Beijing, China

[2]Institut de Recherche en Astrophysique et Planétologie, Université de Toulouse, CNRS, CNES, 31400 Toulouse, France

[3]Tartu Observatory, University of Tartu, Observatooriumi 1, Tõravere, 61602 Tartumaa, Estonia

Abstract. V530 Per is a solar-like member of the young open cluster α Persei, with an ultra-short rotation period (P$\sim$0.32d). We report on two spectropolarimetric campaigns using ESPaDOnS, aimed at characterizing the short-term variability of its magnetic activity and large-scale magnetic field. We used time-resolved spectropolarimetric observations obtained in 2006 and 2018 and reconstructed the brightness distribution and large-scale magnetic field geometry of V530 Per through Zeeman-Doppler imaging. Using the same data sets, we also mapped the spatial distribution of prominences through tomography of Hα emission. We reconstruct, at both epochs, a large, dark spot occupying the polar region of V530 Per while smaller (dark and bright) spots were reconstructed at lower latitudes. The maximal field strength reached $\sim$1 kG. The prominence pattern displayed a stable component that was confined close to the corotation radius. In 2018, we also observed rapidly evolving Hα emitting structures, over timescales ranging from minutes to days. The fast Hα evolution was not linked to any detected photospheric changes in the spot or magnetic coverage.

Keywords. stars: magnetic fields, stars: chromospheres, stars: individual (V530 Per)

1. Introduction

A fraction of young solar analogs in open clusters possess short rotation periods, a puzzling observation indicating that the magnetic braking supposed to act during pre-main sequence evolution was less efficient than expected on these objects, compared to slow rotators observed at the same age. In the most extreme cases, these rapidly rotating stars reach the so-called saturated dynamo regime, and sometimes the supersaturated regime (Pallavicini et al. 1981; Prosser et al. 1996; Wright et al. 2011). We propose here to characterize the magnetic field and related phenomena (e.g., active regions, prominences) of a prototypical saturated star V530 Per (Cang et al. 2020, 2021).

2. Doppler tomography

By applying the Doppler Imaging method (Vogt, Penrod, & Hatzes 1987) using the code of Folsom et al. (2018), we recovered the surface brightness distribution of V530 Per. The reconstruction includes a solar-like differential rotation law, with a roughly solar shear level optimizing our model. We also applied the Zeeman Doppler Imaging

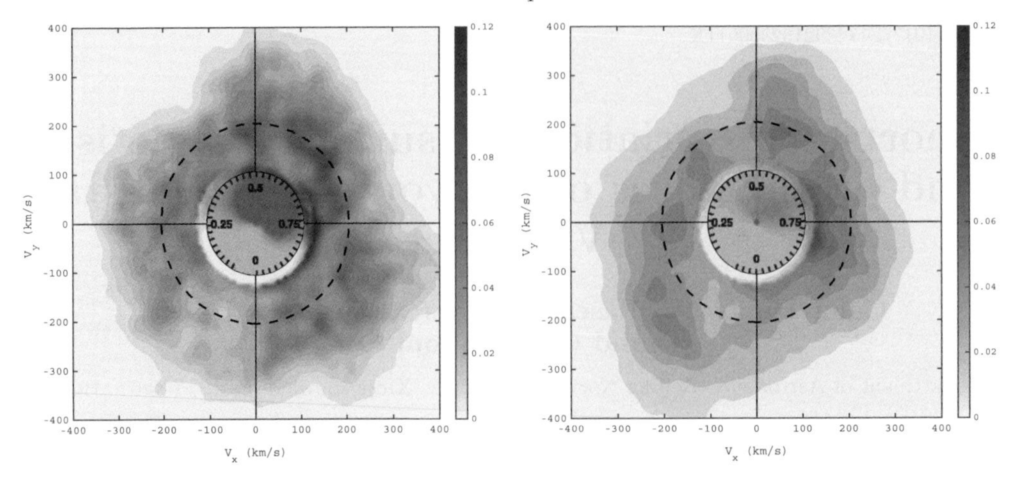

Figure 1. Prominence maps of V530 Per reconstructed from the data of two close nights (22&23 Oct 2018). The inner, filled blue circle represents the stellar surface. Radial ticks inside this circle give the rotational phases of Hα observations. The outer dashed circle is the corotation radius. The color scale depicts the local Hα equivalent width, in units of picometers per 8 km s^{-1} square pixel.

technique (ZDI, Semel 1989; Donati et al. 2006; Folsom et al. 2018) to model the two epochs of Stokes V data, in order to reconstruct the 2D magnetic field distribution of V530 Per. According to the maps, the average surface magnetic field is 177G in 2006 and 222G in 2018, with local peaks slightly above 1kG. About 2/3 of the magnetic energy belongs to the toroidal field component. Moreover, the complexity of the magnetic field is very high, with less than 7% of the magnetic energy stored in the dipolar component. We reconstructed prominence maps of V530 Per for each observational night, unveiling a stable component that was confined close to the co-rotation radius and short-term variations (see Fig.1). The total mass stored as prominences is $\sim 4.6 \times 10^{17}$ kg ($\sim 2.3 \times 10^{-13}$ $M_\odot$). Our observation also suggests that as much as 3.5×10^{16} kg of material has been removed from the system within one day.

References

Cang T.-Q., Petit P., Donati J.-F., Folsom C. P., Jardine M., Villarreal D'Angelo C., Vidotto A. A., et al., 2020, A&A, 643, A39. doi:10.1051/0004-6361/202037693

Cang T.-Q., Petit P., Donati J.-F., Folsom C. P., 2021, A&A, 654, A42. doi:10.1051/0004-6361/202141975

Donati J.-F., Semel M., Carter B. D., Rees D. E., Collier Cameron A., 1997, MNRAS, 291, 658. doi:10.1093/mnras/291.4.658

Donati J.-F., Catala C., Landstreet J. D., Petit P., 2006, ASPC, 358, 362

Donati J.-F., Howarth I. D., Jardine M. M., Petit P., Catala C., Landstreet J. D., Bouret J.-C., et al., 2006, MNRAS, 370, 629. doi:10.1111/j.1365-2966.2006.10558.x

Folsom C. P., Bouvier J., Petit P., Lèbre A., Amard L., Palacios A., Morin J., et al., 2018, MNRAS, 474, 4956. doi:10.1093/mnras/stx3021

Pallavicini R., Golub L., Rosner R., Vaiana G. S., Ayres T., Linsky J. L., 1981, ApJ, 248, 279. doi:10.1086/159152

Prosser C. F., Randich S., Stauffer J. R., Schmitt J. H. M. M., Simon T., 1996, AJ, 112, 1570. doi:10.1086/118124

Semel M., 1989, A&A, 225, 456

Vogt S. S., Penrod G. D., Hatzes A. P., 1987, ApJ, 321, 496. doi:10.1086/165647

Wright N. J., Drake J. J., Mamajek E. E., Henry G. W., 2011, ApJ, 743, 48. doi:10.1088/0004-637X/743/1/48

Winds of Stars and Exoplanets
Proceedings IAU Symposium No. 370, 2023
A. A. Vidotto, L. Fossati & J. S. Vink, eds.
doi:10.1017/S1743921322004355

Water and silicon-monoxide masers monitored towards the "water fountain" sources

H. Imai[1,2], K. Amada[3], J. F. Gómez[4], L. Uscanga[5], D. Tafoya[6], K. Nakashima[7], K.-Y. Shum[7], Y. Hamae[7], R. Burns[8] and G. Orosz[9]

[1]Amanogawa Galaxy Astronomy Research Center, Kagoshima University, Japan

[2]Center for General Education, Kagoshima University

[3]Graduate School of Science and Engineering, Kagoshima University

[4]Instituto de Astrofísica de Andalucía, CSIC, Spain

[5]Departamento de Astronomia, Universidad de Guanajuato, Mexico

[6]Department of Space, Earth and Environment, Chalmers University of Technology, Sweden

[7]Faculty of Science, Kagoshima University

[8]Division of Science, National Astronomical Observatory of Japan

[9]User Support Team, Joint Institute for VLBI in Europe ERIC, Netherlands

Abstract. We have investigated the evolution of 12 "water fountain" sources in real time in the accompanying H_2O and SiO masers through our FLASHING (Finest Legacy Acquisitions of SiO-/H_2O-maser Ignitions by Nobeyama Generation) project. It has been confirmed that these masers are excellent probes of new jet blob ejections, acceleration of the material supplied from the parental circumstellar envelope and entrained by the stellar jets yielding its deceleration. Possible periodic variations of the maser emission, reflecting properties of the central dying stars or binary systems, will be further investigated.

Keywords. masers, stars: AGB and post-AGB, stars: mass loss, stars: winds, outflows

1. Introduction

A "water fountain" source (WF) is classified as an H_2O maser source associated with a high velocity, collimated outflow or jet driven by a dying star in the transition of the AGB phase to the phase of a central star of a planetary nebula. Recent ALMA observations revealed that they are likely experiencing the "common envelope evolution" of low to intermediate-mass binary stars ($M_* \lesssim 4\,M_\odot$) with extremely high mass loss rates (up to $\dot{M}_* \sim 10^{-3}\,M_\odot \mathrm{yr}^{-1}$) for a very short period (<200 yr)(Khouri et al. 2021). Because of such a short-lived event, it has been expected to see spectral and morphological evolutions of the WFs in H_2O and SiO masers over a few decades. We have conducted monitoring observations of these maser sources in the FLASHING project using the Nobeyama 45 m telescope (see Table 1) and ATCA. These observations aim to monitor the spectral evolution of the masers, while interferometric follow-up observations with ATCA and KaVA (KVN and VERA Array) aim to find their morphological evolutions.

Table 1. Specification of the FLASHING observations. The H22 and H40/Z45 receivers have been used for the 22 GHz and 43 GHz bands to observe H_2O and SiO masers, respectively. The pairs of the H22–H40 and H22–Z45 receivers can be used to simultaneously observe these masers (Okada et al. 2020). The latter pair has been operational since 2022 March. The root-mean-square (rms) noise level of H22 will be reduced since 2022 November.

Obs. periods	2018 Dec.—2019 May 2020 Dec.—2021 April 2019 Dec.—2020 April 2021 Dec.—2022 March 2022 Nov.—2023 April (proposed)	
Targets	38 stars (12 WFs, 2 WFCs, 24 AGBs/post-AGBs)	
Receivers	H22 (K-band, RHCP+LHCP)	H40 (Q Band, LHCP) Z45 (Q-band VLP+HLP)
rms noise	σ_{TA}~0.06 Jy (➡ 0.04 Jy)	σ_{TA}~0.1 Jy (H40) σ_{TA}~0.05 Jy (Z45)
Spectral windows	4 x 2 (H22+H40) 2 x 2 + 2 (H22+Z45) Δv≈820 km/s	8 (H22+H40) 5 x 2 (H22+Z45) Δv≈420 km/s
Velocity resolution	0.4 km/s	0.2 km/s
Beam size	~74″ (FWHM, with KQ-optics)	~39″ (with KQ-optics)
Aperture efficiently	~61% (with KQ-optics)	~55% (with KQ-optics)

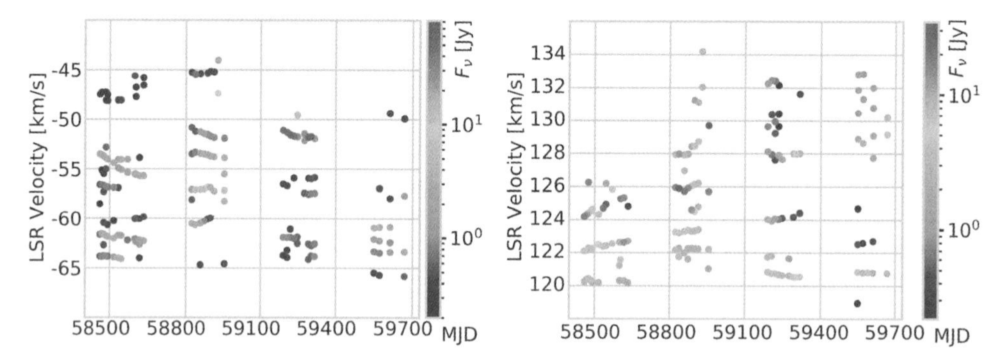

Figure 1. Peaks of H_2O maser spectra of W 43A taken with the NRO 45 m telescope. With respect to the systemic LSR velocity (~35 km s^{-1}), the drifts of the spectral peak velocities indicate further accelerations of the maser clumps, suggesting that the entrained material hosting the masers may be accelerated by the faster jet (~120 km s^{-1}).

2. Progress of the project

We have found new spectral peaks of H_2O masers breaking the records of the top speed of the WF jets by up to 130 km s^{-1} towards IRAS 18286−0959 (Imai et al. 2020) and IRAS 18043−2116 (Uscanga et al. 2022). Due to their too short lifetimes, it should be further investigated whether they exhibit rapid deceleration as predicted (Orosz et al. 2018). For both sources, comparing with the previous H_2O maser distributions (Walsh et al. 2009; Imai et al. 2013a,b) a growth of the maser jet with a very short dynamical timescale (~30 yr) also has been confirmed (Imai et al. 2020; Uscanga et al. 2022).

We also newly found SiO masers associated with IRAS 16552−3050 (Amada et al. 2022), the second case of SiO masers in WFs after W 43A. This was newly yielded by our development of the simultaneous two-band observation system equipped with the Nobeyama 45 m telescope (Okada et al. 2020).

Through the intensive monitoring observations, we have found systematic velocity drifts of H_2O masers in W 43A (see Figure 1). In W 43A, it is suggested that the faster jet traced by CO emission should accelerate the outflow formed in entrained material supplied from the parental circumstellar envelope (Tafoya et al. 2020). The observed maser accelerations support this suggestion.

Periodicity of the maser spectra is expected if the central stellar system is composed of a long period variable such as an OH/IR star (e.g., Imai et al. 2013b) or a binary system (Tafoya et al. 2020; Khouri et al. 2021). This will be confirmed after solving the complexity of the maser spectra affected by chaotic variation and the artificial periodicity due to time gaps of the monitoring program.

References

Amada, K., et al. 2022, AJ, 163, 85
Imai, H., et al. 2020, PASJ, 72, 58
Imai, H., et al. 2013, PASJ, 65, 28
Imai, H., et al. 2013, ApJ, 771, 47
Khouri, T., et al. 2021, Nature Astron., 6, 275
Okada, N., et al. 2020, PASJ, 71, 7
Orosz, G., et al., 2018, MNRAS Lett., 482, L40
Tafoya, D., et al. 2020, ApJL, 890, L14
Uscanga, L., et al. 2022, ApJ/AJ, submitted
Walsh, A. J., et al. 2009, MNRAS, 394, L70

Winds of Stars and Exoplanets
Proceedings IAU Symposium No. 370, 2023
A. A. Vidotto, L. Fossati & J. S. Vink, eds.
doi:10.1017/S174392132200480X

Weakening the wind with ULLYSES: Examining the Bi-Stability Jump

Olivier Verhamme and Jon Olof Sundqvist

Instituut voor Sterrenkunde (IvS), KU Leuven, Celestijnenlaan 200D, 3001 Leuven, Belgium
email: olivier.verhamme@kuleuven.be

Abstract. Radiation-driven mass-loss is an important, but still highly debated, driver for the evolution of massive stars. Current massive star evolution models rely on the theoretical prediction that low luminosity massive stars experience a sudden increase in mass loss below a stellar effective temperature of about 20 000 K. However, novel radiation-driven mass-loss rate predictions show no such bi-stability jump, which effects the post main-sequence evolution of massive stars. The ULLYSES data set provides a unique opportunity to investigate the theoretical bi-stability jump dichotomy and may help to assess the existence of the bi-stability jump in massive star winds. By utilising UV spectra from ULLYSES combined with X-shooter optical data we obtain empirical mass-loss rate constraints, that are no longer degenerate to the effects of wind clumping, and derive novel empirical constraints on the mass-loss behavior across the temperature range of the bi-stability jump. Current preliminary results do not show a clear presence of a bi-stability jump.

Keywords. stars: mass loss, techniques: spectroscopic, stars: winds, outflows

1. Introduction

Winds from hot, massive stars are line driven outflows which can have mass loss rates up to $10^{-5}\frac{M_{\odot}}{yr}$ and terminal velocities which can exceed $2000\frac{km}{s}$. As these these stars lose a very substantial amount of mass, determining correct mass loss rates is important to understand their evolution (figure 1a Björklund et al. (2022)). In current stellar evolution codes the mass loss prescription by Vink et al. (2001) is standardly used to evolve the star, which includes an increase in mass loss rate at the so called bi-stability jump around 20-25K due to the recombination of FeIV-FeIII. However, recent self-consistent models do not predict such a jump in mass loss rate (Björklund et al. 2022). As these two theoretical calculations show drastically different behaviour, it is needed to compare to empirical mass loss rates. This is not the first time mass loss rates of stars around the bi-stability jump have been studied empirically (e.g. Rubio-Díez et al. (2022), Markova & Puls (2008)). However, due to recent observational campaigns with the Hubble space telescope and the VLT X-SHOOTER, we now have access to ample high resolution optical and UV spectra for stars in the B supergiant regime. This allows for the simultaneous systematic fitting of UV-resonance lines and optical recombination lines. As the resonance lines are less sensitive to the clumping factor of the stellar winds it is possible to get a good fit on the mass loss rates and the clumping of the wind at the same time.

2. Method

The goal is to determine the wind, clumping and stellar parameters as well as the stellar parameters in one fitting procedure using a multitude of spectral lines. The high parameter space of the fits requires a special optimization scheme which is able to avoid

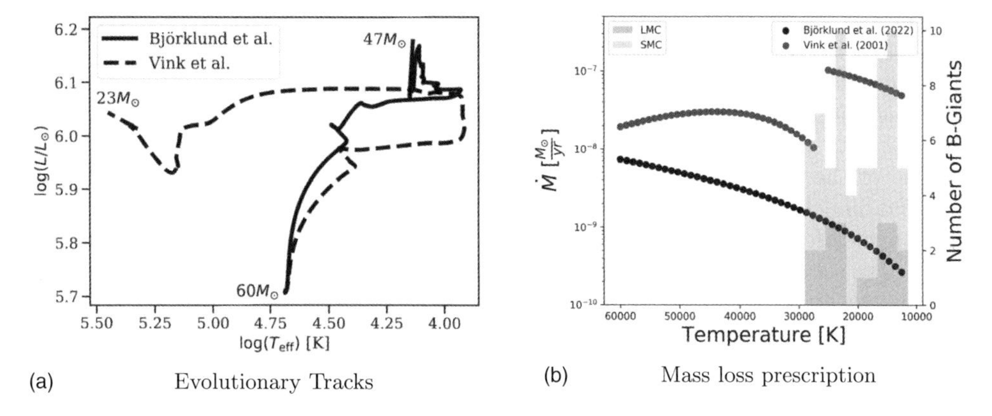

(a) Evolutionary Tracks (b) Mass loss prescription

Figure 1. The left figure shows two MESA models for a massive star with an initial mass of 60 $M_{\odot}$. One model uses the mass loss rate prescription by Vink et al. (2001) while the other uses the prescription of Björklund et al. (2022). The right figure shows the two mass loss rate prescriptions as well as an overlaying histogram showing the amount of B-supergiants in the ULLYSES sample.

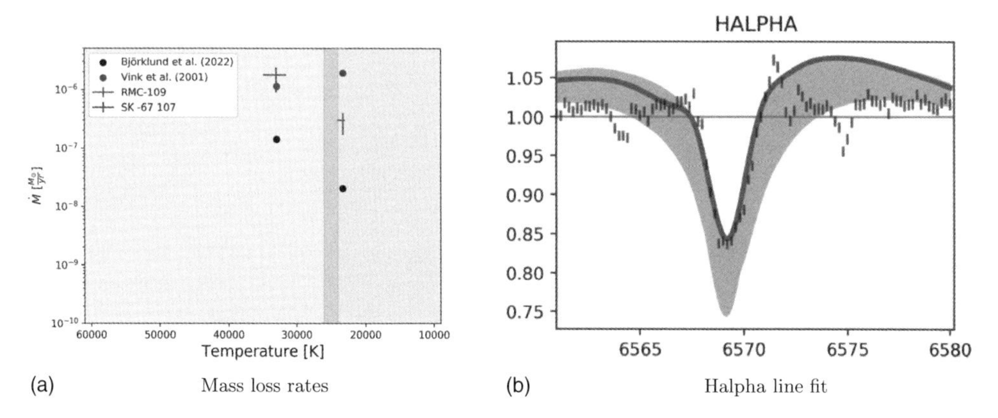

(a) Mass loss rates (b) Halpha line fit

Figure 2. On the left you can see the comparison between the emprical mass loss rates and their equivalent mass loss from the two different mass loss rates prescriptions. The errors given here are the 2-σ errors. On the right you see the H-α fit from RMC-109.

local minima, like a genetic algorithm. Kiwi-GA is a genetic algorithm produced by Brands et al. (2022) and already used for a similar study for stars with higher temperatures. The spectral lines themselves are calculated by Fastwind (Puls et al. 2005), as the GA only decides the parameters of the models which are fitted and computes the goodness of fit. Fastwind is a 1 D stellar atmosphere code which is capapable of modelling clumping of arbitrary optical thickness in the winds (Sundqvist & Puls 2018).

3. Results and Discussion

Kiwi-GA creates around 30.000 FASTWIND models for each fit. As a result, it is possible to use the statistical nature of the fitting procedure to find errors. This procedure is exactly the same as the one described in Brands et al. (2022). Figure 2a shows the mass loss rates of 2 LMC B-supergiants compared to the values given by both the Vink and the Björklund prescriptions. The theoretical models were given the empirically calculated luminosity, spectroscopic mass, effective temperature and for the Vink models the ratio of the escape speed over the terminal velocity. The metalicity was set to 0.5 times solar metalicity as they are LMC objects. Even though figure 2a does not allow for comparison between the 2 stars directly as they do not share the same stellar parameters, it is possible to compare each star individually with the 2 prescriptions. While the star on the hot side of the jump is in relative agreement with the Vink prescriptions, the star on the cool side of the jump has a significantly lower mass loss rate than the Vink prescription. This might point to an overestimation of the mass loss rate of the Vink prescription on the cool side of the jump. The mass loss rates in general are very high compared to the Björklund models and even high compared to the Vink models which are usually seen as high estimates for mass loss. The results here are preliminary and only a small part of a greater data set has been studied so far, as more of the data set is analysed we will be able to give a full description of the wind over the bi-stability jump.

References

Björklund, R., Sundqvist, J. O., Bouret, J. C., Singh, S. M., Puls, J., Najarro, F., 2022, A&A

Brands, Sarah A., de Koter, Alex, Bestenlehner, et al., 2022, A&A, 663, A36

Markova, N., Puls, J., 2008, A&A, 478, 823

Puls, J., Urbaneja, M. A., Venero, R., Repolust, T., Springmann, U., Jokuthy, A., Mokiem M. R., 2005 A&A, 435, 669

Rubio-Díez, M. M., Sundqvist, J. O., Najarro, F.,Traficante, A., Puls, J., et al., 2022, A&A

Sundqvist, J. O., Puls, J., 2018, A&A, 619, 59

Vink, Jorick S., de Koter, A., Lamers, H. J. G. L. M., 2001, A&A, 369, 574

Winds of Stars and Exoplanets
Proceedings IAU Symposium No. 370, 2023
A. A. Vidotto, L. Fossati & J. S. Vink, eds.
doi:10.1017/S1743921323001783

Statistical properties of cold circumstellar envelops observed in NESS–NRO

K. Amada[1], S. Fukaya[1], H. Imai[1,2], P. Scicluna[3,4], N. Hirano[4], A. Trejo-Cruz[4], S. Zeegers[4], F. Kemper[3,5], S. Srinivasan[6], S. Wallström[7,4], T. Dharmawardena[8,4] and H. Shinnaga[1]

[1]Graduate School of Science and Engineering, Kagoshima University, Japan

[2]Center for General Education, Institute for Comprehensive Education, Kagoshima University

[3]European Southern Observatory, Alonso de Cordova Santiago RM, Chile

[4]Institute of Astronomy and Astrophysics, Academia Sinica, Taiwan

[5]European Southern Observatory, Karl-Schwarzschild-Str. 2, Germany

[6]Instituto de Radioastronomía y Astrofísica, UNAM, Mexico

[7]Institute of Astronomy, KU Leuven, Belgium

[8]Max-Planck-Institute for Astronomy, Königstuhl 17, Germany

Abstract. We conducted CO J=1→0 emission line observations for nearby AGB stars using the Nobeyama 45 m telescope. Comparing our results with those from CO J=3→2 observations with JCMT, the circumstellar envelopes observed in CO J=1→0 look more extended than J=3→2. Thus, we could trace the outer, cold parts of the envelopes. We also found four stars in which the CO/^{13}CO ratio changes dramatically outward, but the change implies the effect of selective photodissociation by interstellar ultraviolet radiation, not the third dredge up in the stellar interior. We moreover found two unique stars with aspherical envelope morphology.

Keywords. stars: AGB and post-AGB, stars: mass loss, stars: winds, outflows, stars: carbon

1. Introduction

The asymptotic giant branch (AGB) represents the final stage in the evolution of low- and intermediate-mass stars. Inside the AGB stars, carbon is dredged up from the vicinity of the central core to the stellar surface by the convection (Third Dredge Up (TDU)) caused by thermal pulses (TPs, e.g. Herwig 2005). This repeated TDU process gradually changes the C/^{13}C abundance ratio in the circumstellar envelopes (CSEs). The issues of this TDU process are its efficiency and the mechanism determining the stellar mass-loss rate (MLR). Because the MLR is enhanced by TP, a detached shell-like gas (and dust) distribution with high C/^{13}C ratio is formed in CSEs (e.g. Olofsson et al. 1990). Therefore, these issues are expected to be constrained by determining the CO/^{13}CO ratio over the entire CSEs.

The Nearby Evolved Stars Survey (NESS) project addresses several issues with a volume-complete sample of ∼850 evolved stars within 3 kpc (Scicluna et al. 2022). These stars have been observed in the CO J= 2→1 and 3→2 emission lines, and the sub-mm continuum mainly using the James Clerk Maxwell Telescope (JCMT). In addition, we have conducted staring and mapping observations in CO J= 1→0 emission using the 45 m radio telescope of the Nobeyama Radio Observatory (NRO) to reveal more extended cold gas distribution.

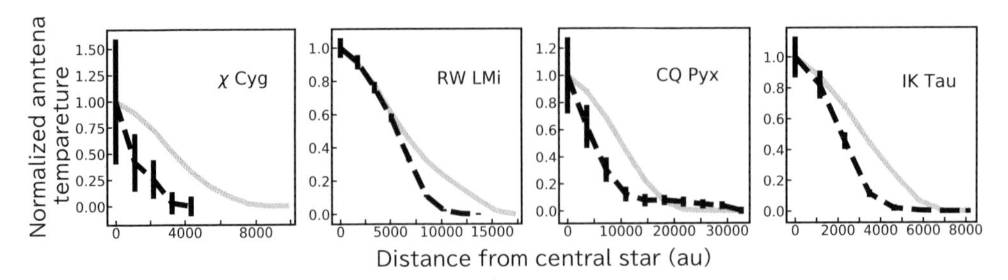

Figure 1. Radial profiles in CO and ^{13}CO J=1→0 emission lines for 4 stars with ^{13}CO intensity decreasing against CO intensity. Grey solid and black dashed lines show CO and ^{13}CO emission lines, respectively.

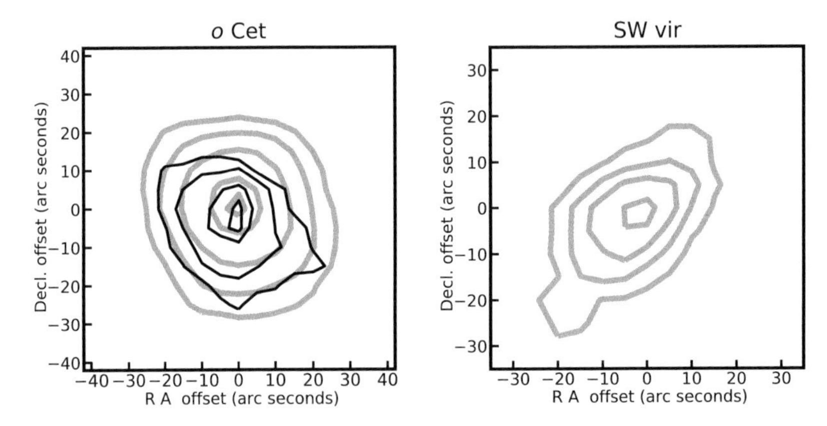

Figure 2. Velocity integrated maps of o Cet (left panel) and SW Vir (right panel). Grey and black lines show the CO and ^{13}CO emission line, respectively. Contour levels in CO and ^{13}CO are peak flux times (10, 20, 40, 80, 95) and (20, 40, 80, 95) in o Cet, (10, 40, 70, 95) in SW Vir.

2. Results of NESS–NRO

We carried out staring observations for 212 stars, and detected CO and ^{13}CO emission lines in 96 and 34 stars, respectively. For the stars detected in CO emission, MLR has been derived from the CO line profile, and the ranged from 10^{-9} to 10^{-5} $M_{\odot}$ yr^{-1}. However, it would be underestimated taking into account the derived MLR/dust-production rate ratio that is ∼5 times smaller than previously derived (Knapp 1985). This small ratio suggests that the beam could not cover the whole CSEs with the beam of the NESS–NRO observations smaller than those CSEs.

We conducted mapping observations for 27 stars, and detected CO and ^{13}CO emission towards 19 and 10 stars, respectively. Comparing the angular sizes of CSEs observed in the CO J=1→0 emission using NRO with those in the CO J=3→2 emission using JCMT (Scicluna et al. 2022), the CO J=1→0 appears to be 2–3 times more extended than the J=3→2 towards all the stars so as to trace outer, cold parts of the envelopes.

Through the radial profiles of the CO and ^{13}CO J=1→0 emission in the CSEs of 10 stars, we found four stars towards which the ^{13}CO intensity, or the CO/^{13}CO intensity ratio, decreases dramatically against the CO intensity outward (see Figure 1). This implies no evidence for TPs or TDUs within recent 2000–7000 yr when developing the present envelope. Note that the radial change in the intensity ratio is caused by a decrease in the ^{13}CO rather than the CO intensity. This suggests the effect of selective photodissociation for CO molecules by intersteller ultraviolet radiation (e.g. Saberi et al. 2020).

We also found asymmetry in the CSEs of SW Vir and o Cet (see Figure 2). The CSE of o Cet appears to be formed aspherically by a bipor outflow and the selective photodissociation. However, the energy origin forming the SW Vir's CSE is still unclear.

References

Herwig, F. 2005, A&ARv, 43, 435
Knapp, G.R. 2014, ApJ, 293, 273
Olofsson, H., et al. 1990, A&A, 230, L13
Scicluna, P., et al. 2021, MNRAS, 512, 1091
Saberi, M., et al. 2020, A&A, 638, A99

Part 3:
Physical ingredients of winds

Winds of Stars and Exoplanets
Proceedings IAU Symposium No. 370, 2023
A. A. Vidotto, L. Fossati & J. S. Vink, eds.
doi:10.1017/S1743921322004598

The origin of planetary winds

Daria Kubyshkina

Space Research Institute, Austrian Academy of Sciences, Schmiedlstrasse 6, A-8042 Graz, Austria

Abstract. Atmospheric escape is a fundamental phenomenon shaping the structure and evolution of planetary atmospheres. Physics of planetary winds range from global processes such as tidal interactions with the host star, through large-scale hydrodynamic outflow, to essentially microphysical kinetic effects, including Jeans-like escape and the interaction of planetary atmospheres with stellar winds and the own magnetic fields of planets. Each of these processes is expected to be most relevant for planets of different properties and at different stages in planetary and stellar evolution. Thus, it is expected that the hydrodynamic outflow guides the evolution of hydrogen-dominated atmospheres of planets having low masses (below that of Neptune) and/or close-in orbits, while the kinetic effects are most important for the long-term evolution of planets with secondary atmospheres, similar to the inner planets in the Solar System. Finally, each of these processes is affected by the interaction with stellar winds.

Keywords. hydrodynamics, atmospheric effects, planets and satellites: general

1. Introduction

Planetary winds, or atmospheric escape, represent the outflow of atmospheric material occurring on different space- and timescales. Depending on the age of the planet and the type of its atmosphere, planetary winds can be driven by a wide range of physical processes. Formally, these processes can be divided into thermal and non-thermal ones, where the former depends directly on the energy deposition in the atmosphere through heating, and the latter includes the processes related to microphysical interactions of the atmospheric species, to the interaction of the atmospheres with stellar winds, and to the effects related to the tidal or magnetic interaction with the host star. In Figure 1, I suggest a simplified classification of these processes.

The most classic form of thermal escape is the kinetic Jeans escape (e.g., Öpik 1963; Chamberlain 1963; Mihalas & Mihalas 1984), which assumes the atmospheres described by the Boltzmann distribution and considers the escape of the most energetic particles from the exobase, i.e., from the level where the atmosphere becomes collisionless. This approach implies that the fraction of particles with energies sufficient to overcome the escape velocity is small, and the whole atmosphere remains in a steady state without significant bulk motion. Such an assumption holds for the compact, terrestrial-like, secondary atmospheres or for the hydrogen-dominated atmospheres of Jupiter-like planets in absence of the significant energy input from an external source. With increasing energy input, the thermal energy of the atmospheric gas grows. At some point, the bulk energy of the atmospheric material overcomes the gravitational well of the planet and forms a continuous bulk outflow (e.g., Watson et al. 1981). This case is considered by the hydrodynamic approach, assuming collisional, fluid-like atmospheres.

In the case of hydrodynamic escape, the outflow can be supported by different energy/heating sources. The most common one is the heating by the high-energy stellar

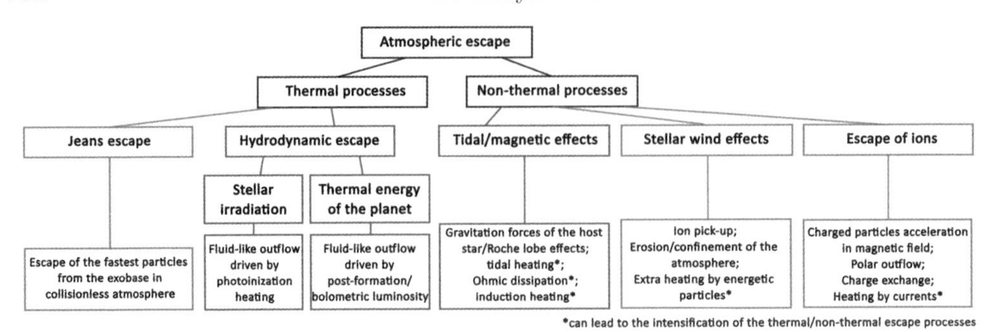

Figure 1. A simplified scheme reflecting the most common types of atmospheric escape, i.e., the mechanisms driving planetary winds. The color of the boxes highlights kinetic processes mainly relevant in the case of secondary atmospheres (orange) and hydrodynamic/global processes that have a major impact on voluminous hydrogen-dominated atmospheres (blue).

irradiation (X-ray + extreme ultraviolet, XUV) absorbed deep in the upper atmosphere. However, the hydrodynamic planetary wind can be also driven by the heat deposited in the planetary core/lower atmosphere, such as the post-formation cooling luminosity of a planet or bolometric heating from the host star.

Non-thermal escape processes, in turn, can be split into three groups: the tidal/magnetic interaction with the host star, interaction with stellar winds, and the ion escape processes at magnetized and non-magnetized planets. The tidal forces of the host star are relevant for the escaping atmospheres of planets in close-in orbits, and their influence can be twofold. First, stellar gravity forces can interact with atmospheric material directly, decreasing the energy needed for the atmospheric particles to leave the gravitation well of the planet. In extreme cases of very small orbital distances, the Roche lobe (the obstacle within which the atmosphere is gravitationally bound to the planet) comes close to the planetary photosphere resulting in the catastrophic Roche lobe overflow (Koskinen et al. 2022). Second, the tidal heating in the lower levels of the atmosphere (e.g., Bodenheimer et al. 2001; Arras & Socrates 2010) can lead to planetary inflation and to the intensification of the thermal escape processes due to the increase in the energy budget of the atmosphere. A similar effect can be caused by the magnetic interaction between the star and a close-in planet, through the induction of currents deep in the atmosphere (e.g., Batygin & Stevenson 2010; Wu & Lithwick 2013; Ginzburg & Sari 2016; Kislyakova et al. 2018). For further detail on these processes, I point the reader to the recent review by Fortney et al. (2021).

Stellar winds represent the hot coronal material (low-density plasma) accelerated to high velocities and carrying stellar magnetic field (see, e.g., Vidotto et al. 2015). Their interaction with planetary atmospheres depends strongly on the presence of planetary magnetic field (see, e.g., Cohen et al. 2022), parameters of the planet (e.g., Carolan et al. 2020) and age of the system (e.g., Kubyshkina et al. 2022). Along with the erosion of planetary atmospheres, extra heating of atmospheres by stellar wind energetic particles, and ion pick-up (drag of ions from the exosphere due to the interaction with the electromagnetic field of the wind), increasing the atmospheric mass loss, stellar winds can confine planetary atmospheres leading to the reduction of the escape if the pressure balance is reached close to the planet (e.g., Christie et al. 2016; Vidotto & Cleary 2020; Carolan et al. 2020, 2021).

The wide range of non-thermal escape processes, including ion pick-up by the stellar wind, relates specifically to the escape of the ionized species, and depends, therefore, on the presence, strength, and configuration of the planetary magnetic field. This includes precipitation of energetic particles/ions from the upper atmosphere, interaction with energized particles from stellar winds or the ionosphere, the processes controlled by the magnetic field of the planet (as polar outflow and acceleration of the particles on closed magnetic lines), and processes driven by photochemistry. All these processes strongly depend on the specific content of planetary atmospheres.

Besides the division into thermal and non-thermal processes, the processes described above can be split into kinetic (indicated by orange boxes in Figure 1) and large-scale hydrodynamic/tidal effects (blue boxes). In general, all of these processes can occur simultaneously and can take place at any planet. Their relative input in driving planetary wind, however, differs depending on the age of the planet (and its host star) and the type of atmosphere. Thus, the hydrodynamic escape (predicting the largest atmospheric mass loss rates among the processes discussed here) plays a major role for primary hydrogen-dominated atmospheres. It drives, therefore, the winds of young planets and of close-in Neptune-like/Jupiter-like planets. The kinetic escape processes, in turn, are relevant for compact secondary (terrestrial-like) atmospheres.

In the following sections, I consider the mechanisms outlined above in more detail. I focus predominantly on the hydrodynamic outflow, and for more detail on the kinetic escape processes, I point the reader to the recent and detailed review by Gronoff et al. (2020). Here, I start with describing the Jeans escape in section 2.1 and give a brief overview of the ion escape from non-magnetized and magnetized planets in sections 2.2 and 2.3. In section 3 I describe the typical approach to the hydrodynamic outflow and discuss in more detail the XUV-driven escape (section 3.1), the hydrodynamic escape driven by the own thermal energy of a planet (section 3.2), and the tidal effects in this context (section 3.3). I describe the influence of the hydrodynamic escape across the wide parameter range in section 3.4 and discuss the influence of magnetic fields on these escape processes in section 4. I summarize my conclusions in Section 5.

2. Kinetic escape processes

In this chapter, I briefly discuss the variety of microphysical/kinetic processes driving atmospheric escape. Though the typical atmospheric escape rates associated with such processes are at least an order of magnitude below those of the hydrodynamic escape, the kinetic processes are crucial for the evolution of planets and their atmospheres. Specifically, the timing and relative input of different thermal and non-thermal processes control the fractionation of chemical elements and their isotopes, and therefore essentially control the atmospheric content. For more information, I point the reader to the recent review on the fractionation of noble gases and other heavy elements including a description of all atmospheric fractionation processes by Lammer et al. (2020b).

2.1. *Jeans escape*

Jeans escape represents the most classic form of thermal escape considering the kinetic escape of most energetic particles in the atmosphere. To evaluate it in the most accurate way, one has to solve the Boltzmann equation (see, e.g., Volkov et al. 2011). However, given the computational costs of such a method, a range of simplifications are commonly applied. In the most common approach, the atmosphere is considered Maxwellian (Mihalas & Mihalas 1984)

$$f(\vec{x},\vec{v}) = n_{\rm i}\left(\frac{1}{\pi u_{\rm th,i}^2}\right)^{3/2} \exp\left(-\frac{v^2}{u_{\rm th,i}^2}\right), \tag{2.1}$$

where $n_{\rm i}$ is the numerical density, and $u_{\rm th,i} = \sqrt{\frac{2k_{\rm b}T}{m_{\rm i}}}$ is a thermal velocity of i-th atmospheric species with the mass $m_{\rm i}$; atmospheric temperature T is assumed to be constant.

The most energetic particles in this distribution, with velocities $v > v_{\rm esc} = \sqrt{\frac{2GM_{\rm pl}}{r}}$ at the exobase (where $M_{\rm pl}$ is the planetary mass and r is the radial distance from the planet center), escape. With such an approach, the flux of escaping particles depends on the relation between the escape velocity and the thermal speed

$$\Phi = n_{\rm i} \left(\frac{u_{\rm th,i}^2}{4\pi} \right)^{1/2} (1 + \lambda_{\rm exo}) \exp(-\lambda_{\rm exo}) \, . \tag{2.2}$$

Parameter $\lambda_{\rm exo} = \frac{v_{\rm esc}^2}{u_{\rm th,i}^2} = \frac{GM_{\rm pl} m_{\rm i}}{k_{\rm b} T_{\rm exo} r_{\rm exo}}$ (Jeans parameter) represents, in a more general sense, the relation of the gravitational energy of the planet to the thermal energy deposited in its atmosphere calculated at exobase. Besides controlling the flux 2.2, or the atmospheric mass loss rate in terms of thermal atmospheric escape, this relation is used to distinguish between the Jeans and the hydrodynamic regimes. As discussed in the introduction, with increasing thermal energy the atmospheric outflow changes qualitatively – the mean energy of the particles becomes high enough to overcome the gravity of the planet itself and the binding due to the collisions between the atmospheric particles, allowing for the planetary wind to engage deep below the exobase level. The border value of $\lambda_{\rm exo}$ was estimated 1-3 upon different assumptions (Hunten 1982; Volkov et al. 2011; Erkaev et al. 2015).

Differently to the hydrodynamic escape, where the dense outflow of light elements (such as hydrogen and helium) can drag away the heavier elements (e.g., Zahnle & Kasting 1986; Hunten et al. 1987; Pepin 1991; Odert et al. 2018; Lammer et al. 2020a; Erkaev et al. 2022), in "collisionless" Jeans escape the escape of different species occurs almost independently. Therefore, in quiet conditions, Jeans escape can describe the escape of lighter particles through the stable heavy-element atmospheres, as, e.g., escape of hydrogen and helium from the present-day Earth, Venus, and Mars (Shizgal & Blackmore 1986; Chamberlain 1969; Hunten 1973). This makes Jeans escape particularly important for the fractionation of light elements in planetary atmospheres.

Summarizing the above, the Jeans escape is most relevant for the atmospheric escape from planets exposed to the thermal escape but remaining under moderate conditions, i.e., with energy input that is non-negligible but also not too large (as in the case of close-in exoplanets or young host stars). The atmospheres of such planets are expected to remain stable. Therefore, Jeans escape typically defines the outflow from compact secondary atmospheres of terrestrial-like planets in evolved (ages of a few Gyr) planetary systems or from cool gas giants with hydrogen-dominated atmospheres close to the hydrostatic equilibrium. In the latter case, and also for $\lambda_{\rm exo}$ close to the border values, the atmospheric mass loss rates given by this approach become comparable to slow hydrodynamic outflow.

The formulation of the Jeans escape presented here has a range of limitations. The real distribution of the atmospheric particles is not quite Maxwellian, especially in the case of strong Jeans escape, and the atmosphere is not isothermal; escape also does not occur only from exobase. The range of corrections exists to account for these facts (see Gronoff et al. 2020, and the references therein) but in the most general case one has to solve the Boltzmann equation to get the most accurate estimate. Also, if one aims at using equation 2.2 as an analytical approximation, one needs to know the position and parameters of the atmosphere at the exobase. In the case of exoplanets, these parameters are typically unknown, while they can vary significantly through the evolution and across planetary parameter space (e.g., Lammer et al. 2008).

2.2. *Ion escape from non-magnetized planets*

The relevance of non-thermal processes in general, as well as the relevance of every specific mechanism, varies between different planets and depends on its basic parameters, such as mass and orbital separation, atmospheric composition, and the age/activity level of the host star. For highly irradiated planets, the atmospheric mass loss rates of non-thermal escape processes described here and in the next section are typically about an order of magnitude lower than thermal escape. However, they remain significant for terrestrial planets, and can even dominate in quiet conditions, as they are expected to be progressively more important with decreasing stellar irradiation and increasing age (e.g., Vidal-Madjar 1978; Shizgal & Lindenfeld 1982). Thus, in the present-day solar system, the dominant escape source is represented by various ion escape processes considered in this section.

The atmospheric species can be ionized through a few channels, of which the dominant one is photoionization by stellar irradiation. Additionally, ions can be produced by collisions with the atmospheric electrons or the energetic particles of the stellar wind. Thus, a large fraction of the species in the upper layers of the planetary atmosphere (exosphere) are represented by ions of different atoms/molecules. In the case of non-magnetized planets, these ions can be removed from the exosphere through ion pick-up (see, e.g., Kislyakova et al. 2013): the ionized particles interact with the magnetic field frozen in the stellar wind plasma and can be dragged away. This process is effective both on terrestrial-like planets and the gas giants; the associated atmospheric mass loss rates remain, however, several times smaller than those of thermal Jeans escape for different planetary types.

Further on, photochemical reactions and collisions with stellar wind particles can also lead to additional heating (and thus, larger thermal escape) and the production of energetic neutral atoms (ENA) with energies sufficient to escape the gravitation well of the planet (e.g., Shizgal & Arkos 1996; Shematovich et al. 1994; Lee et al. 2015) and sputtering of atmospheric species (e.g., Johnson 1994; Johnson et al. 2008).

Among the exothermic photochemical reactions, except the photoionization which will be discussed in more detail in the context of hydrodynamic escape in section 3, the most relevant appear the photodissociation and ion recombination processes. These processes are specifically relevant for the escape of heavy ions, and in the present-day solar system dominate the escape of oxygen, nitrogen, and carbon species (O^+, NO^+, CO^+, CO_2^+) from Earth and Mars (see Gronoff et al. 2020, and references therein).

In terms of fractionation of elemental and isotopic abundances, non-thermal escape processes, similar to Jeans escape, are most relevant for the evolution of secondary atmospheres, after the removal of primordial hydrogen-dominated envelopes by hydrodynamic winds. Differently from Jeans escape, the non-thermal processes are capable of the fast removal of heavy ions, though the former can yet be more effective on longer timescales (e.g., Kulikov et al. 2007; Lammer et al. 2020b).

The study of non-thermal escape processes from exoplanets is complicated by the fact, that the overall picture is strongly dependent on the specific content of the atmosphere, which is, in general, unknown. Therefore, in the ideal case, such studies require testing the stability of the assumed atmosphere in the context of the long-term evolution accounting for the activity history of the host star.

2.3. *Interaction of ions with magnetic fields*

With the inclusion of magnetic fields, the behavior of neutral and ionized species becomes rather different. While the neutrals "do not notice" the magnetic field and escape in a regular way (e.g., through Jeans escape), the motion of the charged particles is bound in a presence of the magnetic field and spiraled/circularized along the field lines. Thus, in the regions where magnetic lines are closed, the charged particles are getting trapped.Instead, in the regions of open magnetic field lines (as polar cusps and reconnection regions), the escape of ionized species intensifies. The escape of plasma from the ionosphere through the polar cusps (so-called polar wind/polar outflow) can reach the mass loss rates comparable to hydrodynamic escape (see, e.g., Carolan et al. 2021).

Further on, in the case of a dipole-like field charged particles can be accelerated in the "magnetic mirror", making oscillatory motion along the line with magnetic field strength increasing towards both ends approaching the planet until they reach the energy sufficient to escape. The charged particles can escape from the line if their pitch angle (between their velocity vector and the magnetic field direction) is sufficiently small. Otherwise, the escape occurs through charge exchange – the collision of an energized ion with a neutral atmospheric atom leading to the production of the energized neutral atom, that can escape, e.g., through Jeans escape, and a regular ion

$$\mathrm{M}^{*+} + M \rightarrow M^{*} + M^{+} . \tag{2.3}$$

The same process can be an effective source of escape also through the collisions of neutrals with energetic ions from the ionosphere and the stellar wind.

Finally, in presence of a magnetic field, numerous current systems appear at the magnetopause and across the ionosphere, which can provide additional heating intensifying the thermal atmospheric escape.

3. Hydrodynamic escape

In contrast to the kinetic Jeans escape discussed in Section 2.1, in the hydrodynamic approach an atmosphere is treated as a collisional, fluid-like medium. Gross (1972) have demonstrated that in case of substantial (hence, dense) hydrogen-dominated atmospheres the temperatures at the exobase can rise to the values order of 10000 K and the values of $\lambda_{\rm exo}$ drop below 1. In this case, the total internal (thermal) energy of the atmospheric gas exceeds the energy of the gravitation well binding the atmospheric material to the planet, and the atmospheric escape represents a bulk outflow rather than the loss of individual particles from the exobase.

Already decades ago, this approach was used to study the early evolution of Earth and Venus (see, e.g., Dayhoff et al. 1967; Sekiya et al. 1980; Watson et al. 1981), and later on it was generalized on the hydrogen-dominated atmospheres of extrasolar planets, specifically close-in Jupiter-like and sub-Neptune-like planets (see, e.g., Lammer et al. 2003; Lecavelier des Etangs et al. 2004; Yelle 2004; Cecchi-Pestellini et al. 2009; Owen & Jackson 2012; Erkaev et al. 2016; Kubyshkina et al. 2018; Caldiroli et al. 2021). For substantial hydrogen-dominated atmospheres of close-in planets, the hydrodynamic escape represents the major source of the atmospheric mass loss, overcoming by an order of magnitude kinetic and non-thermal escape processes discussed in Section 2. Therefore, it is believed to be the main driving mechanism of the atmospheric evolution (e.g., Murray-Clay et al. 2009; Lopez et al. 2012; Chen & Rogers 2016; Kubyshkina et al. 2019, 2020) and therefore to play a major role in shaping the population of low to intermediate mass exoplanets as it is known to date (e.g., Fulton et al. 2017; Fulton & Petigura 2018; Owen & Wu 2017; Gupta & Schlichting 2019; Mordasini 2020).

In general, the hydrodynamic outflow from a planet is described by a set of fluid dynamics equations, which can be written in a 1D form as

$$\frac{\partial \rho}{\partial t} + \frac{\partial(\rho v r^2)}{r^2 \partial r} = S \,, \tag{3.1}$$

$$\frac{\partial \rho v}{\partial t} + \frac{\partial[r^2(\rho v^2 + P)]}{r^2 \partial r} = -\frac{\partial U}{\partial r} + \frac{2P}{r} \,, \tag{3.2}$$

$$\frac{\partial E}{\partial t} + \frac{\partial [vr^2(E+P)]}{r^2 \partial r} = Q + \frac{\partial}{r^2 \partial r}(r^2 \chi \frac{\partial T}{\partial r}) - \frac{\partial (\rho U)}{r^2 \partial r} . \tag{3.3}$$

These three equations describe the mass, momentum, and energy conservation, and the variables are the radial distance from the planet along the substellar line (r), density (ρ), temperature (T), and bulk velocity (v) of the atmosphere, thermal pressure (P), the sum of the kinetic and thermal energies (E), and the gravitational potential U, which can represent the gravitational potential of the central body (planet) or account also for the gravitational forces of the host star (Erkaev et al. 2007). The source term S on the right-hand of equation 3.1 is typically taken as 0, reflecting the assumption that there are no leaks or replenishment of the atmospheric material. Finally, the second term in Equation 3.3 accounts for the thermal conductivity of the neutral gas ($\chi = 4.45 \times 10^4 (\frac{T[\mathrm{K}]}{1000})^{0.7}\,\mathrm{erg\,cm^{-1}\,s^{-1}}$).

The term Q in Equation 3.3 represents the sum of all heating and cooling processes accounted for in the specific model. The number and the type of the specific cooling/heating processes thus depend on the chosen approach and the specific composition of the hydrogen-dominated atmosphere. However, the main heating source is normally the photoionization of the atmospheric species (in primary atmospheres, predominantly, ionization of hydrogen by the XUV radiation of the host star)

$$\mathrm{M} + h\nu \rightarrow \mathrm{M}^+ + e^* . \tag{3.4}$$

In this reaction, the stellar photon of a certain energy ($h\nu$) heats the atmospheric particle M. If the energy of a photon exceeds the ionization threshold of the particle, it leads to the production of an ionized particle and a free electron, that carries away the excess energy (given by the difference between the energy of the photon and the ionization energy). This excess energy can be further lost in the collisions (leading to the heating of the atmosphere) or spent in other, endothermic, reactions, such as recombination or secondary ionization of atmospheric species. Therefore, to accurately define which fraction of the absorbed photon energy is spent on heating, one has to account for all relevant reactions and also for all energy levels of the atmospheric species, as the electron tore away in reaction 3.4 must not be the one with the minimal ionization energy in the general case. The resulting heating efficiency, i.e., the relation of the energy spent on heating to the total absorbed energy, varies with altitude and the wavelength (energy) of the incident radiation, and also depends on the type of a planet (Dalgarno et al. 1999; Yelle 2004; Shematovich et al. 2014; Salz et al. 2015, 2016a,b). Given the application complexity and the numerical costs of such an approach, the real heating efficiency is often reduced to a coefficient η, which is further set to a constant value (as, e.g., in Owen & Jackson 2012; Erkaev et al. 2016; Kubyshkina et al. 2018) or has an analytical dependence on the photon wavelength (as in Murray-Clay et al. 2009; Allan & Vidotto 2019; Kubyshkina et al. 2022). In this case, the volume heating rate of species "i" by the stellar flux of wavelength ν can be written as

$$H_{ion} = \frac{\eta \sigma_{\nu,i} n_{\mathrm{i}}}{2} \int_0^{\frac{\pi}{2}+\arccos(1/r)} I_{\nu,i}(r,\theta) \sin\theta d\theta , \tag{3.5}$$

where n_{i} is the numeric density of species "i", $\sigma_{\nu,i}$ is the wavelength-dependent absorption cross-section, and $I_{\nu,i}(r,\theta)$ describes the spatial variations of the stellar irradiation due to the atmospheric absorption in spherical coordinates.

Other heating and cooling sources are given by other exothermic and endothermic reactions, such as, e.g., collisional excitation of hydrogen (Lyα-cooling)

$$\mathrm{H} + e^* \rightarrow H^* + e . \tag{3.6}$$

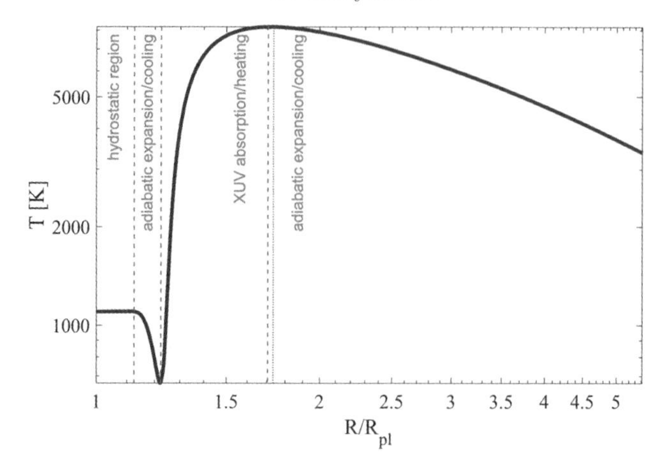

Figure 2. A typical temperature profile of the hot, strongly irradiated sub-Saturn. The mass of the planet is 45.1 $M_{\oplus}$and the radius is 6.0 $R_{\oplus}$. It is located at 0.04 AU from its 0.8 $M_{\odot}$host star (which corresponds to the equilibrium temperature of 1100 K), and receives the XUV flux of $\sim$54000 $erg/s/cm^2$. The mass loss rate estimated in hydrodynamic simulation is 6.8×10^{10} g/s.

The corresponding cooling/heating rates are given by the product of the numerical densities of participating species and the specific reaction rates

$$C_{\mathrm{Ly}\alpha} = 7.5 \times 10^{-19} n_{\mathrm{e}} n_{\mathrm{H}} exp\left(\frac{118348\mathrm{K}}{T}\right). \tag{3.7}$$

To account for these processes (which typically include photodissociation, recombination, collisional ionization, etc.), in addition to the Equations 3.1–3.3 one has to solve the system of equations describing chemical equilibrium. It consists of equations in a form of Equation 3.1 on numerical densities for each individual species included in the specific model, where the source terms represent the sink/replenishment of the species according to chemical reactions.

3.1. *XUV-driven outflow*

The most common and widely studied type of the hydrodynamic escape is an XUV-driven escape, which considers the photoionization heating as the main driving mechanism of the hydrodynamic wind (e.g., Watson et al. 1981; Yelle 2004; Lecavelier des Etangs et al. 2004; Murray-Clay et al. 2009; Owen & Jackson 2012; Shematovich et al. 2014; Salz et al. 2016a; Erkaev et al. 2016; Kubyshkina et al. 2018; Caldiroli et al. 2021). In this approach, the deep layers of otherwise hydrostatic atmosphere are heated up by the stellar irradiation leading to the adiabatic expansion and settling of the transonic planetary wind. In Figure 2, I show a typical temperature profile for the upper atmosphere of the planet experiencing XUV-driven atmospheric escape. The planet is a 45.1 $M_{\oplus}$ and 6.0 $R_{\oplus}$ sub-Saturn orbiting at 0.04 AU around 0.8 $M_{\odot}$ star and receiving the XUV flux of $\sim$54000 $erg/s/cm^2$. According to the hydrodynamic model, it experiences an atmospheric mass loss of 6.8×10^{10} g/s. Here and later in this section, we employ the hydrodynamic model by Kubyshkina et al. (2018) for the examples, unless stated otherwise. It is a 1D model assuming a pure hydrogen atmosphere and accounting for the effects of dissociation, recombination of ions and of the atomic hydrogen to H_2, collisional ionization, Lyα-cooling, and H_3^+-cooling. The stellar irradiation is treated in two bands of 5 nm (X-ray) and 62 nm (EUV), and the constant heating efficiency of 0.15 is assumed in both cases.

In Figure 2, one can see a few distinct regions. The lower boundary ($R = R_{pl}$) of the simulation domain is the observable radius of the planet, i.e., the photosphere. The lowermost region (up to $R \simeq 1.1 R_{pl}$) is gravitationally well bound to the planet and remains hydrostatic, which can be seen by the constant temperature. Higher up, the gravitation force of the planet decreases, and the atmosphere starts to expand adiabatically, and, therefore, to cool (1.1–1.2 R_{pl}). The XUV irradiation of the specific wavelength penetrates the atmosphere layers that are optically thin for this layer and is absorbed at the specific height where the optical depth τ becomes 1. This height lies deeper in the atmosphere for more energetic (short-wavelength) radiation, and higher for the less-energetic part of the spectra. Therefore, the ionizing radiation (energetic enough to tear away the electron from the hydrogen atom) heats the atmosphere in the narrow region somewhat above the photosphere of the planet (1.2–1.7 R_{pl} in Figure 2). Then at some point, the heating is overtaken by adiabatic cooling, and the atmosphere expands up to and beyond the Roche lobe, which is a natural upper boundary of the simulation domain where the gravitation forces of the planet and its host star equilibrate. Above this point, the atmosphere is no more gravitationally bound to the planet.

In terms of other parameters, the outflow represents the upward accelerating wind with density gradually decreasing and ion fraction gradually increasing with radial distance.

The planet discussed above represents the most classic case of the XUV-driven planetary wind. Such a picture is typical for the high gravity planets with well-bound atmospheres. In a more general case, some regions shown in Figure 2 can degenerate. For example, the lowermost hydrostatic region can be absent in the case of the planet with lower gravity or higher thermal energy (see the next section 3.2 for more detail); the inner region of adiabatic cooling (and thus the temperature minimum) can degenerate for massive planets with a very steep density gradient in the inner region; finally, if the Roche lobe is located close to the planet, which is typical for low-mass planets orbiting close to their host stars, the upper region of the adiabatic expansion/cooling can be absent and the heating occurs in the uppermost layers of the upper atmosphere. In the next sections, I consider some extreme cases.

3.2. *Hydrodynamic outflow driven by the own thermal energy of a planet*

Until recently, the XUV irradiation was considered to be the major, and the only relevant, source of energy fuelling the hydrodynamic outflow. Thus, e.g., Sekiya et al. (1980) considers as an additional source of the heating a release of the gravitational energy due to the planetesimal accretion for Earth but concludes that this source is minor in comparison to the stellar high-energy radiation. However, if considering the whole diversity of exoplanets at different stages of evolution, the own thermal energy of the planet can play a crucial role in powering the planetary wind. To account for this type of atmospheric escape, Ginzburg et al. (2016b, 2018) introduced the term "core-powered mass loss" and suggested the post-formation luminosity as the energy source. Later, Gupta & Schlichting (2019) generalized this idea for the bolometric heating from the host star. Given that the post-formation luminosity and the inflation degree of a planet decline with time, the relative input of this atmospheric escape mechanism in comparison to the XUV-driven escape (that occurs on Gyr timescale, see, e.g., King & Wheatley 2021) is expected to be largest for young planets.

For the own thermal energy of the planet to be the major driver of the planetary wind, this energy should dominate the gravitational energy binding the planetary atmosphere. Thus, this type of planetary wind is typical for low-gravity, hot, and/or highly inflated planets. For a prime example, I consider in Figure 3 a hot super-puff type planet of the low mass ($2.1\,M_\oplus$) and a radius of $3.0\,R_\oplus$ orbiting at 0.04 AU around $0.8\,M_\odot$ K-type

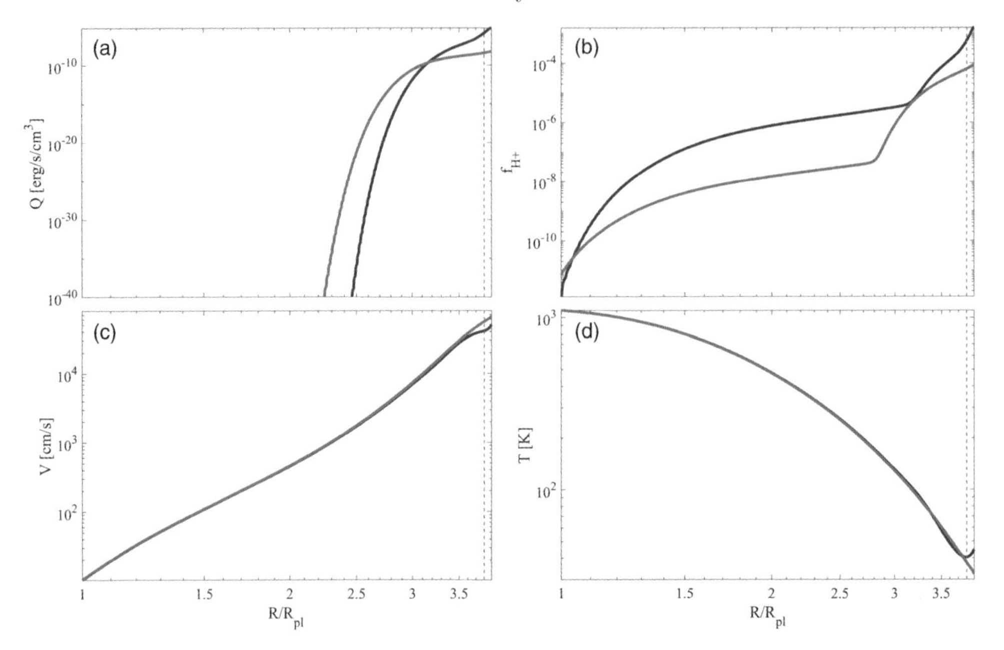

Figure 3. Volume heating rate (a), ion fraction (b), bulk velocity (c), and temperature (d) of the atmosphere of the low gravity planet against radial distance. The red and the blue lines correspond to low ($F_{\rm XUV} = 140$ erg/s/cm^2) and high ($F_{\rm XUV} = 56000$ erg/s/cm^2) XUV irradiation levels, respectively. The mass of the planet is 2.1 $M_\oplus$ and the radius is 3.0 $R_\oplus$. It is located at 0.04 AU from its 0.8 $M_\odot$ host star (which corresponds to the equilibrium temperature of 1100 K). The mass loss rate estimated in hydrodynamic simulations is $\sim 3.5 \times 10^{14}$ g/s for both cases.

star. These parameters correspond to a low density of about 0.4 g/cm^3, suggesting that the planet hosts a large and massive hydrogen-dominated atmosphere (in the case of an evolved planet) or is inflated due to the heat deposited in the planetary core/lower atmosphere (in the case of a young planet). In principle, the hydrodynamic model does not distinguish between the two cases, as the lower boundary of the simulation domain lies at the photosphere. The condition of the lower atmosphere is, therefore, set by the atmospheric parameters fixed at the lower boundary; in the example considered here I adopted the photosphere pressure of ~0.13 bar and the photosphere temperature equal to the equilibrium temperature assuming zero albedos (1100 K) based on the results of the lower atmosphere modeling for the first case (large atmospheric mass fraction Cubillos et al. 2017).

Due to the low gravity of the planet considered in Figure 3, the atmosphere is less gravitationally bound and the density gradient is less steep than at the planets considered in section 3.1, where most of the atmospheric material is concentrated in the inner region. It leads to a dense and relatively slow outflow. The atmospheric density remains high up to the Roche lobe, preventing stellar XUV from penetrating into the deep layers and thus, from effectively heating the atmosphere. In Figure 3, I consider two very different levels of XUV flux exposed at the planet: very low (accounting for the position of the planet) one of 140 erg/s/cm^2 (red lines) and the high one of 56000 erg/s/cm^2 (blue lines). One can see, that despite the very different levels of irradiation, the velocity and temperature profiles (bottom panels of Figure 3) are nearly identical in the two cases. The same holds also for the density profiles not shown here. This is due to the volume heating rate and the ion fraction remaining very low throughout the atmosphere (top panels of Figure 3), and the XUV heating being therefore inefficient, as discussed above.

The inefficiency of the XUV heating does not mean, however, that the planetary wind is absent in this case. Instead, it is driven by the pressure gradient between the lower and the upper atmosphere, and the atmospheric mass loss rates in both considered cases are exceptionally high – $\sim 3.5 \times 10^{14}$ g/s. This suggests, that at this planet (with the wind purely driven by its own thermal energy) the atmosphere is not stable and will escape within a short time order of megayear. Therefore, one would not expect such a planet to be present among the evolved population and such a mechanism to be important out of the initial stages of planetary evolution. However, for less extreme cases, the own thermal energy of the planet can yet be relevant along with the stellar XUV heating and contribute to driving planetary winds on longer timescales.

3.3. *Tidal effects*

Besides the energy (heating) sources considered in sections 3.1 and 3.2, the important role in controlling planetary winds can be played by the tidal interactions with the host star, specifically for close-in planets. As I already mentioned in the Introduction, the interaction of close-in planets with their host stars, magnetic or tidal, can lead to additional heating in the lower layers of the atmospheres and therefore intensify the thermal escape processes, specifically those described in sections 2.1 and 3.2, and it can also affect the ion escape processes. Here, I will discuss a more direct implication of tidal forces onto the atmospheric material relevant in terms of hydrodynamic outflow.

Accounting for the tidal effects, the gravitation potential along the substellar line can be written as (Erkaev et al. 2007)

$$\mathbf{U} = \mathbf{U_0}\left[-\frac{\mathbf{1}}{\zeta} - \frac{\mathbf{1}}{\mu(\xi-\zeta)} - \frac{\mathbf{1}+\mu}{\mathbf{2}\mu\xi^{\mathbf{3}}}\left(\xi\frac{\mathbf{1}}{\mathbf{1}+\mu} - \zeta\right)^{\mathbf{2}}\right]. \tag{3.8}$$

In Equation (3.8), $U_0 = GM_{\rm pl}/R_{\rm pl}$, $\mu = M_{\rm pl}/M_*$, $\xi = a/R_{\rm pl}$, and $\zeta = r/R_{\rm pl}$ (where a is the orbital separation and r is the radial distance from the planet). The second term in the square brackets accounts for the effects of the stellar gravitation forces, and it is easy to see that their role increases with increasing distance from the planet along the substellar line, and, more generally, with decreasing orbital separation and decreasing $M_{\rm pl}/M_*$ ratio. The latter changes from the values order of 10^{-2} in the case of the Jupiter-like planets orbiting M-dwarfs to $\sim 10^{-6}$ for Earth-like planets orbiting G or F-type stars. The term ξ, in turn, reflects the fact that for two identical planets (the same mass, radius, and equilibrium temperature) the tidal effects are larger around the lower-mass host star, as the same temperatures are reached at shorter orbits compared to heavier stars.

For an extreme example, in Figure 4 I compare the atmospheric models for the hot dense Neptune-mass planet orbiting stars of $0.8\,M_\odot$ (at ~ 0.02 AU, blue lines) and $0.4\,M_\odot$ (at ~ 0.004 AU, red lines). The most evident difference between the two cases is the position of the Roche lobe: about $6\,R_{\rm pl}$ in the first case, and only about $1.5\,R_{\rm pl}$ in the second one. Thus, in the first case (more massive host star), the atmospheric outflow represents a typical case of the XUV-driven escape (see the temperature profile in panel (c) of Figure 4), as expected for such a high-gravity planet. The mass loss rate estimated in this case is $\sim 2.0 \times 10^8$ g/s, which is slightly lower than the value predicted for the given XUV flux (391 erg/s/cm^2) by energy-limited approximation adopting the form from Erkaev et al. (2007). It is expected for this type of planet and highlights that the atmospheric outflow is purely XUV-driven, without significant input from the own thermal energy of the planet.

For the planets orbiting $0.4\,M_\odot$ star, the outflow looks completely different. As the Roche lobe is located so close to the photosphere, the gravitation forces of the planet are only effective within $\sim 1\,R_\oplus$ distance from the photosphere of the planet. Moreover,

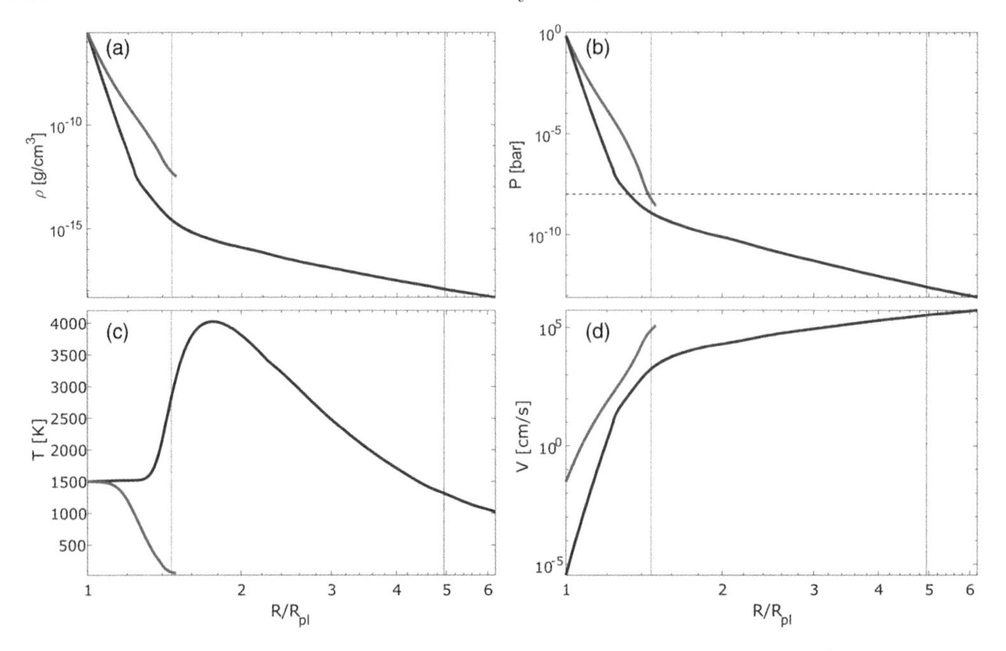

Figure 4. Density (a), pressure (b), temperature (c), and bulk velocity (d) profiles of the two planets of equivalent radius (2.0 $R_\oplus$), mass (16.2 $M_\oplus$), and equilibrium temperature (1500 K), and under similar levels of XUV irradiation (about 400 erg/s/cm^2), but orbiting different stars (which means, that the given $T_{\rm eq}$ is achieved at different orbital distances). The blue solid lines correspond to the planet orbiting 0.8 $M_\odot$star at 0.0194 AU, and the red solid lines to the planet orbiting 0.4 $M_\odot$star at 0.0037 AU. The blue/red dashed vertical lines denote the positions of the sonic points in the respective simulations. The horizontal black dashed line in panel b denotes the 10 nbar level (Roche lobe overflow limit; Koskinen et al. 2022).

comparing the input from the planet into gravitation potential (the first term in the square brackets in Equation 3.8) to the stellar input (the second term in the square brackets), one can see that the stellar gravitational forces remain crucial even within this region. As a result of the shortness of the Roche radius and the impact of the stellar tidal forces, the atmosphere remains very dense up to the Roche lobe (see panel (a) of Figure 4) and the outflow occurs in a manner similar to the case described in section 3.2. However, as the own thermal energy of the planet is relatively small in this case (about 1.5 orders of magnitude below the gravitational one), the outflow is mainly controlled by the tidal forces. The atmospheric mass loss rate estimated in this simulation is $\sim 1.7 \times 10^{12}$ g/s, which is about 3 orders of magnitude above the one given by the energy-limited equation for the XUV flux of 496 erg/s/cm^2. As one can see from panel (b) of Figure 4, the pressure at the upper boundary is consistent with the lower limit predicted for the Roche lobe overflow by Koskinen et al. (2022) (10-100 nbar).

As in the case of the planetary wind driven by the own thermal energy of the planet considered in the previous section, the example presented here is quite extreme (the model planet may even not be physical). This was done to surely isolate the effect of the tidal forces. However, also for more moderate planets, tidal effects can have a significant effect. I discuss it in more detail in the next section.

3.4. *Atmospheric mass loss rates across parameter space*

In this section, I give a brief summary of the mechanisms driving the hydrodynamic planetary winds discussed in sections 3.1–3.3 and consider their influence across the parameter space. To this end, I employ the grid of upper atmosphere models presented in Kubyshkina & Fossati (2021). This grid is based on the hydrodynamic 1D model by Kubyshkina et al. (2018) described in brief at the beginning of section 3 and includes ~ 10000 model planets with masses of 1–108.6 $M_\oplus$ and radii of 1–10 $R_\oplus$ orbiting stars with masses between 0.4 $M_\odot$ and 1.3 $M_\odot$. The orbital distances for each stellar mass cover the range of equilibrium temperatures of 300–2000 K, and the irradiation levels at all orbits vary between the values typical for young stars (near the time of protoplanetary disk dispersal) and for stars at the end of the main sequence (~ 10 Gyr). By construction, the grid is dominated by close-in planets and, further, the non-physical model planets and those where the hydrodynamic approach is not applicable were excluded from consideration. In particular, this includes planets with mean densities lower than 0.03 g/cm^3 or higher than that expected for the rocky core of the given mass, and planets with exobases below the sonic level. The consideration in Kubyshkina & Fossati (2021) also excludes the planets with Roche radii smaller than 1.5 $R_{\rm pl}$. In the present work, however, I include them for completeness.

In the top panel of Figure 5, I show the atmospheric mass loss rates of the grid planets against the generalized Jeans escape parameter $\Lambda = (GM_{\rm pl}m_{\rm H})/(k_{\rm b}T_{\rm eq}R_{\rm pl})$ (the same as parameter $\lambda_{\rm ex}$ in section 2.1, but calculated at the photosphere). To distinguish between the different cases of hydrodynamic outflow, I use the following formalisation. First, to separate the model planets, where the input from the own thermal energy is more significant than the XUV heating, I compare the mass loss rates predicted by hydrodynamic modeling with the ones predicted by energy-limited approximation (Erkaev et al. 2007), assuming that the atmospheric escape is purely driven by stellar XUV

$$\dot{M}_{\rm EL} = \pi \frac{\eta \Phi_{\rm XUV} R_{\rm eff}^2 R_{\rm pl}}{GM_{\rm pl}}. \tag{3.9}$$

Here, $\Phi_{\rm XUV}$ is the XUV flux received by the planets, and the heating efficiency coefficient η is taken 0.15 to consist with the hydrodynamic models. $R_{\rm eff}$ is an effective radius of the XUV absorption, reflecting the narrow interval where the high energy irradiation from the host star is absorbed in the planetary atmosphere (see the discussion in section 3.1), reduced to one specific distance. Taking $R_{\rm eff} = R_{\rm pl}$, as is done in the numerous studies, leads in general to a drastic underestimation of the atmospheric mass loss rates. Adopting $R_{\rm eff}$ defined in the hydrodynamic simulations, in turn, would make the predicted mass-loss rates closer to those in the grid but would make the comparison between the two predictions less meaningful. The effective radii for planets with outflows driven mainly by their own thermal energy, as the one discussed in section 3.2, are formally set at the upper boundary of the simulation domain (see Kubyshkina et al. 2018, for the detail). Thus, the estimates obtained with Equation 3.9 would increase but become physically meaningless. Therefore, to define $R_{\rm eff}$ I adopt here the analytical equation (Chen & Rogers 2016)

$$R_{\rm eff} = R_{\rm pl} + H \ln(\frac{P_{\rm photo}}{P_{\rm XUV}}), \tag{3.10}$$

where H is the atmospheric scale height, $P_{\rm photo}$ is the pressure at the photosphere, and $P_{\rm XUV}$ is the pressure at the XUV absorption level approximated as $(GM_{\rm pl}m_{\rm H})/(\sigma_{\rm 20eV}R_{\rm pl}^2)$. In the latter, $\sigma_{\rm 20eV}$ is the absorption cross-section of hydrogen for a photon energy of 20 eV. Equation 3.10 predicts the effective radii of the XUV absorption similar to those obtained from the hydrodynamic models for planets with the outflow predominantly driven by the stellar XUV heating. Including it in Equation 3.9, therefore, leads to the estimation of the atmospheric mass loss rates comparable to the estimates from the hydrodynamic models for the XUV-driven escape (gray dots in Figure 5). Thus, to distinguish the planets where the winds are mainly driven by their own thermal energy (not accounted for in the Equation 3.9) and the input from the stellar XUV becomes minor, I filter the model planets with atmospheric mass loss rates $\dot{M}$ larger than 10 times the mass loss rates predicted by energy-limited approximation $\dot{M}_{\rm EL}$ (orange dots in Figure 5). Some of these planets can be further affected by the tidal forces, and to avoid mixing up the two effects, I further exclude from this group the planets with the Roche radii smaller than $2R_{\rm pl}$. Additionally, I highlight in Figure 5 the planets in "transitional" state given by the conditions $\dot{M} = 5 - 10 \times \dot{M}_{\rm EL}$ and $R_{\rm roche} > R_{\rm pl}$ (cyan dots).

The Roche radius of the planet is the point on the star-planet line, where the gravitation forces of the planet and the star are balanced. Therefore, its position relative to the photosphere radius is a good proxy of the degree to which the tidal forces of the host star affect the atmospheric outflow. For grid planets, the tidal effects become significant when the Roche lobe is smaller than about $2R_{\rm pl}$ (violet points in Figure 5), and, similarly to the case considered in section 3.3, become dominant when the Roche lobe lies below $\sim 1.5R_{\rm pl}$ (green points in Figure 5). One can see, that the violet points in Figure 5 ($R_{\rm roche} = 1.5 - 2 \times R_{\rm pl}$) represent a transitional state between the tidally driven outflow and planetary winds driven by the internal thermal energy of planets (for $\Lambda \lesssim 20$) or the XUV-driven planetary winds ($\Lambda > 20$).

Speaking about the parameters of planets in the different groups, the group planets with the outflow driven by their own thermal energy should be dominated by planets with low gravity and high temperature, as, in this case, the relation of the gravitational energy to the thermal energy (given, in essence, by the parameter Λ) is small. Indeed, most of the group of orange points in Figure 5 is concentrated at $\Lambda < 20$. The occurrence rate of different equilibrium temperatures of planets in this group is strongly biased by the construction of the grid, so in Figure 5 I only demonstrate the distribution of planets in this group across planetary masses (bottom left panel). One can see, that the majority of planets in this group have masses below $10\,M_\oplus$, and their fraction in the total number of grid planets with $R_{\rm roche} > 2R_{\rm pl}$ (shown by the empty bins) is consistently increasing with decreasing mass. However, the number of such planets remains significant up to 20–$30\,M_\oplus$ and some single planets are present up to $50\,M_\oplus$. The shape of the distribution for all grid planets with $R_{\rm roche} > 2R_{\rm pl}$ is mainly defined by the density cut described at the beginning of this section.

Formally, the potential number of planets whose outflow is dominated by tidal effects increases with decreasing stellar mass, as is shown by the empty histogram bins in the bottom right panel of Figure 5. To plot these bins, I calculated the Roche lobes for all the planets in the grid parameter space, applied the density filter described above, and filtered the planets with $R_{\rm roche} < 1.5R_{\rm pl}$. However, a significant fraction of these planets have the Roche lobes that lie below the photosphere level – and this fraction also increases towards lower stellar masses. Therefore, the actual number of "tidally dominated" planets existing in the grid (which, in fact, only includes planets with $R_{\rm roche} > 1.2R_{\rm pl}$) maximizes at the stellar mass of $0.6\,M_\odot$.

4. Magnetic field and the escape of planetary atmospheres

The effect of a magnetic field, if present at the planet, is essential in describing escaping atmospheres and their interaction with magnetized stellar winds, and therefore, in the interpretation of the observations (see, e.g., Matsakos et al. 2015; Villarreal D'Angelo et al. 2018; Carolan et al. 2021; Cohen et al. 2022). However, the effect of the magnetic field on planetary winds, and specifically on atmospheric mass loss rates, can be ambiguous. While for compact atmospheres and strong enough fields, the upper atmosphere does not interact with the stellar wind directly anymore, reducing the effects of erosion and ion pick-up, some processes discussed in Section 2.3 can lead to additional heating or intensification of ion escape in specific regions. Thus, at

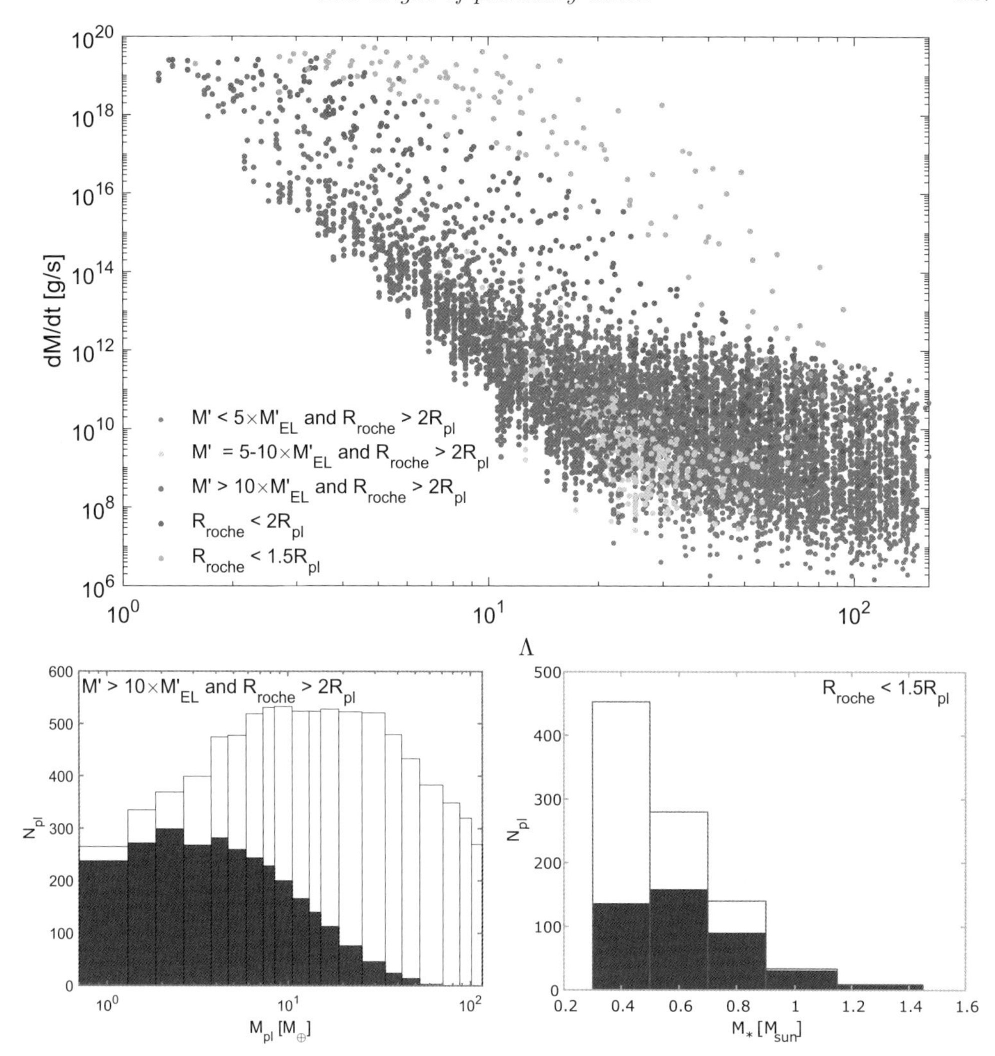

Figure 5. Top panel: atmospheric mass loss rates against the gravitational Jeans parameter calculated at the photosphere. Color code reflects the presumable major driver of the atmospheric escape. Gray points represent XUV-dominated planets, orange points highlight the planets with dominating internal thermal energy, and green points show the planets dominated by tidal effects. Cyan and violet points correspond to the transitional regimes. See the detailed discussion in the text. Bottom left panel: number of planets in different planetary mass bins in the grid of upper atmosphere models. The empty bins show the total number of planets in the grid with $R_{\rm roche} > 2R_{\rm pl}$, and the blue bins correspond to the planets with atmospheric mass loss dominated by their own thermal energy of the planet. Bottom right panel: number of planets with Roche radius below 1.5 of the photosphere radius. The empty bins show the total number of such planets within the parameter space of the grid and the blue bins reflect the actual number of planets in the grid. Note, that a significant fraction of planets shown by the empty bins were in general not considered, as the Roche lobe lies below the photosphere level. All plots are based on the grid of upper atmosphere models presented in Kubyshkina & Fossati (2021).

strongly magnetized planets, polar outflow can represent one of the major escape sources both in the context of hydrodynamic (e.g., Carolan et al. 2021) and non-thermal (e.g., Zhang et al. 2022) escape.

For terrestrial planets in Solar System (dominated by non-thermal escape processes), it was shown that a weak magnetic field can intensify the escape, while the strong field is, in general, expected to be protective (Gunell et al. 2018; Sakai et al. 2018; Egan et al. 2019; Ramstad & Barabash 2021). Further on, for the crustal magnetic field at Mars, the overall effect on the atmospheric mass loss depends on the configuration and inclination angle of the magnetic field (Weber et al. 2021; Li et al. 2022; Curry et al. 2022).

For Hot Jupiters, where planetary winds are driven by hydrodynamic processes, models predict significant suppression of the atmospheric mass losses for magnetic field strength above 0.1–3 G (see Owen & Adams 2014; Trammell et al. 2014; Khodachenko et al. 2015; Arakcheev et al. 2017; Khodachenko et al. 2021). For more moderate planets hosting hydrogen-dominated atmospheres, the effect remains unclear. E.g., Carolan et al. (2021) predict for 0.7 $M_{\rm jup}$ planet a small increase in escape for magnetic field strength increasing from 0 to 5 G.

To summarize, the actual effect of magnetic fields on planetary winds depends on the type of planetary atmospheres, parameters and position of the planet, and configuration of the field itself.

5. Conclusions

The processes driving planetary winds are extremely various, and spread from microphysical kinetic processes to global hydrodynamic flows. Relevant mechanisms vary in dependence on the properties of the planet, its orbit, type of the atmosphere, and the age and mass of the host star. Thus, escape from evolved terrestrial-like planets near the habitable zone is, in general, dominated by non-thermal escape processes, including ion escape and the interaction with stellar winds. However, at young ages, when the stellar flux is high even at relatively far orbits, the planetary wind is expected to be driven by thermal processes. If such escape is hydrodynamic or kinetic (Jeans escape) is defined by the parameters of a planet and the energy budget of its atmosphere.

Planetary winds from Neptune-like to Jupiter-like planets are, in general, hydrodynamic. Their atmospheric mass loss rates dominate, typically, over those of kinetic processes and spread over a few orders of magnitude depending on planetary and stellar parameters. The mass loss rates maximize for low-gravity hot planets, where the outflow is driven by their own thermal energy rather than stellar irradiation. For the majority of planets, however, the hydrodynamic outflow is driven by stellar XUV radiation. The mass loss rates, in this case, span between 10^7–10^{12} g/s. Finally, for close-in planets, the tidal forces of the host star can play a significant role. Their influence increases for planets orbiting low-mass stars due to closer orbital distances.

Given the wide range of possible driving mechanisms, modeling planetary winds represents a non-trivial task, specifically for planets outside of Solar System. Even though some mechanisms depend mainly on general parameters of the planet, such as its mass, radius, and temperature (which are, to some extent, measurable), in general case one has to deal with parameters that can be poorly constrained. In particular, the content of exoplanetary atmospheres, crucial for non-thermal processes is typically unknown; therefore, modeling these processes should, in the best case, include the analysis of specific atmospheres for stability in the evolutionary context. The latter is, however, further complicated by the uncertain evolution history of the host star (Kubyshkina et al. 2019). The same problem holds for magnetic fields at exoplanets, which cannot, to date, be directly measured. Therefore, the presence of magnetic fields can potentially only be constrained by the effect it causes on planetary outflow and its interaction with stellar wind and, hence, on observations (e.g., Lyα). However, given the wide range of parameters affecting these processes, the task appears difficult for most of the planets.

Besides the intrinsic parameters of the planet, its environment plays a crucial role. It includes, in the first place, the parameters of the host star and the orbital distance of the planet. In general, the latter can be constrained from the observations with good accuracy, while some of the stellar parameters remain unknown and are set on the basis of theoretical and empirical models. This includes, in particular, stellar EUV radiation and parameters of stellar winds (see, e.g., Vidotto 2021).

References

Allan, A. & Vidotto, A. A. 2019, MNRAS, 490, 3760. doi:10.1093/mnras/stz2842

Arakcheev, A. S., Zhilkin, A. G., Kaigorodov, P. V., et al. 2017, Astronomy Reports, 61, 932. doi:10.1134/S1063772917110014

Arras,, P. & Socrates, A. 2010, ApJ, 714, 1. doi:10.1088/0004-637X/714/1/1

Batygin, K. & Stevenson, D. J. 2010, ApJ Letters, 714, L238. doi:10.1088/2041-8205/714/2/L238

Bodenheimer, P., Lin, D. N. C., & Mardling, R. A. 2001, ApJ, 548, 466. doi:10.1086/318667

Caldiroli, A., Haardt, F., Gallo, E., et al. 2021, A&A, 655, A30. doi:10.1051/0004-6361/202141497

Caldiroli, A., Haardt, F., Gallo, E., et al. 2022, A&A, 663, A122. doi:10.1051/0004-6361/202142763

Carolan, S., Vidotto, A. A., Hazra, G., et al. 2021, MNRAS, 508, 6001. doi:10.1093/mnras/stab2947

Carolan, S., Vidotto, A. A., Plavchan, P., et al. 2020, MNRAS, 498, L53. doi:10.1093/mnrasl/slaa127

Cecchi-Pestellini, C., Ciaravella, A., Micela, G., et al. 2009, A&A, 496, 863. doi:10.1051/0004-6361/200809955

Chamberlain, J. W. 1963, Planetary and Space Science, 11, 901. doi:10.1016/0032-0633(63)90122-3

Chamberlain, J. W. 1969, ApJ, 155, 711. doi:10.1086/149905

Chen, H. & Rogers, L. A. 2016, ApJ, 831, 180. doi:10.3847/0004-637X/831/2/180

Christie, D., Arras, P., & Li, Z.-Y. 2016, ApJ, 820, 3. doi:10.3847/0004-637X/820/1/3

Cohen, O., Alvarado-Gómez, J. D., Drake, J. J., et al. 2022, ApJ, 934, 189. doi:10.3847/1538-4357/ac78e4

Cubillos, P., Erkaev, N. V., Juvan, I., et al. 2017, MNRAS, 466, 1868. doi:10.1093/mnras/stw3103

Curry, S. M., Tatum, P., Mitchell, D., et al. 2022, MNRAS, 517, L121. doi:10.1093/mnrasl/slac099

Dalgarno, A., Yan, M., & Liu, W. 1999, ApJS, 125, 237. doi:10.1086/313267

Dayhoff, M. O., Eck, R. V., Lippincott, E. R., et al. 1967, Science, 155, 556. doi:10.1126/science.155.3762.556

Egan, H., Jarvinen, R., Ma, Y., et al. 2019, MNRAS, 488, 2108. doi:10.1093/mnras/stz1819

Erkaev, N. V., Kulikov, Y. N., Lammer, H., et al. 2007, A&A, 472, 329. doi:10.1051/0004-6361:20066929

Erkaev, N. V., Lammer, H., Odert, P., et al. 2015, MNRAS, 448, 1916. doi:10.1093/mnras/stv130

Erkaev, N. V., Lammer, H., Odert, P., et al. 2016, MNRAS, 460, 1300. doi:10.1093/mnras/stw935

Erkaev, N., Scherf, M., Herbort, O., et al. 2022, arXiv:2209.14691 (accepted for publication in MNRAS)

Fortney, J. J., Dawson, R. I., & Komacek, T. D. 2021, Journal of Geophysical Research (Planets), 126, e06629. doi:10.1029/2020JE006629

Fulton, B. J., Petigura, E. A., Howard, A. W., et al. 2017, AJ, 154, 109. doi:10.3847/1538-3881/aa80eb

Fulton, B. J. & Petigura, E. A. 2018, AJ, 156, 264. doi:10.3847/1538-3881/aae828

Ginzburg, S., Schlichting, H. E., & Sari, R. 2016, ApJ, 825, 29. doi:10.3847/0004-637X/825/1/29

Ginzburg, S. & Sari, R. 2016, ApJ, 819, 116. doi:10.3847/0004-637X/819/2/116

Ginzburg, S., Schlichting, H. E., & Sari, R. 2018, MNRAS, 476, 759. doi:10.1093/mnras/sty290

Gronoff, G., Arras, P., Baraka, S., et al. 2020, Journal of Geophysical Research (Space Physics), 125, e27639. doi:10.1029/2019JA027639

Gross, S. H. 1972, Journal of Atmospheric Sciences, 29, 214. doi:10.1175/1520-0469(1972)029<0214:OTETOH>2.0.CO;2
Gunell, H., Maggiolo, R., Nilsson, H., et al. 2018, A&A, 614, L3. doi:10.1051/0004-6361/201832934
Gupta, A. & Schlichting, H. E. 2019, MNRAS, 487, 24. doi:10.1093/mnras/stz1230
Hazra, G., Vidotto, A. A., Carolan, S., et al. 2022, MNRAS, 509, 5858. doi:10.1093/mnras/stab3271
Hunten, D. M. 1973, Journal of Atmospheric Sciences, 30, 1481. doi:10.1175/1520-0469(1973)030<1481:TEOLGF>2.0.CO;2
Hunten, D. M. 1982, Planetary and Space Science, 30, 773. doi:10.1016/0032-0633(82)90110-6
Hunten, D. M., Pepin, R. O., & Walker, J. C. G. 1987, Icarus, 69, 532. doi:10.1016/0019-1035(87)90022-4
Johnson, R. E. 1994, SSR, 69, 215. doi:10.1007/BF02101697
Johnson, R. E., Combi, M. R., Fox, J. L., et al. 2008, SSR, 139, 355. doi:10.1007/s11214-008-9415-3
Khodachenko, M. L., Shaikhislamov, I. F., Lammer, H., et al. 2015, ApJ, 813, 50. doi:10.1088/0004-637X/813/1/50
Khodachenko, M. L., Shaikhislamov, I. F., Lammer, H., et al. 2021, MNRAS, 507, 3626. doi:10.1093/mnras/stab2366
King, G. W. & Wheatley, P. J. 2021, MNRAS, 501, L28. doi:10.1093/mnrasl/slaa186
Kislyakova, K. G., Lammer, H., Holmström, M., et al. 2013, Astrobiology, 13, 1030. doi:10.1089/ast.2012.0958
Kislyakova, K. G., Fossati, L., Johnstone, C. P., et al. 2018, ApJ, 858, 105. doi:10.3847/1538-4357/aabae4
Koskinen, T. T., Lavvas, P., Huang, C., et al. 2022, ApJ, 929, 52. doi:10.3847/1538-4357/ac4f45
Kubyshkina, D., Fossati, L., Erkaev, N. V., et al. 2018, A&A, 619, A151. doi:10.1051/0004-6361/201833737
Kubyshkina, D., Cubillos, P. E., Fossati, L., et al. 2019, ApJ, 879, 26. doi:10.3847/1538-4357/ab1e42
Kubyshkina, D., Vidotto, A. A., Fossati, L., et al. 2020, MNRAS, 499, 77. doi:10.1093/mnras/staa2815
Kubyshkina, D. I. & Fossati, L. 2021, Research Notes of the American Astronomical Society, 5, 74. doi:10.3847/2515-5172/abf498
Kubyshkina, D., Vidotto, A. A., Villarreal D'Angelo, C., et al. 2022, MNRAS, 510, 2111. doi:10.1093/mnras/stab3594
Kulikov, Y. N., Lammer, H., Lichtenegger, H. I. M., et al. 2007, SSR, 129, 207. doi:10.1007/s11214-007-9192-4
Lammer, H., Selsis, F., Ribas, I., et al. 2003, ApJ Letters, 598, L121. doi:10.1086/380815
Lammer, H., Kasting, J. F., Chassefière, E., et al. 2008, SSR, 139, 399. doi:10.1007/s11214-008-9413-5
Lammer, H., Leitzinger, M., Scherf, M., et al. 2020, Icarus, 339, 113551. doi:10.1016/j.icarus.2019.113551
Lammer, H., Scherf, M., Kurokawa, H., et al. 2020, SSR, 216, 74. doi:10.1007/s11214-020-00701-x
Lecavelier des Etangs, A., Vidal-Madjar, A., McConnell, J. C., et al. 2004, A&A, 418, L1. doi:10.1051/0004-6361:20040106
Lee, Y., Combi, M. R., Tenishev, V., et al. 2015, GRL, 42, 9015. doi:10.1002/2015GL065291
Li, S., Lu, H., Cao, J., et al. 2022, ApJ, 931, 30. doi:10.3847/1538-4357/ac6510
Lopez, E. D., Fortney, J. J., & Miller, N. 2012, ApJ, 761, 59. doi:10.1088/0004-637X/761/1/59
Matsakos, T., Uribe, A., & Königl, A. 2015, A&A, 578, A6. doi:10.1051/0004-6361/201425593
Mihalas, D. & Mihalas, B. W. 1984, New York, Oxford University Press, 1984, 731 p.
Mordasini, C. 2020, A&A, 638, A52. doi:10.1051/0004-6361/201935541
Murray-Clay, R. A., Chiang, E. I., & Murray, N. 2009, ApJ, 693, 23. doi:10.1088/0004-637X/693/1/23

Odert, P., Lammer, H., Erkaev, N. V., et al. 2018, Icarus, 307, 327. doi:10.1016/j.icarus.2017.10.031
Owen, J. E. & Jackson, A. P. 2012, MNRAS, 425, 2931. doi:10.1111/j.1365-2966.2012.21481.x
Owen, J. E. & Adams, F. C. 2014, MNRAS, 444, 3761. doi:10.1093/mnras/stu1684
Owen, J. E. & Wu, Y. 2017, ApJ, 847, 29. doi:10.3847/1538-4357/aa890a
Öpik, E. J. 1963, Geophysical Journal, 7, 490. doi:10.1111/j.1365-246X.1963.tb07091.x
Pepin, R. O. 1991, Icarus, 92, 2. doi:10.1016/0019-1035(91)90036-S
Ramstad, R. & Barabash, S. 2021, SSR, 217, 36. doi:10.1007/s11214-021-00791-1
Sakai, S., Seki, K., Terada, N., et al. 2018, GRL, 45, 9336. doi:10.1029/2018GL079972
Salz, M., Schneider, P. C., Czesla, S., et al. 2015, A&A, 576, A42. doi:10.1051/0004-6361/201425243
Salz, M., Schneider, P. C., Czesla, S., et al. 2016, A&A, 585, L2. doi:10.1051/0004-6361/201527042
Salz, M., Czesla, S., Schneider, P. C., et al. 2016, A&A, 586, A75. doi:10.1051/0004-6361/201526109
Sekiya, M., Nakazawa, K., & Hayashi, C. 1980, Progress of Theoretical Physics, 64, 1968. doi:10.1143/PTP.64.1968
Shematovich, V. I., Bisikalo, D. V., & Gerard, J. C. 1994, JGR, 99, 23217. doi:10.1029/94JA01769
Shematovich, V. I., Ionov, D. E., & Lammer, H. 2014, A&A, 571, A94. doi:10.1051/0004-6361/201423573
Shizgal, B. & Lindenfeld, M. J. 1982, JGR, 87, 853. doi:10.1029/JA087iA02p00853
Shizgal, B. & Blackmore, R. 1986, Planetary and Space Science, 34, 279. doi:10.1016/0032-0633(86)90133-9
Shizgal, B. D. & Arkos, G. G. 1996, Reviews of Geophysics, 34, 483. doi:10.1029/96RG02213
Trammell, G. B., Li, Z.-Y., & Arras, P. 2014, ApJ, 788, 161. doi:10.1088/0004-637X/788/2/161
Vidal-Madjar, A. 1978, GRL, 5, 29. doi:10.1029/GL005i001p00029
Vidotto, A. A., Fares, R., Jardine, M., et al. 2015, MNRAS, 449, 4117. doi:10.1093/mnras/stv618
Vidotto, A. A. & Cleary, A. 2020, MNRAS, 494, 2417. doi:10.1093/mnras/staa852
Vidotto, A. A. 2021, Living Reviews in Solar Physics, 18, 3. doi:10.1007/s41116-021-00029-w
Villarreal D'Angelo, C., Esquivel, A., Schneiter, M., et al. 2018, MNRAS, 479, 3115. doi:10.1093/mnras/sty1544
Volkov, A. N., Johnson, R. E., Tucker, O. J., et al. 2011, ApJ Letters, 729, L24. doi:10.1088/2041-8205/729/2/L24
Watson, A. J., Donahue, T. M., & Walker, J. C. G. 1981, Icarus, 48, 150. doi:10.1016/0019-1035(81)90101-9
Weber, T., Brain, D., Xu, S., et al. 2021, Journal of Geophysical Research (Space Physics), 126, e29234. doi:10.1029/2021JA029234
Wu, Y. & Lithwick, Y. 2013, ApJ, 763, 13. doi:10.1088/0004-637X/763/1/13
Yelle, R. V. 2004, Icarus, 170, 167. doi:10.1016/j.icarus.2004.02.008
Zahnle, K. J. & Kasting, J. F. 1986, Icarus, 68, 462. doi:10.1016/0019-1035(86)90051-5
Zhang, H., Fu, S., Fu, S., et al. 2022, ApJ, 937, 4. doi:10.3847/1538-4357/ac8a93

Winds of Stars and Exoplanets
Proceedings IAU Symposium No. 370, 2023
A. A. Vidotto, L. Fossati & J. S. Vink, eds.
doi:10.1017/S174392132300025X

Stellar wind from low-mass main-sequence stars: an overview of theoretical models

Munehito Shoda

Department of Earth and Planetary Science, School of Science, The University of Tokyo, 7-3-1 Hongo, Bunkyo-ku, Tokyo, 113-0033, Japan

Abstract. The stellar wind from low-mass stars affects the evolution of the whole stellar system in various ways. To better describe its quantitative contributions, we need to understand the theoretical aspects of stellar wind formation. Here, we present an overview of the theoretical models of stellar wind. The classical thermally-driven wind model fails in reproducing the anti-correlation between the coronal temperature and wind speed observed in the solar wind, thus needs modification with magnetic-energy injection. Specifically, energy input by Alfvén wave is likely to be important. Indeed, a number of solar-wind observations are well reproduced by the Alfvén-wave models, although it could be risky to directly apply the Alfvén-wave models to general low-mass stars. For a better description of stellar wind from low-mass stars with a variety of activity levels, the hybrid model would be better, in which we consider the effect of flux emergence as well as Alfvén wave.

Keywords. solar wind, stars: winds, outflows, stars: mass loss

1. Introduction

Stellar wind plays several important roles in the long-term evolution of the stellar system. For example, stellar rotational evolution in the main sequence is regulated by the angular momentum loss by the magnetized stellar wind (magnetic braking, Kraft 1967; Weber & Davis 1967; Sakurai 1985; Kawaler 1988). Stellar wind also affects the evolution of planets in the form of enhanced/suppressed atmospheric loss (Vidotto & Cleary 2020; Mitani et al. 2022) and water (hydrogen) supply (Daly et al. 2021). To better understand these processes, we need to know how the stellar-wind properties depend on the stellar fundamental parameters, such as luminosity, radius, mass, metallicity, and age (Reimers 1975; Schröder & Cuntz 2005; Cranmer & Saar 2011; Suzuki et al. 2013).

One problem underlying the stellar wind studies is the observational difficulty. Since the direct observation of the stellar wind from low-mass stars is impossible (except the solar wind), we need to infer the stellar wind parameters from model-dependent indirect methods, including the excessive absorption of the Lyman-alpha line (Wood et al. 2005, 2014, 2021), slingshot prominence (Jardine & Collier Cameron 2019), and the planetary response to the stellar wind (Vidotto & Bourrier 2017; Kavanagh et al. 2019). In spite of the various methods proposed, the number of samples is still insufficient to derive statistical properties (see Figure 8 of Vidotto 2021).

Given the observational difficulty, improving theoretical models is essential to promote our knowledge of the stellar wind. In this article, we overview the current status of the stellar-wind models. We first briefly review the classical thermally-driven wind model and its limitation. More sophisticated "magnetically-driven wind model" is discussed in the subsequent section, with a particular focus on the Alfvén-wave driven models. The summary and future prospect are given in the final section.

2. Parker's thermally-driven wind model

Motivated by the discovery of the million-kelvin envelop of the solar atmosphere (solar corona, Edlén 1943) and the double-tale structure of the comet (Biermann 1957), Parker (1958) proposed a model of the solar (and stellar) wind, which is now called the thermally-driven wind model.

2.1. *Model description*

In the thermally-driven wind model, we assume that the solar wind is described by the hydrodynamic equations. Substituting the energy equation with the polytropic equation, the governing equations of the solar wind are therefore given as follows.

$$\frac{\partial}{\partial t}\rho + \nabla \cdot (\rho \boldsymbol{v}) = 0, \tag{2.1}$$

$$\rho \frac{\partial}{\partial t}\boldsymbol{v} + \rho (\boldsymbol{v} \cdot \nabla) \boldsymbol{v} = -\nabla p - \rho \frac{GM_\odot}{r^3}\boldsymbol{r}, \tag{2.2}$$

$$\frac{\partial}{\partial t}\left(\frac{p}{\rho^\alpha}\right) + \boldsymbol{v} \cdot \nabla \left(\frac{p}{\rho^\alpha}\right) = 0, \tag{2.3}$$

where α is the polytropic index. For simplicity, we further assume the time independence and spherical symmetry, which yield

$$\frac{d}{dr}\left(\rho v_r r^2\right) = 0, \tag{2.4}$$

$$v_r \frac{dv_r}{dr} + \frac{dp}{dr} + \rho \frac{GM_\odot}{r^2} = 0, \tag{2.5}$$

$$\frac{d}{dr}\left(\frac{p}{\rho^\alpha}\right) = 0. \tag{2.6}$$

These equations are reduced to the equation of Mach number $M = v_r/c_s$ (where $c_s = \sqrt{\alpha p/\rho}$ is the sound speed) given by (Kopp & Holzer 1976)

$$\frac{M^2-1}{2M^2}\frac{dM^2}{dr} = \left[1 + \left(\frac{\alpha - 1}{2}\right) M^2\right]\left[\frac{2}{r} - \frac{1}{2}\left(\frac{\alpha+1}{\alpha-1}\right)\frac{GM_\odot/r^2}{E + GM_\odot/r}\right], \tag{2.7}$$

where

$$E = \frac{1}{2}v_r^2 + \frac{\alpha}{\alpha - 1}\frac{p}{\rho} - \frac{GM_\odot}{r} \tag{2.8}$$

is the constant of motion (constant in r). We note that the sign of the right-hand-side of Eq. (2.7) is positive in $r < r_\mathrm{c}$ and negative in $r > r_\mathrm{c}$, where r_c is the critical radius given by

$$r_\mathrm{c} = \frac{GM_\odot}{E}\frac{5 - 3\alpha}{4(\alpha - 1)}. \tag{2.9}$$

When the corona is adiabatic ($\alpha = 5/3$), $r_\mathrm{c} = 0$, which means that the Mach number is everywhere a decreasing function of r. Thus, the solar wind is never supersonic without non-adiabatic processes. In reality, due to thermal conduction and in-situ plasma heating, the effective value of α is smaller than 5/3, thus allowing M to increase in r.

When $\alpha < 5/3$, Eq. (2.7) has two types of solutions that exhibit the subsonic outflow in the vicinity of the Sun (Figure 1). One is the subsonic (breeze) solution that is always $M < 1$, with maximum M at $r = r_\mathrm{c}$. The other is the transonic (wind) solution that satisfies $M = 1$ at $r = r_\mathrm{c}$, in which M always increases in r. Given the sufficiently large pressure gap between the corona and the interstellar medium, the transonic solution is

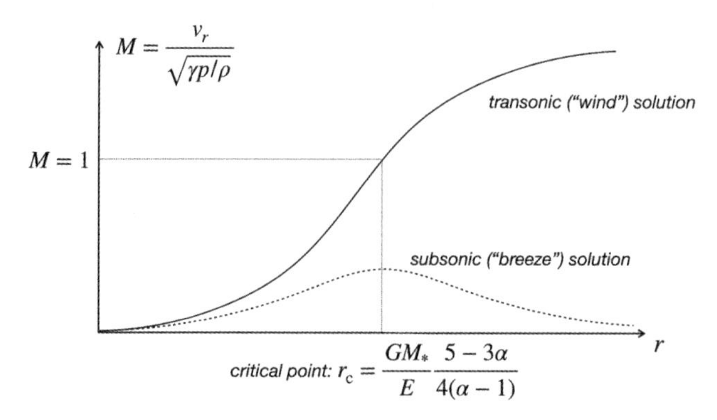

Figure 1. Mathematical solutions of the Mach-number equation Eq. (2.7) with $\alpha < 5/3$.

the only possible stationary state (Parker 1958; Velli 1994). Thus, the thermal expansion of the coronal plasma leads to the supersonic outflow.

2.2. *Limitation*

Although the thermally-driven wind model explains the emergence of the supersonic outflow as a natural consequence of coronal heating, several observations disagree with what the model predicts. For example, the solar wind is found to become slower as the coronal temperature increases, as inferred from the ionization state of the solar wind ions (Geiss et al. 1995; von Steiger et al. 2000). This is contradictory to the thermally-driven model, which predicts faster solar wind for higher coronal temperature (Parker 1958).

Another example of disagreement is the relation between the expansion factor of the solar wind and the wind speed. Fast solar wind emanates from the magnetic flux tube with a small expansion factor, and vice versa (Wang & Sheeley 1990; Arge & Pizzo 2000). According to the thermally-driven wind model, the wind tends to be faster with larger flux-tube expansion (Kopp & Holzer 1976), which is in the opposite sense to observation.

One missing physics in the thermally-driven wind model is the role of magnetic field. The solar corona is the magnetically-dominated region (Gary 2001; Iwai et al. 2014), and thus, magnetic field should dominate the energetics of the solar wind. Since the coronal thermal energy originates from the dissipation of magnetic energy (Alfvén 1947; Parker 1972, 1988; Rappazzo et al. 2008), the ideal model of the solar/stellar wind is such that solves the coronal heating and wind acceleration via the magnetic energy deposition. In the following, we call such a model "magnetically-driven wind model".

3. Magnetically-driven wind models

3.1. *Various forms of magnetic energy injection*

In the magnetically-driven wind models, we assume that the energy of the solar/stellar wind is supplied by magnetic field. Since the magnetic energy is transported via the Poynting flux, the wind models are categorized according to how the Poynting flux is generated. Under the ideal magnetohydrodynamic approximation, the vertical component of the Poynting flux is written (in cgs Gauss unit) as follows.

$$S_z = \frac{c}{4\pi}\boldsymbol{E}\times\boldsymbol{B}\Big|_z = \left(\frac{B_\perp^2}{4\pi}\right)v_z - \frac{B_z}{4\pi}\left(\boldsymbol{v}_\perp\cdot\boldsymbol{B}_\perp\right), \tag{3.1}$$

where X_z and $\boldsymbol{X}_\perp$ denote the vertical (z) and horizontal (x and y) components of $\boldsymbol{X}$, respectively.

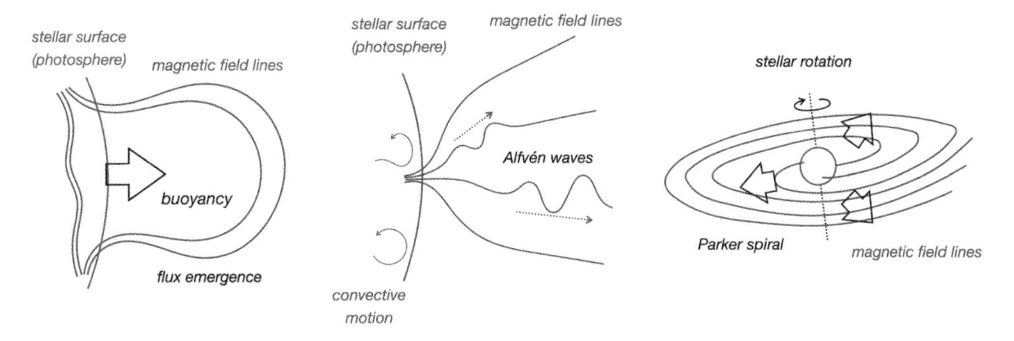

Figure 2. Various forms of Poynting-flux transport. Left: energy transport by the rising-up motion of the horizontal field (flux emergence). Center: energy transport by the transverse oscillations of the velocity and magnetic field (Alfvén wave). Right: energy transport by the magnetic rotation of the stellar wind (Parker spiral).

The first term in the right hand side of Eq. (3.1) represents the energy transfer by the rising-up motion of horizontal magnetic field, that is, flux emergence. The second term corresponds to the energy transfer by the horizontal displacements in velocity and magnetic field. Alfvén wave (Alfvén 1942) is an example of this type of energy transfer, but we note that the stellar rotation coupled with magnetic field (Parker spiral, Parker 1958; Weber & Davis 1967) also contributes to the energy transfer via the second term. The schematic images of the three processes are summarized in Figure 2.

Flux emergence, Alfvén wave and Parker spiral have different roles in the solar/stellar wind acceleration. Energy transferred by flux emergence is likely to be deposited below the critical point, because the typical height of the coronal loop is lower than the critical point. Since the energy input below the critical point leads to the increased mass flux (Hansteen & Velli 2012), the flux emergence should contribute to the larger mass-loss rate. Meanwhile, energy injection by Parker spiral occurs above the critical point, thus Parker spiral contributes to larger wind speed without increasing mass flux (Holzwarth & Jardine 2007; Shoda et al. 2020). Alfvén wave deposits energy both below and above critical points, depending on the shape of flux tube (Cranmer & van Ballegooijen 2005; Verdini & Velli 2007; Perez & Chandran 2013; Chandran & Perez 2019).

3.2. *Alfvén-wave models*

Since the effect of Alfvén wave is the easiest to incorporate in the solar wind models (using, for example, WKB approximation, Alazraki & Couturier 1971; Belcher 1971; Jacques 1977), the Alfvén-wave model is the most widely studied in the magnetically-driven wind models. It is capable of reproducing various observational aspects of the solar wind, including the radial structure of the solar wind (Suzuki & Inutsuka 2005; Cranmer et al. 2007; Verdini et al. 2010; Matsumoto & Suzuki 2012, 2014; Shoda et al. 2018, 2019; Matsumoto 2021), the anti-correlation between the flux-tube expansion and the wind speed (Suzuki & Inutsuka 2006; Cranmer et al. 2007; Shoda et al. 2022), and the global (full-sphere) structure of the solar wind (van der Holst et al. 2014; Usmanov et al. 2018; Réville et al. 2020). We show in Figure 3 the result of a direct numerical simulation of the Alfvén-wave-powered solar wind, which directly demonstrates the performance of the Alfvén-wave model.

In spite of the great performance of the Alfvén-wave model in reproducing the solar wind, it could be risky to assume that the Alfvén-wave model is applicable to any low-mass main-sequence stars. Indeed, Shoda et al. (2020) modeled the stellar wind from

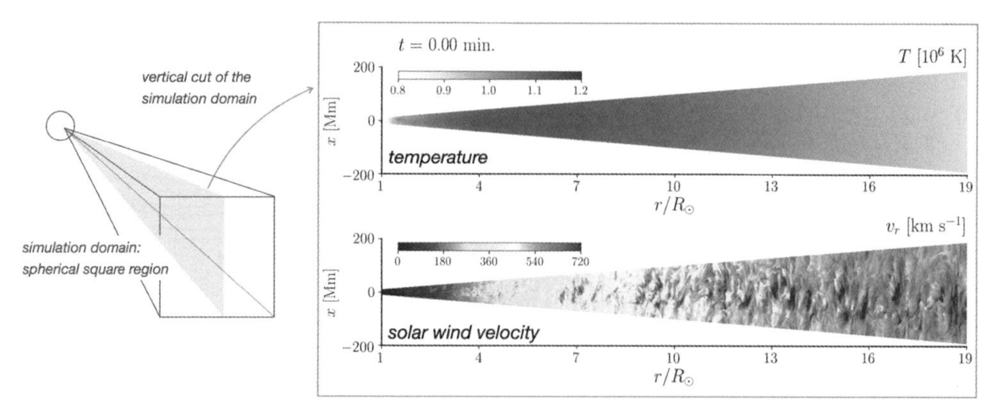

Figure 3. 3D simulation result of the solar wind based on the Alfvén-wave secario (Shoda et al. 2019). The two panels on the right show the profiles of temperature (top) and solar wind velocity (bottom) on the two-dimensional cut (indicated by the left picture).

Sun-like stars under the Alfvén-wave (+ Parker-spiral) scenario, predicting that the mass-loss rate of fast rotators is significantly smaller than the observed values (Wood et al. 2005; Jardine & Collier Cameron 2019; Vidotto 2021). Stellar wind from M-type stars are also predicted to be much weaker than observation in the Alfvén-wave models (Cranmer & Saar 2011; Sakaue & Shibata 2021). These results indicate that Alfvén waves (and Parker spiral) are insufficient to power the stellar wind from magnetically active stars.

3.3. *Flux emergence as a feeding mechanism to the stellar wind*

Flux emergence is a process in which the magnetic bundle embedded in the convection zone emerges above the surface due to magnetic buoyancy (Cheung & Isobe 2014). Flux emergence is often followed by the formation of coronal loops that exhibit large thermal and magnetic energies. Once the magnetic reconnection between the coronal loop and open field line (interchange reconnection, Crooker et al. 2002) occurs, the solar wind can be powered by the coronal-loop energy. In terms of energy flux, it means that the solar wind is accelerated by (a fraction of) energy flux carried by flux emergence. The solar wind models based on the flux emergence and interchange reconnection are called reconnection/loop-opening (RLO) models (Fisk et al. 1999; Fisk 2003; Antiochos et al. 2011; Rappazzo et al. 2012).

Although RLO models are more difficult to numerically demonstrate than Alfvén-wave models, several observations support the RLO scenario as a driving mechanism of the solar wind (Sakao et al. 2007; Harra et al. 2008; Brooks et al. 2015; Bale et al. 2022). Recent observations of magnetic switchback (Bale et al. 2019; Kasper et al. 2019) also indicate the frequent occurrence of interchange reconnection at the base of the corona (Fisk & Kasper 2020; Zank et al. 2020; Drake et al. 2021). With this background, the role of flux emergence and subsequent interchange reconnection has been attracting more interest in these years.

With the current capability of the magnetic-field observation on the Sun, the interchange reconnection is estimated to feed an insufficient amount of energy to the solar wind (Cranmer & van Ballegooijen 2010). However, the existence of small-scale coronal loops without corresponding root magnetic fields on the surface indicates that we might significantly underestimate the rate of flux emergence due to the lack of resolution, in particular in the open-field regions (Wang 2020). Recent observations of the small-scale coronal loops (campfires, Berghmans et al. 2021) support this hypothesis.

Assuming that the rate of flux emergence in the coronal hole is the same as that in the quiet Sun, Wang (2020) estimated the energy flux by flux emergence (measured at the coronal base) as follows.

$$F^{\rm FE} \sim 2.9 \times 10^5 \left(\frac{B_0}{10 \text{ G}}\right) \times \left(\frac{h}{6.4 \text{ Mm}}\right) \left[\frac{E}{2.25 \times 10^{-3} \text{ Mx cm}^2 \text{ s}^{-1}}\right] \text{ erg cm}^{-2} \text{ s}^{-1}, \tag{3.2}$$

where B_0 is the coronal magnetic field, h is the typical vertical scale of the magnetic loop, and E is the rate of flux emergence having total unsigned flux $\gtrsim 1 \times 10^{18}$ Mx. Suppose that h and E do not vary with the magnetic activity, Eq. (3.2) means that the energy flux from flux emergence is proportional to the coronal magnetic field, which is nearly equivalent to the unsigned magnetic flux on the surface. Thus, the flux emergence possibly dominates the energetics in the solar active region and magnetically active stars.

4. Summary and conclusion

In this article, we have reviewed models of the stellar wind from low-mass stars. The thermally-driven wind model naturally explains the origin of the supersonic outflow with the polytropic index $\alpha < 5/3$. Several observations show, however, qualitative disagreements with the thermally-driven wind model, thus prompting us to explicitly consider the effect of magnetic field. The Alfvén-wave model is a representative of magnetically-driven wind models, which is found to perform well in the solar wind. However, the Alfvén wave could be insufficient to power the stellar wind from magnetically-active stars, and thus, it might be necessary to consider the additional effects. Recent studies of the solar flux emergence and solar wind indicate that the flux emergence possibly playes a significant role when the magnetic activity is sufficiently strong. Our conclusion is that, to precisely model the stellar wind from stars with a range of magnetic activity, we might need a hybrid model, in which we consider both Alfvén wave and flux emergence.

Acknowledgement

Munehito Shoda is supported by a Grant-in-Aid for Scientific Research from MEXT/JSPS of Japan, 22K14077.

References

Alazraki, G. & Couturier, P. 1971, A&A, 13, 380
Alfvén, H. 1942, Nature, 150, 405
Alfvén, H. 1947, MNRAS, 107, 211
Antiochos, S. K., Mikić, Z., Titov, V. S., Lionello, R., & Linker, J. A. 2011, ApJ, 731, 112
Arge, C. N. & Pizzo, V. J. 2000, JGR, 105, 10465
Bale, S. D., Badman, S. T., Bonnell, J. W., et al. 2019, Nature, 576, 237
Bale, S. D., Drake, J. F., McManus, M. D., et al. 2022, arXiv e-prints, arXiv:2208.07932
Belcher, J. W. 1971, ApJ, 168, 509
Berghmans, D., Auchère, F., Long, D. M., et al. 2021, arXiv e-prints, arXiv:2104.03382
Biermann, L. 1957, The Observatory, 77, 109
Brooks, D. H., Ugarte-Urra, I., & Warren, H. P. 2015, Nature Communications, 6, 5947
Chandran, B. D. G. & Perez, J. C. 2019, Journal of Plasma Physics, 85, 905850409
Cheung, M. C. M. & Isobe, H. 2014, Living Reviews in Solar Physics, 11, 3
Cranmer, S. R. & Saar, S. H. 2011, ApJ, 741, 54
Cranmer, S. R. & van Ballegooijen, A. A. 2005, ApJS, 156, 265
Cranmer, S. R. & van Ballegooijen, A. A. 2010, ApJ, 720, 824

Cranmer, S. R., van Ballegooijen, A. A., & Edgar, R. J. 2007, ApJS, 171, 520
Crooker, N. U., Gosling, J. T., & Kahler, S. W. 2002, Journal of Geophysical Research (Space Physics), 107, 1028
Daly, L., Lee, M. R., Hallis, L. J., et al. 2021, Nature Astronomy, 5, 1275
Drake, J. F., Agapitov, O., Swisdak, M., et al. 2021, A&A, 650, A2
Edlén, B. 1943, ZA, 22, 30
Fisk, L. A. 2003, Journal of Geophysical Research (Space Physics), 108, 1157
Fisk, L. A. & Kasper, J. C. 2020, ApJ Letters, 894, L4
Fisk, L. A., Schwadron, N. A., & Zurbuchen, T. H. 1999, JGR, 104, 19765
Gary, G. A. 2001, Solar Physics, 203, 71
Geiss, J., Gloeckler, G., & von Steiger, R. 1995, Space Science Reviews, 72, 49
Hansteen, V. H. & Velli, M. 2012, Space Science Reviews, 172, 89
Harra, L. K., Sakao, T., Mandrini, C. H., et al. 2008, ApJ Letters, 676, L147
Holzwarth, V. & Jardine, M. 2007, A&A, 463, 11
Iwai, K., Shibasaki, K., Nozawa, S., et al. 2014, Earth, Planets and Space, 66, 149
Jacques, S. A. 1977, ApJ, 215, 942
Jardine, M. & Collier Cameron, A. 2019, MNRAS, 482, 2853
Kasper, J. C., Bale, S. D., Belcher, J. W., et al. 2019, Nature, 576, 228
Kavanagh, R. D., Vidotto, A. A., Ó. Fionnagáin, D., et al. 2019, MNRAS, 485, 4529
Kawaler, S. D. 1988, ApJ, 333, 236
Kopp, R. A. & Holzer, T. E. 1976, Solar Physics, 49, 43
Kraft, R. P. 1967, ApJ, 150, 551
Matsumoto, T. 2021, MNRAS, 500, 4779
Matsumoto, T. & Suzuki, T. K. 2012, ApJ, 749, 8
Matsumoto, T. & Suzuki, T. K. 2014, MNRAS, 440, 971
Mitani, H., Nakatani, R., & Yoshida, N. 2022, MNRAS, 512, 855
Parker, E. N. 1958, ApJ, 128, 664
Parker, E. N. 1972, ApJ, 174, 499
Parker, E. N. 1988, ApJ, 330, 474
Perez, J. C. & Chandran, B. D. G. 2013, ApJ, 776, 124
Rappazzo, A. F., Matthaeus, W. H., Ruffolo, D., Servidio, S., & Velli, M. 2012, ApJ Letters, 758, L14
Rappazzo, A. F., Velli, M., Einaudi, G., & Dahlburg, R. B. 2008, ApJ, 677, 1348
Reimers, D. 1975, Memoires of the Societe Royale des Sciences de Liege, 8, 369
Réville, V., Velli, M., Panasenco, O., et al. 2020, ApJS, 246, 24
Sakao, T., Kano, R., Narukage, N., et al. 2007, Science, 318, 1585
Sakaue, T. & Shibata, K. 2021, ApJ, 919, 29
Sakurai, T. 1985, A&A, 152, 121
Schröder, K. P. & Cuntz, M. 2005, ApJ Letters, 630, L73
Shoda, M., Iwai, K., & Shiota, D. 2022, ApJ, 928, 130
Shoda, M., Suzuki, T. K., Asgari-Targhi, M., & Yokoyama, T. 2019, ApJ Letters, 880, L2
Shoda, M., Suzuki, T. K., Matt, S. P., et al. 2020, ApJ, 896, 123
Shoda, M., Yokoyama, T., & Suzuki, T. K. 2018, ApJ, 853, 190
Suzuki, T. K., Imada, S., Kataoka, R., et al. 2013, PASJ, 65, 98
Suzuki, T. K. & Inutsuka, S.-i. 2005, ApJ Letters, 632, L49
Suzuki, T. K. & Inutsuka, S.-I. 2006, Journal of Geophysical Research (Space Physics), 111, A06101
Usmanov, A. V., Matthaeus, W. H., Goldstein, M. L., & Chhiber, R. 2018, ApJ, 865, 25
van der Holst, B., Sokolov, I. V., Meng, X., et al. 2014, ApJ, 782, 81
Velli, M. 1994, ApJ Letters, 432, L55
Verdini, A. & Velli, M. 2007, ApJ, 662, 669
Verdini, A., Velli, M., Matthaeus, W. H., Oughton, S., & Dmitruk, P. 2010, ApJ Letters, 708, L116
Vidotto, A. A. 2021, Living Reviews in Solar Physics, 18, 3

Vidotto, A. A. & Bourrier, V. 2017, MNRAS, 470, 4026
Vidotto, A. A. & Cleary, A. 2020, MNRAS, 494, 2417
von Steiger, R., Schwadron, N. A., Fisk, L. A., et al. 2000, JGR, 105, 27217
Wang, Y. M. 2020, ApJ, 904, 199
Wang, Y. M. & Sheeley, N. R., J. 1990, ApJ, 355, 726
Weber, E. J. & Davis, Leverett, J. 1967, ApJ, 148, 217
Wood, B. E., Müller, H.-R., Redfield, S., & Edelman, E. 2014, ApJ Letters, 781, L33
Wood, B. E., Müller, H.-R., Redfield, S., et al. 2021, ApJ, 915, 37
Wood, B. E., Müller, H. R., Zank, G. P., Linsky, J. L., & Redfield, S. 2005, ApJ Letters, 628, L143
Zank, G. P., Nakanotani, M., Zhao, L. L., Adhikari, L., & Kasper, J. 2020, ApJ, 903, 1

Winds of Stars and Exoplanets
Proceedings IAU Symposium No. 370, 2023
A. A. Vidotto, L. Fossati & J. S. Vink, eds.
doi:10.1017/S1743921322004719

The Driving of Hot Star Winds

Andreas A.C. Sander

Zentrum für Astronomie der Universität Heidelberg, Astronomisches Rechen-Institut,
Mönchhofstr. 12-14, 69120 Heidelberg, Germany
email: andreas.sander@uni-heidelberg.de

Abstract. In the regime of hot stars, winds were not seen as a common thing until the era of UV astronomy. Since we have access to the UV wavelength range, it has become clear that winds are not an exotic phenomenon limited to some special objects, but actually ubiquitous among hot and massive stars. The opacities due to spectral lines are the decisive ingredient that allows hot, massive stars to launch powerful winds. While the fundamental principles of these so-called line-driven winds have been realized decades ago, their proper quantitative prediction is still a major challenge today. Established theoretical and empirical descriptions have allowed us to make major progress on all astrophysical scales. However, we are now reaching their limitations as we still lack various fundamental insights on the nature of hot star winds, thereby hampering us from drawing deeper conclusions, not least when dealing with stellar or sub-stellar companions. This has spawned a new generation of researchers searching for answers with a yet unprecedented level of detail in observational and new theoretical approaches.

In these proceedings, the fundamental principles of driving hot star winds will be briefly reviewed. Starting from the classical CAK theory and its extensions, over Monte Carlo and recent comoving-frame-based simulations, the different methods to describe and model the acceleration of hot star winds will be introduced. The review continues with briefly discussing instabilities as well as qualitative and quantitative insights for OB- and Wolf-Rayet-star winds. Moreover, the challenges of companions and their impact on radiation-driven winds are outlined.

Keywords. radiative transfer, stars: atmospheres, stars: early-type, stars: mass loss, stars: winds, outflows, stars: Wolf-Rayet, binaries: general

1. Introduction

While hot stars ($T_{\rm eff} > 10\,\mathrm{kK}$) appear blue to the human eye, their actual flux maximum is located at ultraviolet (UV) wavelengths. The onset of UV astronomy at the end of the 1960s (Morton 1967; Lucy & Solomon 1967) revealed that this highly energetic flux does not simply escape from the stars, but can give rise to strong stellar winds due to the absorption of the star's radiation in spectral lines. Albeit the photon is eventually re-emitted, this re-emission happens in an arbitrary direction, while most of the absorption occurs in radial direction. This results in a radial net transfer of momentum from the photons to the matter in the outermost layers of the star, giving rise to a stellar wind.

The idea of radiation pressure being powerful enough to remove material from a stellar surface had already been proposed a generation earlier for known (optical) emission-line stars (Milne 1926; Beals 1929), but the ubiquitous occurrence of radiation-driven winds in hot stars was only discovered with the accessibility of the UV spectral range, revealing P Cygni profiles in many hot star spectra. Albeit hot star winds are subject to inherent instabilities (e.g., Lucy & Solomon 1970, see also Sect. 3), the overall shape of the wind-affected line profiles is largely constant over time, thereby justifying also a stationary, i.e. time-independent, description. In this limit, the radiative acceleration for a spherically

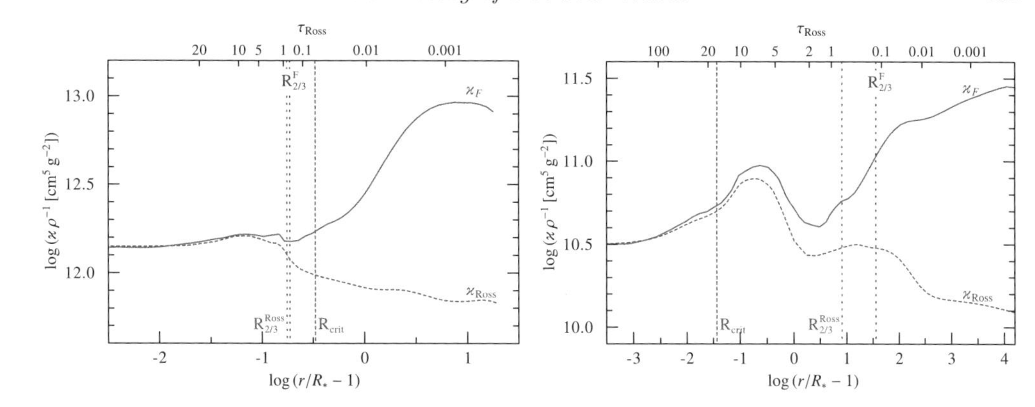

Figure 1. Comparison of flux-weighted (red, solid) and Rosseland mean opacity (blue, dashed) for dynamically-consistent atmosphere models of a B supergiant (left panel) and a hydrogen-free WN star (right panel).

symmetric star can be written as

$$a_{\rm rad}(r) = \frac{1}{c}\int\limits_0^\infty \varkappa_\nu(r)F_\nu(r)\mathrm{d}\nu = \varkappa_F(r)\frac{L}{4\pi c r^2}. \tag{1.1}$$

It is usually convenient to express $a_{\rm rad}$ not as an absolute value, but normalize it to the gravitational acceleration that tries to pull material back to the star. One can thus define the quantity

$$\Gamma_{\rm rad}(r) := \frac{a_{\rm rad}(r)}{g(r)} = \varkappa_F(r)\frac{L}{4\pi cGM}. \tag{1.2}$$

From Eq. (1.2), we immediately obtain that the strength of a radiatively driven wind depends mainly on three quantities: the stellar luminosity L, its mass M, and the *flux-weighted mean opacity* $\varkappa_F(r)$. The latter opacity dependence is what makes the description of radiatively-driven winds complicated as $\varkappa_F(r)$ has a multitude of inherent dependencies on the physical conditions in the outermost layers of a star.

The flux-weighted mean opacity $\varkappa_F$ is crucial to understand how stellar winds can escape a star, even if they are considerably below the classical Eddington Limit of $\Gamma_{\rm e} = 1$, defining the traditional structural stability limit for a star. The definition of $\Gamma_{\rm e}$ is similar to Eq. (1.2), but using only the Thomson opacity $\varkappa_{\rm Thomson}\rho(r) = \sigma_{\rm e}n_{\rm e}(r)$ describing the scattering of free electrons instead of the full $\varkappa_F$ Although $\varkappa_{\rm Thomson}$ usually makes a large contribution to the radiative acceleration $a_{\rm rad}$ of a stellar wind, it would (in most cases) not be sufficient to overcome gravity on its own. Instead, the aforementioned absorption in spectral lines plays a key role as a second important component. Bound-bound and bound-free opacities can contribute as well, resulting in a total opacity of

$$\varkappa = \varkappa_{\rm bound-free} + \varkappa_{\rm free-free} + \varkappa_{\rm Thomson} + \varkappa_{\rm lines} \tag{1.3}$$

for hot star winds where components such as molecular transitions or dust are absent due to the high $T_{\rm eff}$-regime.

Given that the calculation of the components beyond $\varkappa_{\rm Thomson}$ in Eq. (1.3) requires a significant numerical effort, the use of tabulated opacities such as OPAL (Iglesias & Rogers 1996) is a common ingredient in astrophysical simulations, e.g., in stellar evolution modelling or time-dependent (magneto-)hydrodynamical simulations.

Such tables provide the Rosseland opacity

$$\varkappa_{\mathrm{Ross}}^{-1} = \frac{\int_0^\infty \varkappa_\nu^{-1} \frac{\partial B_\nu}{\partial T} \mathrm{d}\nu}{\int_0^\infty \frac{\partial B_\nu}{\partial T} \mathrm{d}\nu} \tag{1.4}$$

as a function of temperature and density. As evident from comparing Eq. (1.4) with the implicit definition of $\varkappa_F$ in Eq. (1.1), the definition of the Rosseland opacity $\varkappa_{\mathrm{Ross}}$ and the flux-weighted mean opacity $\varkappa_F$ is not identical. As long as we are in an optically thick regime that fulfils the conditions of local thermodynamic equilibrium (LTE), the values of $\varkappa_F$ can be approximated with $\varkappa_{\mathrm{Ross}}$. The winds of hot stars, however, are not in LTE. In Fig. 1, the difference between the two quantities is illustrated for two different types of hot stars. In the inner, hydrostatic regime where the star is optically thick, the two quantities align. Further out in the stellar wind, $\varkappa_F$ considerably exceeds the values of $\varkappa_{\mathrm{Ross}}$, revealing a completely different trend due to the Doppler-shift of the spectral lines in the wind, giving rise to additional absorption (and thus opacity) at blueshifted wavelengths. Consequently, simulations relying (only) on Rosseland opacity tables cannot produce a realistic stellar wind situation.

2. Methods to calculate the radiative acceleration

With the discovery of the UV P Cygni profiles, the principles of radiative driving were quickly inferred, providing first formulations of hot star mass-loss and theoretical considerations for radiation-driven winds (e.g. Lucy & Solomon 1970; Castor 1974). These efforts eventually culminated in the work by Castor, Abbott, & Klein (1975) giving an analytic approximation for a_{rad} and the resulting mass-loss rate $\dot{M}$. In what is nowadays called the *CAK theory* – abbreviated after their initials – one assumes that the contributions from bound-free and free-free opacity are negligible. Hence, Eq. (1.3) reduces to:

$$\varkappa \approx \varkappa_{\mathrm{Thom}} + \varkappa_{\mathrm{lines}} = (1+\mathcal{M})\varkappa_{\mathrm{Thom}}. \tag{2.1}$$

In the second formulation of Eq. (2.1), the line opacities are expressed in the form of the Thomson opacity, multiplied with a so-called *force multiplier* $\mathcal{M}$. The computation of $\mathcal{M}$ can be further simplified by considering only photons emerging radially from the stellar disk ("radial streaming approximation") and using the Sobolev approximation (Sobolev 1960). Neglecting line overlap and splitting strictly into optically thin and thick cases, $\mathcal{M}$ can be calculated as

$$\mathcal{M}(t) = \frac{1}{c} \sum_{\mathrm{i}=1}^{N_{\mathrm{lines}}} \left(\Delta\nu_{\mathrm{D}} F_\nu\right)_i \begin{cases} \varkappa_{\mathrm{line}}/\varkappa_{\mathrm{Thom}} & \text{for optically thin lines} \\ t^{-1} & \text{for optically thick lines} \end{cases} \tag{2.2}$$

with t denoting the so-called Sobolev optical depth $t = \varkappa_{\mathrm{e}}\rho v_{\mathrm{th}} \left|\frac{\mathrm{d}v}{\mathrm{d}r}\right|^{-1}$, which is independent of a particular line opacity. The resulting curve for $\mathcal{M}(t)$ is then approximated as

$$\mathcal{M} = k t^{-\alpha} \left(10^{-11}\mathrm{cm}^3 \frac{n_{\mathrm{e}}(r)}{W(r)}\right)^\delta. \tag{2.3}$$

The expression (2.3) for $\mathcal{M}$ employs already the extended notation with three coefficients k, α, δ, the latter including the geometrical dilution factor $W(r)$. While α and k go back to the original CAK work, the δ-parameter was introduced by Abbott (1982) to account for ionization changes in the wind. The CAK descriptions were then further extended in the following two decades (e.g. Friend & Abbott 1986; Pauldrach et al. 1986; Kudritzki et al. 1989; Puls et al. 2000), most notably to remove the radial streaming approximation and to interpret the connection to the underlying line-strength distribution. It is the simplicity of requiring only a power-law to describe $a_{\rm rad}$ – and a set of coefficients – that explains the success of the (modified) CAK theory until today. The solution of the equation of motion using the CAK description for $a_{\rm rad}$ also motivates the β-velocity law

$$v(r) = v_\infty \left(1 - \frac{R_*}{r}\right)^\beta \tag{2.4}$$

to describe the velocity stratification of a hot star wind. In the original CAK calculation with negligible gas pressure ("zero sound-speed approximation"), one obtains $\beta = 0.5$. With further considerations and later extensions, values of $\beta \approx 0.8$ are nowadays seen as more representative. In the quantitative spectroscopy of hot stars with model atmospheres, β is often a free parameter that can be indirectly constrained from reproducing UV line profiles together with other spectral features.

An alternative approaches to obtain the $a_{\rm rad}$ is to compute the radiative transfer using a Monte Carlo (MC) approach. For hot star winds, this method was first implemented by Abbott & Lucy (1985). Technically, the interaction of so-called "photon packages" are tracked throughout the wind, which is described as a series of layers (1D shells). In each shell, photons can potentially transfer momentum and energy to the wind material. In contrast to CAK, this method can account for multiple scatterings of each photon, which becomes important in more dense winds that are characterized by higher mass-loss rates. The dense winds of Wolf-Rayet (WR) stars, which cannot be explained in the framework of CAK, have thus been one of the focus applications of MC radiative transfer codes (e.g. Schmutz 1993; de Koter et al. 1997; Vink & de Koter 2005). The known loss of energy and thus luminosity between the onset of the wind and the outer boundary provides a way to compute the mass-loss rate $\dot{M}$ via the relation

$$\frac{1}{2}\dot{M}\,(v_\infty - v_{\rm esc}) = L(R_*) - L(r \to \infty) \tag{2.5}$$

if the terminal wind velocity v_∞ is known, e.g. by assuming a β-law. The widely used mass-loss recipe for OB stars from Vink et al. (2000); Vink et al. (2001) makes use of this, assuming a scaling of v_∞ with $v_{\rm esc}$ and a β-law. Later, Müller & Vink (2008) extended the method to also predict v_∞. The same technique is used in more recent MC calculations such as in Vink & Sander (2021). Combined with some approximations for the atmosphere, the 1D MC calculations are relatively fast on modern computers and have thus been used to probe a wider parameter space with various grids of models. A major outcome of these calculations was the association of an observed change in the terminal velocities around $T_{\rm eff} \approx 21\,{\rm kK}$ (Lamers et al. 1995) with a change in mass-loss rates (Vink et al. 1999). This increase in $\dot{M}$ towards cooler temperatures is termed the *bi-stability jump*, using a terminology introduced originally in modelling efforts for the B hypergiant P Cyg (Pauldrach & Puls 1990). MC models further helped to shape our understanding about the different role of iron (and iron-group elements) compared to CNO and intermediate-mass elements in O-star winds with the first group determining the conditions of the inner wind (and thus $\dot{M}$) and the second group providing the acceleration in the outer wind, thereby setting the terminal velocity v_∞ (e.g. de Koter et al. 1997; Vink et al. 1999).

Given its straight-forward principles, the MC concept can also be used for multi-dimensional simulations. While this is computationally more costly and thus has not been used for large model grids so far, Šurlan et al. (2012, 2013) have used the concept to model the effect of optically thick clumps on the spectral imprint on UV resonance lines in a hot star wind.

Despite the success of the MC approach, it has also limitations. From Eq. (2.5), there is no guarantee that the hydrodynamic equation of motion

$$a_{\text{rad}} - \frac{1}{\rho}\frac{\mathrm{d}P_{\text{gas}}}{\mathrm{d}r} = v\frac{\mathrm{d}v}{\mathrm{d}r} + g(r) \tag{2.6}$$

describing the balance of radiation and gas pressure versus gravity and inertia is actually fulfilled locally, i.e. at every point in the stellar wind. Instead, only the global energy budget, which can be obtained from integrating Eq. (2.6), is consistent. The local consistency can be improved by reformulating the equation of motion in a way that the velocity field can be described by the LambertW function (see, e.g., Müller & Vink 2008; Gormaz-Matamala et al. 2021), but requires the assumption of an isothermal wind. In particular more dense winds are not well approximated by a constant temperature, thereby limiting the validity of the MC results, in particular for the crucial regime of the wind launching region, which sets the mass-loss rate $\dot{M}$.

To obtain a locally consistent description of the radiative acceleration, a third approach has emerged that has traditionally been used in quantitative spectroscopy. The solution of the radiative transfer in the co-moving frame (CMF) goes back to a series of papers starting with Mihalas et al. (1975), describing what is essentially a numerical solution of the integral in Eq. (1.1). While the opacities and emissivities do not remain isotropic in the observer's frame, they do so in the CMF, providing a significant advantage in programming and computing. This method has been widely applied in expanding stellar atmosphere codes for quantitative spectroscopy such as CMFGEN (e.g. Hillier & Miller 1998), PoWR (e.g. Hamann et al. 1991), FASTWIND (e.g. Puls et al. 2020), or PHOENIX (e.g. Baron & Hauschildt 1998). With their focus on spectral analysis, the elemental coverage was traditionally limited to elements that either were visible in the observable part of the spectrum or necessary for obtaining a correct temperature and ionization equilibrium. Nonetheless, their local solution of $a_{\text{rad}}(r)$ make CMF models a suitable input for a local solution of the hydrodynamic equation of motion (2.6).

While such hydrodynamically-consistent CMF calculations were envisioned already in the 1980s (see Pauldrach et al. 1986), their success only became possible with the capabilities to include many different elements and solve initial stability problems. The first successful model was presented by Gräfener & Hamann (2005), reproducing the spectrum of a prototypical early-type WC star. Later, a series of hydrogen-rich, late-type WN stars (Gräfener & Hamann 2008) yielded a first $\dot{M}$ recipe based on dynamically-consistent CMF atmosphere models. The method has since also been applied to O- and B-type stars with different codes varying in the implementation details (e.g. Krtička & Kubát 2010; Sander et al. 2017; Sundqvist et al. 2019). For OB-type wind predictions, the CMF-based calculations (e.g. Krtička & Kubát 2017, 2018; Björklund et al. 2021) yield a considerable reduction in mass-loss rates compared to the Vink et al. (2001) description. First comparisons with observations (e.g. Hawcroft et al. 2021; Ramachandran et al. 2019) indicate that the inferred rates from OB-type models could be too low, at least in some parameter regimes. Another issue, e.g. seen in OB-type models by (Björklund et al. 2021), are the often too high terminal velocities obtained in the CMF models. Identifying the underlying reasons is a matter of ongoing research and crucial to obtain better wind predictions. Recent models for Wolf-Rayet stars (Sander et al. 2020; Sander & Vink 2020) do not show the same pattern. One possible ingredient could be the uncertain turbulent velocities at the base of the wind, which can have a strong influence on the derived $\dot{M}$ (e.g. Lucy 2010; Krtička & Kubát 2017; Björklund et al. 2021) and likely differ between OB and WR-type stars. Moreover, there might be unaccounted processes in the 1D models, especially in the case of optically thin winds, resulting from multi-D effects. While a full CMF-based 3D wind treatment is computationally not feasible, ongoing efforts exist towards more realistic 3D simulations (e.g. Moens et al. 2022; Poniatowski et al. 2022) which could provide the necessary constraints and descriptions for 1D models with a more detailed radiative transfer.

The three methods to calculate the radiative acceleration (mCAK, MC, CMF) provide complementary toolsets, reaching from fast, but more approximate, to very detailed, but numerically expensive treatments. Their insights have enhanced our view on radiative driving, but have also prompted new questions. The recent progress in the CMF-based models is promising, but eventually the need for a CAK-like, parametrized description will be necessary for a realistic, but also efficient treatment of hot star winds in time-dependent and large-scale simulations.

3. Instabilities and wind clumping

The acceleration mechanism of line-driven winds is subject to inherent instabilities. This was expected already by Lucy & Solomon (1970) when introducing their steady-state formalism to explain the observed OB-star winds. It took about a decade until multiple approaches aimed to describe the time-dependent behaviour and the resulting instabilities (e.g. Holzer 1977; MacGregor et al. 1979; Carlberg 1980). These were later combined in the line-force perturbation analysis work of Owocki & Rybicki (1984). The basic idea behind the instability of line-driven acceleration stems from the picture that a velocity-dependent line-shift enables new absorption capabilities. Due to the wind, a particular line transition can absorb photons of different wavelengths with bluer wavelengths being reached for higher wind velocities. Consequently, a positive perturbation in the velocity field exposes material to "fresh" flux. This in turn leads to additional absorption and thus further wind acceleration, thereby again enhancing the wind velocity.

Various simulation efforts have been performed over the decades (e.g. Feldmeier et al. 1997; Owocki & Puls 1999; Sundqvist & Owocki 2013), including extensions to multiple dimensions (Dessart & Owocki 2005; Driessen et al. 2022) or the consideration of magnetic fields (ud-Doula & Owocki 2002; Driessen et al. 2021). To avoid extensive numerical costs or obtain (semi-)analytic descriptions, the use of the CAK theory and its later reformulations and extensions (e.g., Gayley 1995; Owocki & Puls 1999) is common. Still, it is important to underline that while the absolute magnitude of the instabilities in line-driven winds is sensitive to its physical approximations and numerical treatments, their existence as such is not a product of any simplifications. The so-called *line-driven instability* or *line-deshadowing instability*, abbreviated LDI, is expected from fundamental considerations.

The LDI provides a mechanism for breaking up homogeneous wind structures, causing porosity (i.e. "clumping") in physical and velocity space. Wind-intrinsic X-rays in hot star winds are commonly attributed to shocks resulting from colliding clumps (Feldmeier et al. 1997). Yet, it is unclear whether the LDI alone can fully explain the observed clumping and X-rays. There are observational (e.g., Lépine & Moffat 2008) and structural indications (e.g., Cantiello et al. 2009) that inhomogeneities exist already beneath the regime that is susceptible to the LDI. Consequently, the LDI might only enhance effects that occur already in the near-sonic layers of hot stars.

4. The complex contribution of the different elements and ions

Beside the acceleration due to free electrons, Castor et al. (1975) included only C III in their original paper. Due to their prominent P Cygni profiles in the UV, the CNO elements were initially seen as the main contributors for driving the winds of hot stars. Later extensions to more elements (e.g., Abbott 1982) then revealed that this picture was too simplified and also elements like iron (Fe) were playing an important role with

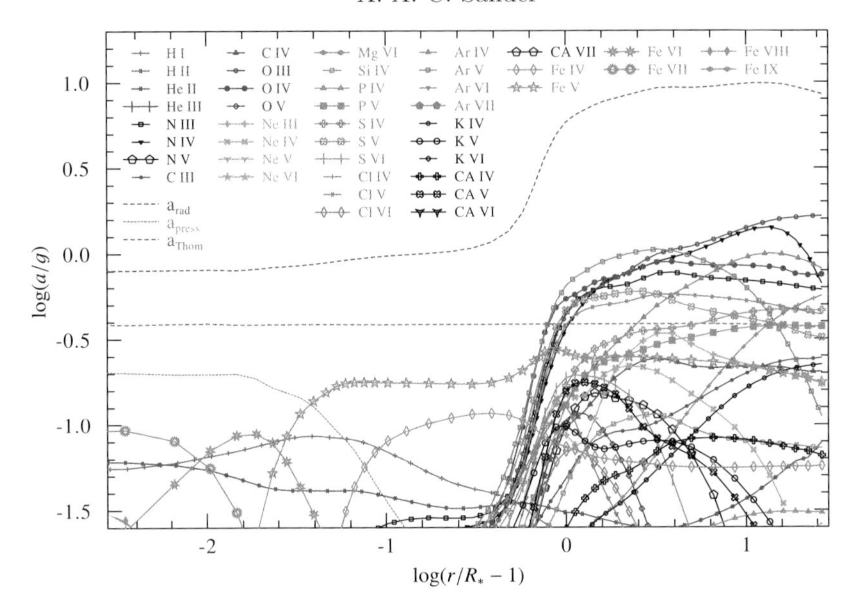

Figure 2. Contributions to the radiative acceleration in an O supergiant with $T_{\rm eff} = 42\,\rm kK$. The underlying model to produce this plot is taken from Sander et al. (2017).

different elements and ions dominating the wind driving in different temperature regimes. Moreover, the CAK-inherent approximation that bound-free and free-free opacities do not significantly contribute to $a_{\rm rad}$ turned out to be invalid for more dense winds.

The contributions of the individual elements to the total radiative acceleration is not trivial. Early calculations for ζ Pup and τ Sco by Pauldrach (1987) demonstrated that the contributions from different elements to the total (modified) CAK force multiplier $\mathcal{M}$ are substantially changing between different types of stars and different parts of the wind. Moreover, elements not visible in the observable spectra like Ne and Ar can be efficient wind drivers despite their low abundances.

With the detailed CMF radiative transfer we can nowadays obtain an even more detailed picture, which we illustrate in Figs. 2, 3 and 4 for three different types of hot stars. In these plots, we only separate Thomson scattering and gas pressure, while the individual ion contributions include both line and (true) continuum contributions. The comparison of the three plots underlines the importance of Fe for driving stellar winds. As briefly mentioned above, iron is particularly crucial for launching a stellar wind in the first place and thereby indirectly determining the mass-loss rate $\dot{M}$ (e.g. Pauldrach et al. 1993; de Koter et al. 1997; Vink et al. 1999). In the outer wind, a variety of elements is responsible for the further wind acceleration, thereby setting v_∞. Especially the latter includes elements that are usually not detectable in the spectra, such as argon in the case of the depicted O supergiant model (Fig. 2). In general, the elemental abundance is not a good indicator for its contribution to the wind driving. In the B supergiant model (Fig. 3), ions of Fe, Si, and S dominate the acceleration of the outer wind, despite the more abundant CNO elements. Hydrogen and helium do not contribute notably to the line acceleration. However, their bound-free contribution can be significant if their ionization stage changes in the wind. This is the case in the depicted WR model (Fig. 4), where the contributions of He II and H I in the outer wind are clearly visible.

The more dense winds of WR stars are still quite enigmatic, but the example in Fig. 4 illustrates that for He-burning stars the wind launching region is much further inward than in the case of O and B stars. The launching of the wind is then fully tied to the

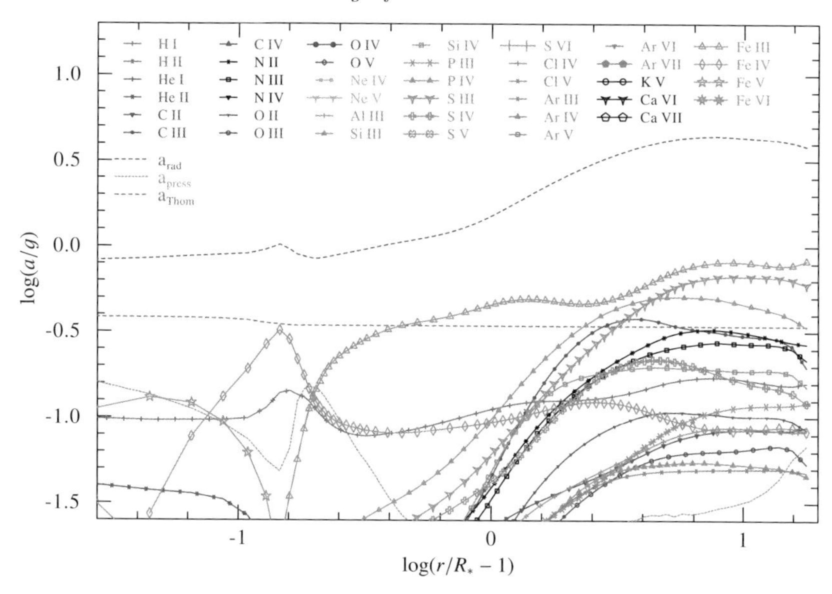

Figure 3. Contributions to the radiative acceleration in a B supergiant with $T_{\rm eff} = 25\,$kK. The underlying model to produce this plot is taken from Sander et al. (2018).

iron opacities (see also Gräfener & Hamann 2005; Vink & de Koter 2005). The observational evidence for WR stars appearing preferably in higher metallicity environments (e.g. Shenar et al. 2020) is thus nicely backed by wind driving studies. This means that also the mass-loss of WR stars scales (mainly) with iron, which has severe consequences for stellar evolution in the early Universe (e.g. Vink et al. 2021). In older population synthesis (e.g. Hurley et al. 2000), WR winds were treated as metallicity-independent due to the WC stage where self-produced carbon reaches the surface of the stars. However, as indicated by the modelling results in Sander et al. (2020), the higher ionization stages of carbon do not yield significant line opacities and thus cannot compensate the declining impact of iron opacities on the wind mass-loss rates at lower metallicities.

Beside the classical, He-burning WR stars, the WR phenomenon also occurs at the upper end of the main sequence. Due to their high L/M-ratio, very massive stars (VMS) with $M_{\rm ini} \approx 100\,M_\odot$ and higher already show WN-type spectra during their central hydrogen burning stage. Such objects, which exist for example in NGC 3603 and the R136 cluster in the LMC (e.g. Crowther & Dessart 1998) or the Galactic Center region (e.g. Figer et al. 2002), have typically less dense winds than the bulk of the He-burning WR stars (if compared at the same metallicity). In the last 15 years, these stars have gained substantial attention, not least due to their influence on stellar populations (e.g. Vink et al. 2015; Senchyna et al. 2021). Yet, major differences exist in modelling their mass-loss rates (e.g. Gräfener & Hamann 2008; Vink et al. 2011; Gräfener 2021) and the resulting evolutionary paths (e.g., Sabhahit et al. 2022, see also these proceedings).

5. The influence of companions

Multiplicity is common among massive stars. Consequently, a significant fraction of massive stars has one or more stellar or compact companion. Depending on the configuration of the system – concerning both the properties of the objects as well as their orbital parameters – the (radiative) driving of the stellar winds can be considerably affected. In any case, the presence of a companion breaks the spherical symmetry of the system. While this is also the case for rotating individual stars (e.g. Müller & Vink 2014), the

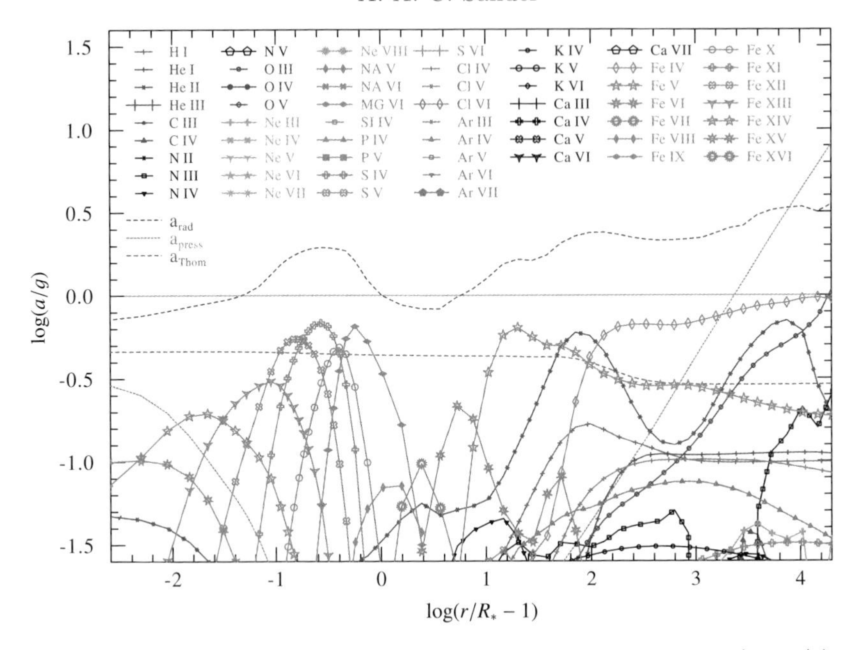

Figure 4. Contributions to the radiative acceleration in a WN star with $T_{\rm eff}(\tau = 2/3) = 32\,\mathrm{kK}$, $T_{\rm eff}(\tau_{\rm sonic} = 27) = 122\,\mathrm{kK}$, and $X_{\rm H} = 0.2$. The underlying model to produce this plot is taken from Sander et al. (2022, submitted).

geometry is not as complex as in case of a companion. With additional radiation coming from non-radial directions or radiation missing from particular angles, typical assumptions about the radiation field inherent to most 1D model efforts are invalid. 2D and 3D calculations become necessary, but are numerically much more costly unless other modelling aspects are significantly simplified. For practical reasons, a full 3D radiative transfer is typically limited to specific applications (e.g. Šurlan et al. 2013; Hennicker et al. 2018, 2020) without performing a full non-LTE model atmosphere calculations, expect for first test cases (Hauschildt & Baron 2014). Multi-dimensional radiative driving studies incorporating the influence of companions therefore need to make significant approximations either the radiative acceleration or for the influence of the companion. Without claiming completeness, the following aspects can lead to major changes in radiation-driven winds of hot stars:

Binary evolution: Mass transfer in a binary system considerably changes the stellar parameters of both components. In particular in cases where the donor star is already evolved, the loss of surface material leads to a configuration that fosters stronger stellar winds. Directly, the higher L/M-ratio of the donor star, which is now "overluminous" compared to a single star of the same mass, brings the donor closer to the Eddington limit and thus increases the mass-loss rate. In addition, the mass transfer can prevent the donor from leaving the regime of radiation-driven winds due to avoiding cooler surface temperatures.

Roche Lobe modification: The existence of a stellar wind with $\Gamma_{\rm rad} > 0$ modifies the effective gravity $g_{\rm eff} = g(1 - \Gamma_{\rm rad})$ of a star. Hot star winds eventually reach supersonic speeds and thus $\Gamma_{\rm rad} > 1$, meaning that the effective gravity becomes negative and there is no longer a Roche Lobe for this star (e.g. Dermine et al. 2009). Unless the subsonic layers of the star alone fill their Roche Lobe, also the term "Roche Lobe overflow" (RLOF) becomes meaningless. Due to the absence of a Roche Lobe, material can escape in any direction. Still, the gravitational potential of the companion provides a preference for matter overflow, leading to a so-called "focussed wind" scenario (Friend & Castor 1982; Gies & Bolton 1986; Hirai & Mandel 2021). Recently, the term "wind RLOF" has also been used in the context of hot star winds to describe supersonic winds focussed towards an accretor (El Mellah et al. 2019a,b). This terminology was originally introduced by Mohamed & Podsiadlowski (2007) for AGB and RSG stars with strong winds. There, the stars as such do not fill their Roche Lobe. However, as RSG and AGB winds are characterized by subsonic velocities, the wind material can fill the star's Roche Lobe in a similar way than the (hydrostatic) star itself. The material is then removed mainly through the inner Lagrangian point. Unless the acceleration is extremely shallow, a similar situation of nearly complete redirection onto the accretor does not occur for hot star winds. El Mellah et al. (2019b) calculated that a maximum of approximately 20% of the donor wind could be accreted by the (compact) companion in an X-ray binary.

Wind-wind collision: In typical cases, both companions in a massive binary system are hot stars, leading e.g. to systems of type O+O or WR+O. The collision region of the winds from the two components then gives rise to a variety of phenomena, most notably (additional) X-ray emission arising from the shock region and non-thermal radio emission (e.g. Prilutskii & Usov 1976; Stevens et al. 1992; De Becker 2007). The additional X-ray flux can ionize the wind and thus alter the radiative acceleration. In most cases, the higher ionization stages will provide fewer line transitions, thus leading to a reduction of the wind driving. A further reduction of the wind driving arises due to the existence of the UV-intensive companion itself. The absorption of the photons from the other hot star can lead to the so-called *radiative breaking* (e.g. Stevens & Pollock 1994; Gayley et al. 1997) as the absorbed momentum has a component opposite to the radial wind direction. In extreme cases, this effect could lead to a collapse of the wind-colliding region, enabling temporary accretion of material from one star onto the other (e.g. Soker 2007). More frequently, however, are other wind collision effects, such as (optical) excess line emission in WR+O binaries (e.g. Lührs 1997; Hill et al. 2002) or dust production, in particular in environments where the wind from a carbon-rich WR star (WC star) meets a hydrogen-rich O-star wind (e.g. Allen et al. 1972; Williams et al. 1990; Lau et al. 2021).

Irradiation: Massive binaries including a compact object (i.e. a neutron star or black hole) as one of their components, can be a major X-ray source due to the accretion of material onto the compact object. In these so-called High Mass X-ray binaries (HMXBs), the ionization structure and thus also the driving of the stellar wind of the non-degenerate star can be significantly altered if the X-ray source is close enough. Orbital modulations of characteristic UV wind profiles were first detected in the late 1970s by Hatchett & McCray (1977), providing observational evidence for the phase-dependent change of the donor star winds. Similar to the X-ray effect in colliding winds, the higher ionization leads to a reduction or termination of the radiative driving (e.g. MacGregor & Vitello 1982; Stevens & Kallman 1990). More detailed calculations of the radiative driving reveal that contrary to the general reduction, there can also be an enhancement of the radiative force in some cases if the X-ray luminosity remains moderate (e.g. Krtička et al. 2012; Sander et al. 2018). The lower terminal wind velocities resulting from X-ray irradiation can foster the accretion of material onto the compact object, thereby affecting the further evolution of the system (e.g. Krtička et al. 2022; Ramachandran et al. 2022).

6. Summary & Conclusions

The outermost layers of hot stars provide an inherent non-LTE environment. The winds of these stars are mainly driven by radiative acceleration. The most important contributors to this acceleration are Thomson scattering and the absorption of photons in spectral lines. As the latter corresponds to a strongly radius-dependent opacity and the launch of the winds is usually impossible without considering this additional contribution, hot star winds are often simply termed *line-driven*. In reality, additional acceleration components can contribute, such as free-free and bound-free opacities, but their importance is usually much weaker than the line and free electron opacities. Gas pressure only has a substantial impact in the subsonic layers.

The line driving opacity itself consists of a plethora of contributing elements and ions with individual contributions depending strongly on the particular stellar parameters, most notably T_{eff}. The iron group elements play a crucial role in hot star winds due to their millions of available line transitions and the resulting supply of opacity (and thus acceleration). In the onset region of hot star winds, lighter elements are often too highly ionized to contribute significant line opacities. Thus, it is the iron opacity that largely determines the mass-loss rate of hot star winds. Overall, one can distinguish between two types of line-driven hot star winds:

OB-type winds are optically thin with individual optically thick lines. The imprint of the stellar wind is mainly seen in the UV with additional, smaller features in the optical and (near) IR for higher mass-loss rates (e.g. in supergiants). OB-type winds occur in most hot, massive stars, ranging from the main sequence over the supergiant regime to hydrogen-stripped stars that are not sufficiently close to the Eddington limit to launch WR-type winds.

WR-type winds are optically thick up to large radii. Their optical spectra are characterized by emission lines. Caused by the proximity to the Eddington limit, WR-type winds occur not only in evolved hydrogen-depleted stars (classical WR stars), but also in very massive (hydrogen-rich) stars and (some) luminous blue variables such as AG Car, η Car, or P Cyg.

Radiatively driven winds are commonly described and simulated with either the semi-analytic (modified) CAK description or more comprehensive Monte Carlo or CMF calculations. The latter two allow for a more realistic physical treatment. In particular, the CMF modelling can provide a local consistency, resulting in complex shapes for the flux-weighted mean opacity $\varkappa_F$ that cannot be sufficiently approximated by using Rosseland opacities. While recipes for the wind mass-loss rate $\dot{M}$ based on such detailed $\varkappa_F$-models become more and more available, the aim for an analytic description usually requires compromises. Consequently, no current $\dot{M}$-formula captures the full $\varkappa_F$-complexity.

Companions of hot stars can have a severe influence on their radiatively driven stellar winds. Via mass-transfer in binary systems, they can cause stronger stellar winds for one of the stars. X-rays due to wind-wind collision or accretion can significantly alter the ionization stratification, usually weakening the radiative acceleration. The supersonic nature of hot star winds further modifies the effective gravity of the stars, making it difficult or even impractical to define a Roche lobe.

Radiatively-driven winds are subject to inherent instabilities, requiring time-dependent modelling to study their effects. While considerable imprints of time-dependent processes, e.g. clumping and shocks, are seen in such simulations, they also show that the bulk of the matter outflow can be sufficiently described by stationary models. Quantitative spectroscopy of hot stars makes use of this result in so-called unified model atmospheres where the hydrostatic and the supersonic layers are treated consistently to obtain synthetic spectra of hot stars including their winds. Observational evidence indicates that both OB- and WR-type winds are inhomogeneous ("clumped"). The origin of the clumps is debated and might even reach down into the subsonic layers. The existence of clumps requires approximations in (1D) model atmospheres and considerably impacts the diagnosis of wind mass-loss rates.

Acknowledgements

AACS is funded by the Deutsche Forschungsgemeinschaft (DFG, German Research Foundation) in the form of an Emmy Noether Research Group – Project-ID 445674056 (SA4064/1-1, PI Sander)

References

Abbott, D. C. 1982, *ApJ*, 259, 282
Abbott, D. C. & Lucy, L. B. 1985, *ApJ*, 288, 679
Allen, D. A., Swings, J. P., & Harvey, P. M. 1972, *A&A*, 20, 333
Baron, E. & Hauschildt, P. H. 1998, *ApJ*, 495, 370
Beals, C. S. 1929, *MNRAS*, 90, 202
Björklund, R., Sundqvist, J. O., Puls, J., et al. 2021, *A&A*, 648, A36
Carlberg, R. G. 1980, *ApJ*, 241, 1131
Castor, J. L. 1974, *MNRAS*, 169, 279
Castor, J. I., Abbott, D. C., & Klein, R. I. 1975, *ApJ*, 195, 157
Cantiello, M., Langer, N., Brott, I., et al. 2009, *A&A*, 499, 279
Crowther, P. A. & Dessart, L. 1998, *MNRAS*, 296, 622
De Becker, M. 2007, *The Astronomy and Astrophysics Review*, 14, 171
de Koter, A., Heap, S. R., & Hubeny, I. 1997, *ApJ*, 477, 792
Dermine, T., Jorissen, A., Siess, L., et al. 2009, *A&A*, 507, 891
Dessart, L. & Owocki, S. P. 2005, *A&A*, 437, 657
Driessen, F. A., Kee, N. D., & Sundqvist, J. O. 2021, *A&A*, 656, A131
Driessen, F. A., Sundqvist, J. O., & Dagore, A. 2022, *A&A*, 663, A40
El Mellah, I., Sander, A. A. C., Sundqvist, J. O., et al. 2019, *A&A*, 622, A189
El Mellah, I., Sundqvist, J. O., & Keppens, R. 2019, *A&A*, 622, L3
Feldmeier, A., Puls, J., & Pauldrach, A. W. A. 1997, *A&A*, 322, 878
Figer, D. F., Najarro, F., Gilmore, D., et al. 2002, *ApJ*, 581, 258
Friend, D. B. & Castor, J. I. 1982, *ApJ*, 261, 293
Friend, D. B. & Abbott, D. C. 1986, *ApJ*, 311, 701
Gayley, K. G. 1995, *ApJ*, 454, 410
Gayley, K. G., Owocki, S. P., & Cranmer, S. R. 1997, *ApJ*, 475, 786
Gormaz-Matamala, A. C., Curé, M., Hillier, D. J., et al. 2021, *ApJ*, 920, 64
Gies, D. R. & Bolton, C. T. 1986, *ApJ*, 304, 389
Gräfener, G. & Hamann, W.-R. 2005, *A&A*, 432, 633
Gräfener, G. & Hamann, W.-R. 2008, *A&A*, 482, 945
Gräfener, G. 2021, *A&A*, 647, A13
Hamann, W.-R., Duennebeil, G., Koesterke, L., et al. 1991, *A&A*, 249, 443
Hatchett, S. & McCray, R. 1977, *ApJ*, 211, 552
Hauschildt, P. H. & Baron, E. 2014, *A&A*, 566, A89
Hawcroft, C., Sana, H., Mahy, L., et al. 2021, *A&A*, 655, A67
Hennicker, L., Puls, J., Kee, N. D., et al. 2018, *A&A*, 616, A140
Hennicker, L., Puls, J., Kee, N. D., et al. 2020, *A&A*, 633, A16
Holzer, T. E. 1977, *JGR*, 82, 23
Hill, G. M., Moffat, A. F. J., & St-Louis, N. 2002, *MNRAS*, 335, 1069
Hillier, D. J. & Miller, D. L. 1998, *ApJ*, 496, 407
Hirai, R. & Mandel, I. 2021, *PASA*, 38, e056
Hurley, J. R., Pols, O. R., & Tout, C. A. 2000, *MNRAS*, 315, 543
Iglesias, C. A. & Rogers, F. J. 1996, *ApJ*, 464, 943
Krtička, J. & Kubát, J. 2010, *A&A*, 519, A50
Krtička, J., Kubát, J., & Skalický, J. 2012, *ApJ*, 757, 162
Krtička, J. & Kubát, J. 2017, *A&A*, 606, A31
Krtička, J. & Kubát, J. 2018, *A&A*, 612, A20
Krtička, J., Kubát, J., & Krtičková, I. 2022, *A&A*, 659, A117

Kudritzki, R. P., Pauldrach, A., Puls, J., et al. 1989, *A&A*, 219, 205
Lamers, H. J. G. L. M., Snow, T. P., & Lindholm, D. M. 1995, *ApJ*, 455, 269
Lépine, S. & Moffat, A. F. J. 2008, *AJ*, 136, 548
Lau, R. M., Hankins, M. J., Kasliwal, M. M., et al. 2021, *ApJ*, 909, 113
Lucy, L. B. & Solomon, P. M. 1967, *AJ*, 72, 310
Lucy, L. B. & Solomon, P. M. 1970, *ApJ*, 159, 879
Lucy, L. B. 2010, *A&A*, 524, A41
Lührs, S. 1997, *PASP*, 109, 504
MacGregor, K. B., Hartmann, L., & Raymond, J. C. 1979, *ApJ*, 231, 514
MacGregor, K. B. & Vitello, P. A. J. 1982, *ApJ*, 259, 267
Milne, E. A. 1926, *MNRAS*, 86, 459
Mihalas, D., Kunasz, P. B., & Hummer, D. G. 1975, *ApJ*, 202, 465
Moens, N., Poniatowski, L. G., Hennicker, L., et al. 2022, *A&A*, 665, A42
Mohamed, S. & Podsiadlowski, P. 2007, *15th European Workshop on White Dwarfs*, 372, 397
Morton, D. C. 1967, *ApJ*, 150, 535
Müller, P. E. & Vink, J. S. 2008, *A&A*, 492, 493
Müller, P. E. & Vink, J. S. 2014, *A&A*, 564, A57
Owocki, S. P. & Rybicki, G. B. 1984, *ApJ*, 284, 337
Owocki, S. P. & Puls, J. 1999, *ApJ*, 510, 355
Pauldrach, A., Puls, J., & Kudritzki, R. P. 1986, *A&A*, 164, 86
Pauldrach, A. 1987, *A&A*, 183, 295
Pauldrach, A. W. A. & Puls, J. 1990, *A&A*, 237, 409
Pauldrach, A. W. A., Feldmeier, A., Puls, J., et al. 1993, *Space Science Reviews*, 66, 105
Prilutskii, O. F. & Usov, V. V. 1976, *Soviet Astronomy*, 20, 2
Poniatowski, L. G., Kee, N. D., Sundqvist, J. O., et al. 2022, *A&A*, 667, A113, arXiv:2204.09981
Puls, J., Pauldrach, A. W. A., Kudritzki, R.-P., et al. 1993, *Reviews in Modern Astronomy*, 6, 271
Puls, J., Springmann, U., & Lennon, M. 2000, *A&A Supplement*, 141, 23
Puls, J., Najarro, F., Sundqvist, J. O., et al. 2020, *A&A*, 642, A172
Ramachandran, V., Hamann, W.-R., Oskinova, L. M., et al. 2019, *A&A*, 625, A104
Ramachandran, V., Oskinova, L. M., Hamann, W.-R., et al. 2022, *A&A*, 667, A77, arXiv:2208.07773
Sabhahit, G. N., Vink, J. S., Higgins, E. R., et al. 2022, *MNRAS*, 514, 3736
Sander, A. A. C., Hamann, W.-R., Todt, H., et al. 2017, *A&A*, 603, A86
Sander, A. A. C., Fürst, F., Kretschmar, P., et al. 2018, *A&A*, 610, A60
Sander, A. A. C., Vink, J. S., & Hamann, W.-R. 2020, *MNRAS*, 491, 4406
Sander, A. A. C. & Vink, J. S. 2020, *MNRAS*, 499, 873
Schmutz, W. 1993, *Space Science Reviews*, 66, 253
Senchyna, P., Stark, D. P., Charlot, S., et al. 2021, *MNRAS*, 503, 6112
Shenar, T., Gilkis, A., Vink, J. S., et al. 2020, *A&A*, 634, A79
Sobolev, V. V. 1960, *Harvard University Press*
Soker, N. 2007, *ApJ*, 661, 482
Stevens, I. R. & Kallman, T. R. 1990, *ApJ*, 365, 321
Stevens, I. R., Blondin, J. M., & Pollock, A. M. T. 1992, *ApJ*, 386, 265
Stevens, I. R. & Pollock, A. M. T. 1994, *MNRAS*, 269, 226
Sundqvist, J. O. & Owocki, S. P. 2013, *MNRAS*, 428, 1837
Sundqvist, J. O., Björklund, R., Puls, J., et al. 2019, *A&A*, 632, A126
Šurlan, B., Hamann, W.-R., Kubát, J., et al. 2012, *A&A*, 541, A37
Šurlan, B., Hamann, W.-R., Aret, A., et al. 2013, *A&A*, 559, A130
ud-Doula, A. & Owocki, S. P. 2002, *ApJ*, 576, 413
Vink, J. S., de Koter, A., & Lamers, H. J. G. L. M. 1999, *A&A*, 350, 181
Vink, J. S., de Koter, A., & Lamers, H. J. G. L. M. 2000, *A&A*, 362, 295
Vink, J. S., de Koter, A., & Lamers, H. J. G. L. M. 2001, *A&A*, 369, 574
Vink, J. S. & de Koter, A. 2005, *A&A*, 442, 587

Vink, J. S., Muijres, L. E., Anthonisse, B., et al. 2011, *A&A*, 531, A132
Vink, J. S., Heger, A., Krumholz, M. R., et al. 2015, *Highlights of Astronomy*, 16, 51
Vink, J. S., Higgins, E. R., Sander, A. A. C., et al. 2021, *MNRAS*, 504, 146
Vink, J. S. & Sander, A. A. C. 2021, *MNRAS*, 504, 2051
Williams, P. M., van der Hucht, K. A., The, P. S., et al. 1990, *MNRAS*, 247, 18P

Winds of Stars and Exoplanets
Proceedings IAU Symposium No. 370, 2023
A. A. Vidotto, L. Fossati & J. S. Vink, eds.
doi:10.1017/S1743921322003507

Slingshot Prominences, Formation, Ejection and Cycle Frequency in Cool Stars

S. Daley-Yates and M. M. Jardine

School of Physics and Astronomy, University of St Andrews, North Haugh, St Andrews, Fife, Scotland KY16 YSS, UK

Abstract. Stars lose mass and angular momentum during their lifetimes. Observations of H-alpha absorption of a number of low mass stars, show prominences transiting the stellar disc and being ejected into the extended stellar wind. Analytic modelling have shown these M-dwarf coronal structures growing to be orders of magnitude larger than their solar counterparts. This makes prominences responsible for mass and angular momentum loss comparable to that due to the stellar wind. We present results from a numerical study which used magnetohydrodynamic simulations to model the balance between gravity, magnetic confinement, and rotational acceleration. This allows us to study the time dependent nature of prominence formation. We demonstrate that a prominence, formed beyond the co-rotation radius, is ejected into the extended stellar wind in the slingshot prominence paradigm. Mass, angular momentum flux and ejection frequency have been calculated for a representative cool star, in the so-called Thermal Non-Equilibrium (TNE) regime.

Keywords. Sun: prominences, stars: magnetic fields, methods: numerical

1. Introduction

In the context of solar prominences the prominent is viewed off limb, tend to be close to the surface and can be carried away by CME But typically rain back onto the stellar surface. As such they do not make a significant contribution to the mass loss of the sun. In contrast the stellar prominences, slingshot prominences, these escape the gravitational field of the star, into interstellar space. They do contribute a significant amount to the mass loss of the star.

For the kind of stars that we are interested in, they typically have stronger magnetic fields than the sun and rotate faster and are less than 1.4 solar masses. Prominences on M-dwarfs form further away from the surface than those on the sun. This relates to the terminology of slingshot prominences, where a prominence forms so far away from the stellar surface, that it is beyond co-rotation, where the effective gravity that the prominence feels is away from the star. This is due to rotation, gravitation and magnetic tension and forms a channel for these prominences to escape into space.

To enlarge on this notion, we borrow a concept from the massive stars community, the notion of different types of magnetospheres. This boils down to the position of the Kepler co-rotation radius and the Alfven radius and their position relative to each other. In the case where we have the co-rotation surface inside the Alfven surface, the last closed magnetic field line is inside the co-rotation. This allows material to move back and forth but there is no net acceleration of material away from the star for magnetically supported material. In the other case where you have the Alfven surface inside the co-rotation radius, material suspended in the magnetic field feels a net force away from

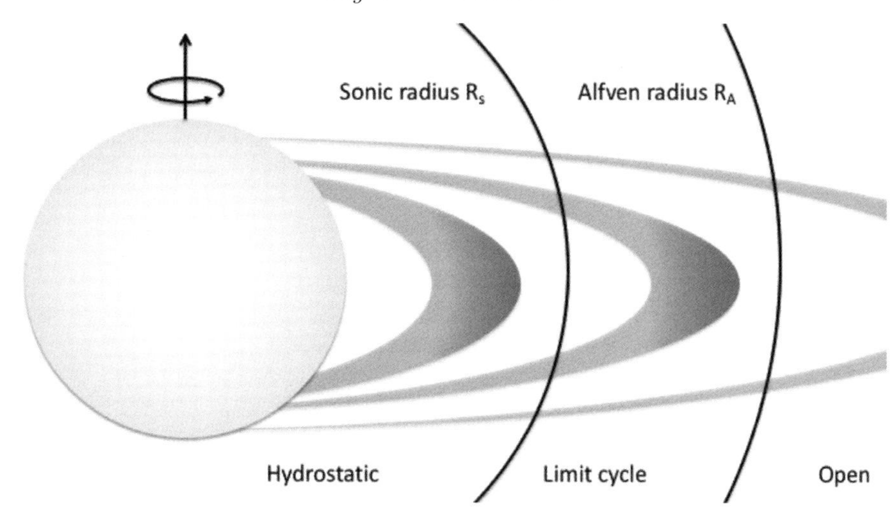

Figure 1. Prominence position relative to different characteristic radii. Figure reproduced from Jardine & Cameron (2019).

the star. Slingshot prominences are those that form in this stable point between the co-rotation and Alfven surfaces. See diagram in Fig. 1

For massive stars, material in the stable point is shock heated. For cool stars the material is subsonic and not shock heated. It can reach densities where it is susceptible to the cooling instability, where gas in the closed magnetosphere undergoes rapid cooling where, as the gas reduces in temperature, the cooling becomes more efficient leading to a runaway cooling. This results in cool (10^4 K) quasi-neutral material which we can see in observations.

2. Observations

F Some of the most compelling observations are in H_{alpha}. The example of Speedy mic, a fast rotating dwarf star (period 0.5 days), we can see features in H_{alpha} velocity shifts indicative of absorbing material passing across the stellar disk. From these observations, it was determined that there were about 25 prominences and they were stable over 13 stellar rotations (Dunstone et al. 2006). More recently, radio observations by Climent et al. (2020) of the star AB Dor at 8.4 GHz showed two sites for emission offset from the star. The distance from the star is of the order of 10 stellar radii. While this is at the upper end of the range for distance from the stellar surface that has been predicted by analytic models, it is not inconceivable. There is a wealth of observational evidence for prominences spanning several decades (Cameron & Woods 1992; Barnes et al. 1998, 2000; Skelly et al. 2008, 2009, 2010; Leitzinger et al. 2016).

3. Modelling

Analytic models of prominences have been carried. The first of which were concerned with 1D models which integrate the equations of mass, energy and momentum conservation from the stellar surface to find stable points in the corona. These early models demonstrated the validity of the slingshot prominences principle but were limited to consider highly idealised settings and did not account for the deformation of the magnetic field by a beard prominence (Cameron & Robinson 1989; Unruh & Jardine 1997). More recent and sophisticated modelling accounts for this effect by deriving an equation that describes the path of a prominence bearing field line through the corona by incorporating the Lorentz force and therefore that action of the prominence material on the

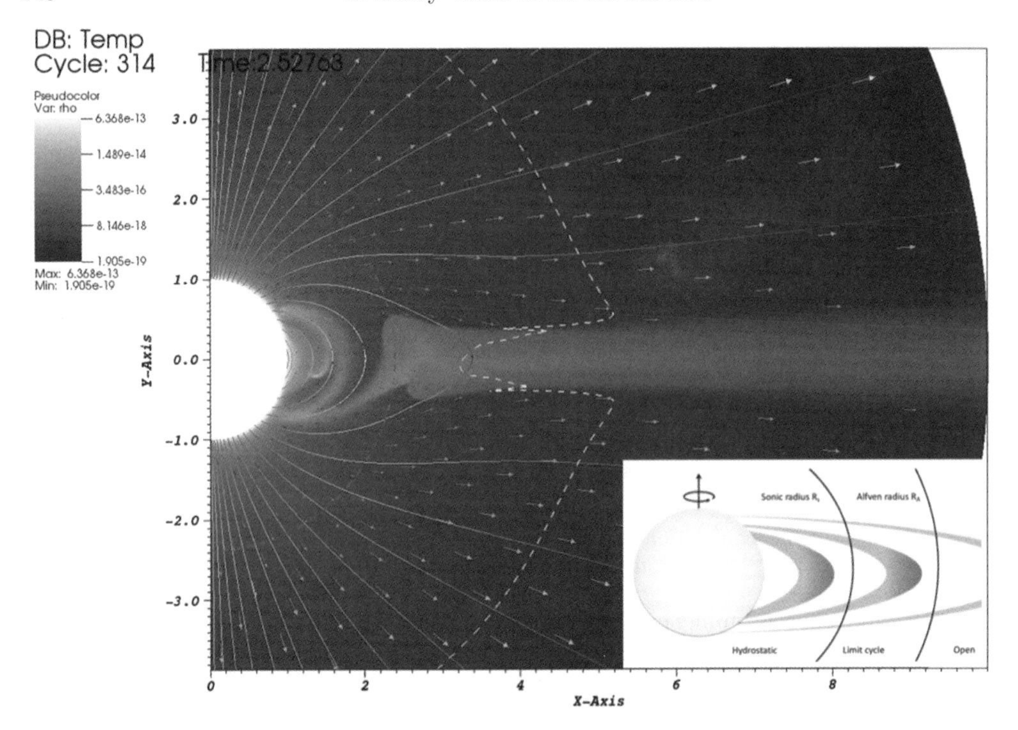

Figure 2. 2D global simulation illustrating the formation of a slingshot prominence. Blue and red dashed lines indicate the Kepler co-rotation and Alfvén radii respectively and the continuous lines and arrows indicate the magnetic and velocity fields. Inset: diagram showing the separate regions of wind behavior (image from Jardine & Cameron (2019)).

field (Waugh & Jardine 2019; Waugh et al. 2022). This tells you the deformation of the field line based on its extent beyond the co-rotation radius. This allows for the production of steady state prominence morphology. These models have been used effectively by Faller & Jardine (2022) to predict the prominence position and occurrence rate. They were able to show that field strength and geometry are central to determining the mass a prominence can support and the rate at which they are ejected.

4. Simulations

The analytic modelling that has been conducted so far lacks time dependence. This means that the models only capture an idealised snapshot, which doesn't tell you how long it takes to form a prominence nor its escape time or how frequently this happens. Knowledge of these processes requires time dependent, dynamic simulations. This introduces additional physics for example thermal conduction and radiatively thin cooling. However, the underlying principles remain the same. This in principle should allow for the analytic models to act as inputs to the simulations. However this has proved to be challenging as the analytic models produce starting numerical grids that are far from equilibrium. This is because the analytic models are typically not force free when the additional physics of a dynamic simulation are accounted for. This means that a simulation that captures the formation of prominences is necessary. Fig. 2 shows a work in progress simulation that lacks realistic energetics, however does communicate the concept of dynamic prominence formed in a cool star corona, whose ejected material is embedded in a free streaming stellar wind. The simulation illustrates the possible antisymmetric nature of prominence formation that the analytic modelling is not capable of. The simulation shows a cyclic behavior of the prominence switching its connection to the surface. Either to the northern or southern hemisphere.

Caption: 2D simulation with the star on the left and the magnetic field lines are illustrated by the stream lines and the velocity vectors with the arrows. We can see magnetic channelling of material and a formation of a disk like structure. There is also a stable point between the co-rotation and Alfven radii. This mirrors the results of similar simulations of massive stars conducted by ud-Doula et al. (2008).

5. Conclusions

Centrifugally ejected prominences, known as slingshot prominences, form a major contributor to cool star mass and angular momentum loss. They form the only observable component of cool star winds. Passed analytic modelling tells us the formation sites and steady state properties of prominences, they can not tell us the dynamic properties such as formation and ejection time scales. This makes time dependent simulations the next vital step in understanding prominences and to have a complete picture of stellar mass and angular momentum loss.

References

Barnes J. R., Cameron A. C., Unruh Y. C., Donati J.-F., Hussain G. A. J., 1998, MNRAS, 299, 904

Barnes J. R., Cameron A. C., James D. J., Donati J.-F., 2000, MNRAS, 314, 162

Cameron A. C., Robinson R., 1989, MNRAS, 236, 57

Cameron A. C., Woods J. A., 1992, MNRAS, 258, 360

Climent J. B., Guirado J. C., Azulay R., Marcaide J. M., Jauncey D. L., Lestrade J.-F., Reynolds J. E., 2020, A&A, 641, A90

Daley-Yates S., Padioleau, T., Tremblin, P., Kestener, P., Mancip, M., 2021, A&A, 653, 13

Dunstone N. J., Barnes J. R., Collier Cameron A., Jardine M., 2006, MNRAS, 365, 530

Faller J. S., Jardine M. M., 2022, MNRAS, 513, 4

Waugh R. F. P., Jardine M., 2019, MNRS, 483, 1513

Waugh R. F. P., Jardine M., 2022, MNRS, 514, 5465

Jardine M., Cameron A. C., 2019, MNRS, 482, 2853

Leitzinger M., Odert P., Zaqarashvili T. V., Hanslmeier R. G. A., Lammer H., 2016, MNRAS, 463, 965

Unruh Y. C., Jardine M., 1997, A&A, 321, 177

ud-Doula A., Owocki S. P., Townsend R. H. D., 2008, MNRAS, 385, 97

Skelly M., Unruh Y., Cameron A. C., Barnes J., Donati J.-F., Lawson W., Carter B., 2008, MNRAS, 385, 708

Skelly M. B., Unruh Y. C., Barnes J. R., Lawson W. A., Donati J.-F., Cameron A. C., 2009, MNRAS, 399, 1829

Skelly M. B., Donati J. F., Bouvier J., Grankin K. N., Unruh Y. C., Artemenko S. A., Petrov P., 2010, MNRAS, 403, 159

Winds of Stars and Exoplanets
Proceedings IAU Symposium No. 370, 2023
A. A. Vidotto, L. Fossati & J. S. Vink, eds.
doi:10.1017/S1743921322004963

Effect of stellar flares and coronal mass ejections on the atmospheric escape from hot Jupiters

Gopal Hazra[1], Aline A. Vidotto[2], Stephen Carolan[3], Carolina Villarreal D'Angelo[4] and Ward Manchester[5]

[1]Dept. of Astrophysics, University of Vienna, Türkenschanzstrasse 17, A-1180 Vienna, Austria

[2]Leiden Observatory, Leiden University, NL-2300 RA Leiden, the Netherlands

[3]School of Physics, Trinity College Dublin, Dublin 2, Ireland

[4]Instituto de Astronomía Teórica y Experimental (IATE-CONICET). Laprida 854, Córdoba, Argentina

[5]Department of Climate and Space Sciences and Engineering, University of Michigan, Ann Arbor, MI 48109, USA

Abstract. Spectral observations in the Ly-α line have shown that atmospheric escape is variable and for the exoplanet HD189733b, the atmospheric evaporation goes from undetected to enhanced evaporation in a 1.5 years interval. To understand the temporal variation in the atmospheric escape, we investigate the effect of flares, winds, and CMEs on the atmosphere of hot Jupiter HD189733b using 3D self-consistent radiation hydrodynamic simulations. We consider four cases: first, the quiescent phase including stellar wind; secondly, a flare; thirdly, a CME; and fourthly, a flare followed by a CME. We find that the flare alone increases the atmospheric escape rate by only 25%, while the CME leads to a factor of 4 increments, in comparison to the quiescent case. We also find that the flare alone cannot explain the observed high blue-shifted velocities seen in the Ly-α. The CME, however, leads to an increase in the velocity of escaping atmospheres, enhancing the blue-shifted transit depth.

Keywords. stars: activity, stars: flares, stars:wind, outflows

---

1. Introduction

Atmospheric escape from the hot Jupiters is well observed from various spectroscopic observations (e.g., Vidal-Madjar et al. 2003; Lecavelier des Etangs et al. 2012). The radiation from host star ionises the upper atmospheres of these planets and drives a photoionized planetary outflow (e.g., Murray-Clay et al. 2009; Hazra et al. 2020). As a result atmosphere escapes from the planet. This escaping planetary outflow further interacts with the stellar environments (e.g., stellar wind, stellar coronal mass ejections (CMEs) and stellar magnetic field). Depending upon the pressure balance between strength of the stellar wind and planetary outflow, atmospheric escape gets enhanced or confined by the stellar wind (see Carolan et al. (2021) for details). The transients events (CMEs and flares) also play a crucial role on the atmospheric escape. Generally, CMEs change the properties of stellar wind plasma and flares enhance the energy deposition in the upper atmospheres of the planet affecting the dynamics of planetary atmosphere.

The transit observations of the classic hot Jupiter HD189733b showed that the atmosphere of HD189733b is undergoing through atmospheric loss due to atmospheric escape

(Lecavelier des Etangs et al. 2012). Not only it goes through an atmospheric loss, it has a temporal variation in the atmospheric escape rate most likely due to the transients events on the host star. Two transit observations of HD189733b on two epochs using the hubble space telescope (HST) showed a temporal variation in the evaporating atmosphere. While in the first epoch of observations (April, 2010) evaporation was not detected, a transit absorption depths of $14.4\% \pm 3.6\%$ in Ly-α was observed in the second epoch (September 2011). Just 8-hours before the second transit observation, an x-ray flare was observed in the host stars supporting the idea that the atmosphere of HD189733b went through an enhancement of escape caused by a transient event of the host star (Lecavelier des Etangs et al. 2012).

In this work, we study the effect of a flare, a CME and both simultaneously on the atmospheres of HD189733b to understand how these transient events affect the atmospheric escape rate. We develop a radiation driven 3D atmospheric escape model where the photoionisation due to stellar radiation in the upper atmospheres is considered self-consistently by solving radiation hydrodynamic equations. This self-consistent model gives us a unique opportunity to study the effect of stellar radiation, stellar outflows on the atmosphere of hot Jupiters and corresponding transit signature in the Ly-α line.

2. 3D atmospheric escape model and planetary outflow

We have used the BATS-R-US code to model atmospheric escape and its interaction with the stellar wind for the star-planet system HD189733. We have developed a new module in the BATS-R-US framework to solve radiation hydrodynamic equations in the rotating frame of the planet in a 3D Cartesian box keeping the planet at the origin. In the new module, we include photoionisation, collisional ionisation and corresponding heating and cooling terms. For details of the model, see Section-2 of Hazra et al. (2022).

As the X-ray and Extreme Ultra Violet (EUV) part of stellar spectra is responsible for photoionisation, we need information about XUV radiation from the host star HD189733A. Because of not availability of EUV spectra of the star from observations currently, we calculate that part of the spectra from empirical relation given in equation (3) of Sanz-Forcada et al. (2011) from measured X-ray luminosity in the quiescent phase of the star. The estimated value of XUV (X-ray and EUV) luminosity during the queiscent phase of the star is $L_{\rm xuv} = 1.3 \times 10^{29}$ erg s^{-1}. Although, our model is capable of incorporating multi-species, we assume a hydrogen dominated atmosphere for simplicity.

We show the basic features of planetary outflow due to self-consistent deposition of stellar radiation from our escape model in figure 1. In the figure 1(a), the total density of outflowing planetary materials is shown in the orbital plane. As it is clear from the figure that the planetary material moves in a clockwise direction after it is able to escape from the planetary gravity due to the Coriolis force. The velocity streamlines are shown in black contours. The neutral density is shown in figure 1(b). For tidally locked planet, the night side does not receive enough stellar radiation and the night side material is no longer ionised leading to the formation of a planetary tail of neutral material as seen in figure 1(b).

Figure 1(c) and figure 1(d) show the temperature and velocity of the planetary materials. A shock is formed when the oppositely deflected material meets due to interplay between tidal force and Coriolis force, which is found in the high temperature region i.e., the tangential between quadrants I and IV and quadrants II and III in figure 1(c).The sonic surface is shown in black contour in figure 1(d). All of these basic features of planetary outflow are very consistent with previous studies (e.g., McCann et al. 2019). The atmospheric escape rate from the planet is 6.0×10^{10} g s^{-1}, which is consistent with previous models (Salz et al. 2016, and reference therein) in spite of the different implementation of various physical processes.

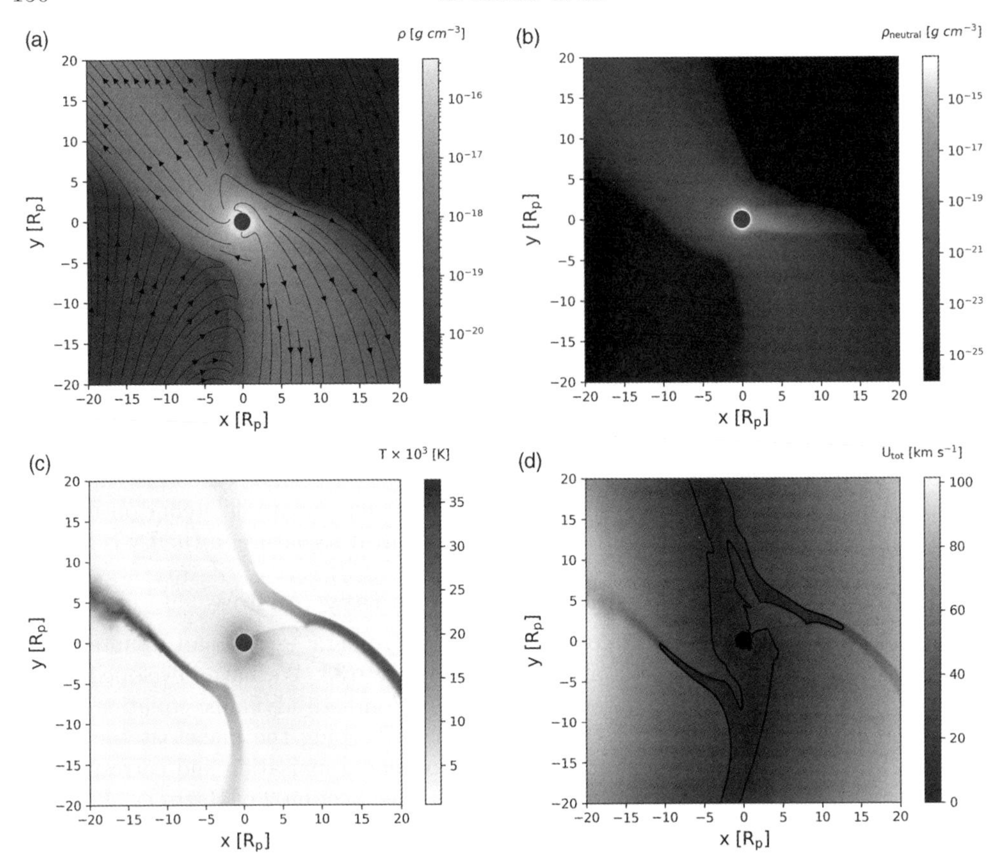

Figure 1. Basic features of planetary outflow. (a) Total density with velocity streamlines, (b) neutral density, (c) temperature distribution and (d) Total velocity with sonic surface in black contour. All plots are in the orbital plane.

3. Effect of stellar transient events on the planetary atmosphere

To study the effect of flares and CMEs on the atmospheres of the hot Jupiter HD189733b, we consider three instances of transient events. In first instance, we consider only a flare, then we consider a CME and finally we consider the effect of a flare associated with a CME. As we mentioned earlier, flares enhance the stellar radiation and CME changes the stellar wind plasma density. Hence we change stellar radiation and stellar wind parameters according to the case we study. We consider four cases to capture the full stellar environments as given below.

3.1. *Case-I: quiescent phase*

In the quiescent phase of the star, we consider the XUV radiation at quiet phase and the background stellar wind. The stellar wind in this case interacts with the XUV driven planetary outflow as presented in Section 2. In our model, stellar wind is modelled as a 1D isothermal Parker wind and injected from the negative x boundary of our simulation box. The properties of the stellar wind temperature and mass-loss rate is taken from stellar wind simulation of K dwarf HD189733A (Kavanagh et al. 2019). The final properties of planetary outflow is depicted in the first column of the figure 2. The total density, neutral density, temperature and total velocity are shown from top to bottom rows of figure 2 respectively in the first column. The black contours in the top row show the velocity

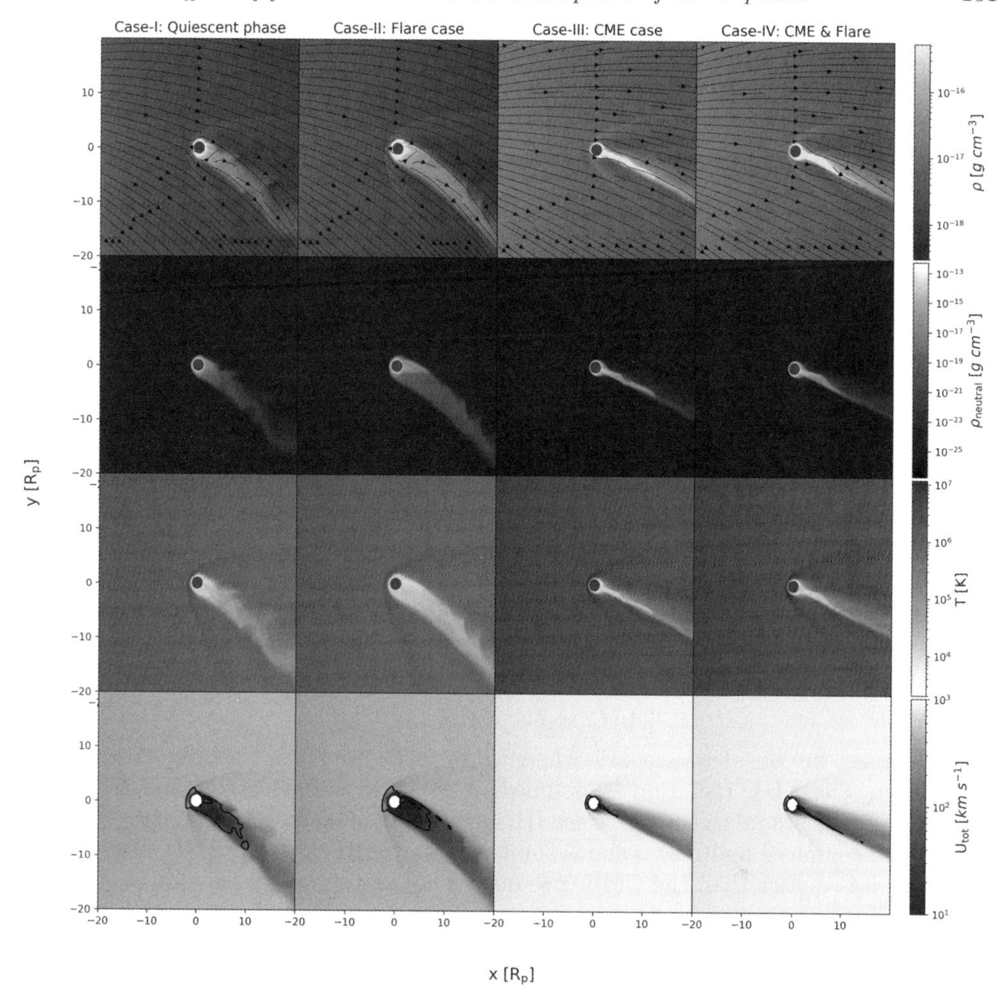

Figure 2. Total density, density of neutral material, temperature and total velocity are shown in first column from top to bottom rows respectively for the quiescent phase. Second, third and fourth columns show same quantity as first column but for a flare case, a CME case and for a flare and a CME, respectively. Black streamlines in the first row show velocity. The black contour in the bottom row show sonic surface for different cases.

streamlines. As it is clear from total density plot, the stellar wind pushes the day-side planetary material in a tail like structure. The atmospheric escape rate from the planet is 7.6×10^{10} g s^{-1}, which is 27% higher than the escape rate of planetary outflow without stellar wind (Section 2).

3.2. *Case-II: flare case*

For the flare case, we have considered the flare event that happened 8 hours before the transit event in september 2011 for HD189733A (Lecavelier des Etangs et al. 2012). The observed peak X-ray flux during the flaring event is 1.3×10^{-12} erg cm^{-2} s^{-1}, which gives an XUV radiation of $F_{xuv} = 1.5 \times 10^5$ erg cm^{-2} s^{-1}. The stellar wind parameters are same as Case-I. The fundamental properties total density, neutral density, temperature and total velocity of planetary materials in this case is shown in the second column of figure 2. The enhanced XUV radiation due to the flare lifts up more planetary material and ionises more neutral material than quiescent phase leading higher atmospheric escape rate. The atmospheric escape rate is this case is 1.0×10^{11} g s^{-1}

3.3. *Case-III: CME only case*

We model CME as a dense and faster stellar wind. To simulate a reasonable CME, we have assumed a density of 2.1×10^{-17} g cm^{-3} and a speed of 755 km s^{-1}, which are a factor of 4 and 2.4 larger than the quiescent stellar wind respectively. The temperature of the CME is 5.4×10^{4} K. Since we do not consider a flare in this case, XUV radiation is taken same as quiescent phase of the star.

As CME is much faster and denser than stellar, the ram pressure associated with CME is stronger too. As a result, when the CME interacts with planetary atmosphere it totally confines the day side planetary outflow. This is clearly seen in the column 3 of figure 2. In the CME case, the planetary tail is more inclined towards the line joining star and planet compared to the stellar wind case. This is because the planetary tail follows the resultant CME velocity direction after its get affected by the orbital motion of the planet and as CME is faster than stellar wind, the tail gets less affected. The atmospheric escape rate in this case is 3.2×10^{11} g s^{-1}, which 5.3 times higher than planetary outflow only (Section 2).

3.4. *Case-IV: CME and flare*

In this case, we consider a scenario where a flare is followed by a CME directing towards the planet. The XUV radiation is assumed to be same as flare case (Case-II) and the CME properties are taken same as Case-III. The overall planetary atmospheric behaviour (see fourth coumn of figure 2) remains similar as the CME case (Case-III). However, the neutral tail is denser than the CME case only. The atmospheric escape rate is this case 4.0×10^{11} g s^{-1}, which is highest among the all cases considered here.

4. Comparison with transit observation: Ly-α line

We have calculated the synthetic transit spectra for all our four cases to compare with the available Ly-α transit observations for HD189733b. The details of how we calculate synthetic transit spectra is given in Hazra et al. (2022). In figure 3(a), the theoretical transit spectra at mid-transit for all cases including no stellar wind case are shown. The higher transit depths in the high-velocity material show the fact that the stellar CME (Case-III and Case-IV, magenta and orange lines) affects the high-velocity blue-shifted materials in compare to the stellar wind cases (Cases -I and II, blue and green lines).

To compare directly with reported observed values, we convolve our theoretical transit profiles with G140M grating of STIS. In figure 3(b), the convolved lines are shown. The color lines showed the Ly-α absorption after it gets absorbed by the planetary atmospheres in each case we studied here. The out-of-transit Ly-α line is also shown in black for comparison. A zoom in version of the blue wing [(-230) - (-140)] km s^{-1}] and red wing [60-110 km s^{-1}] are shown in figure 3(c) and (d) respectively. Both of the cases, where we include CME (Case-III and IV) show a deeper transit depth but the CME only case (Case-III) shows largest integrated blue wing absorption of 5.1%. Although CME case produces larger transit depth in the blue wing among all the model we considered here, none of our model is able to explain the large absorption of $14.4 \pm 3.6\%$ seen in the blue-wing observation.

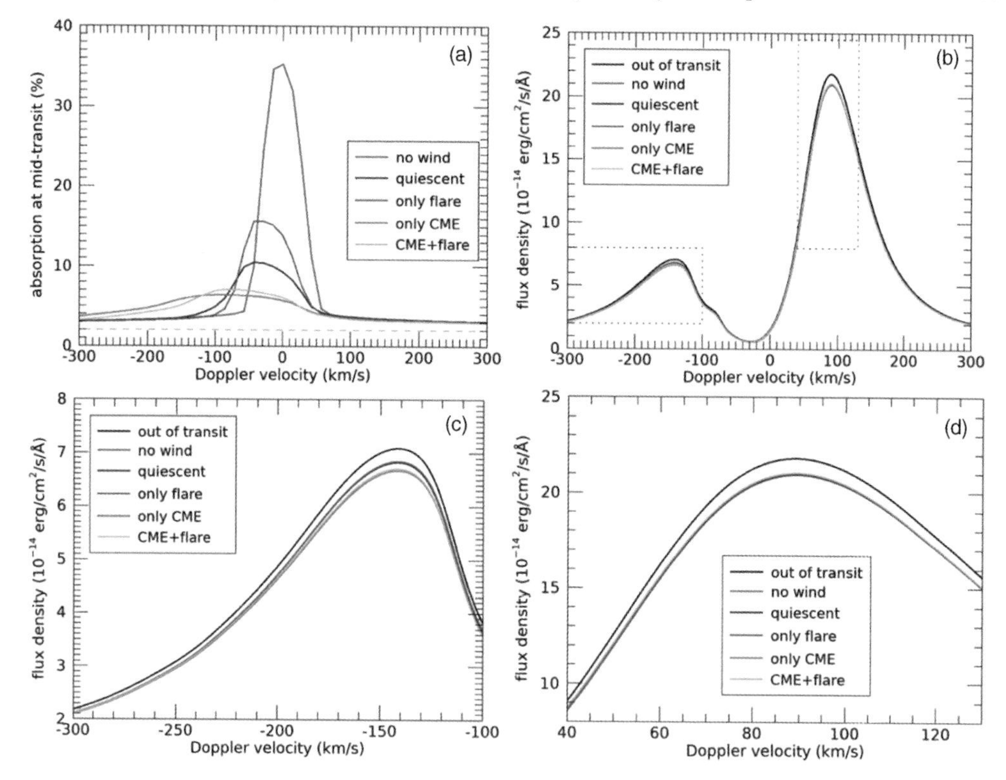

Figure 3. Synthetic Ly-α line profile at mid transit. (a) theoretical line profile; (b) predicted observed profile, convolved with G140M grating mode; (c) same as b, but for the blue wing of the line; (d) same as b, but for the red wing of the line.

5. Conclusions

In this work, we have presented a 3D radiation driven atmospheric escape model and the interaction between radiation driven planetary outflow and stellar transient events. Our model included both neutral and ionised hydrogen (multi-species) and considered photoionisation, collisional ionisation and radiation recombination.

We found that the atmospheric escape rate (mass-loss rate) is higher when we consider CME in our simulations. The CME case with flare enhances the escape rate significantly giving the highest mass-loss rate of 4.0×10^{11} g s^{-1} among all the cases considered here. We also calculated the transit spectra for all cases. We found that the quiescent and flare cases are not able to reproduce the strong absorption in the blue-wing of Ly-α. The CME cases with or without flare (Case-III and Case-IV) show larger absorption of blue-shifted material in compared to the quiescent and flare cases (Case-I and Case-II). Hence, the observed temporal variation seen in blue wing of Ly-α line is more likely to be a consequence of the CME affecting the planetary atmospheres, than the flare changes the energy deposition in the atmospheres of the hot Jupiters.

References

A. Vidal-Madjar, A. Lecavelier des Etangs, J. M. Désert, G. E. Ballester, R. Ferlet, G. Hébrard, and M. Mayor. An extended upper atmosphere around the extrasolar planet HD209458b. *Nature*, 422(6928):143–146, Mar 2003. doi: 10.1038/nature01448.

A. Lecavelier des Etangs, V. Bourrier, P. J. Wheatley, H. Dupuy, D. Ehrenreich, A. Vidal-Madjar, G. Hébrard, G. E. Ballester, J. M. Désert, R. Ferlet, and D. K. Sing. Temporal variations in the evaporating atmosphere of the exoplanet HD 189733b. *A&A*, 543:L4, July 2012. doi: 10.1051/0004-6361/201219363.

Ruth A. Murray-Clay, Eugene I. Chiang, and Norman Murray. Atmospheric Escape From Hot Jupiters. *ApJ*, 693(1):23–42, Mar 2009. doi: 10.1088/0004-637X/693/1/23.

Gopal Hazra, Aline A. Vidotto, and Carolina Villarreal D'Angelo. Influence of the Sun-like magnetic cycle on exoplanetary atmospheric escape. *MNRAS*, 496(3):4017–4031, August 2020. doi: 10.1093/mnras/staa1815.

S. Carolan, A. A. Vidotto, C. Villarreal D'Angelo, and G. Hazra. Effects of the stellar wind on the Ly α transit of close-in planets. *MNRAS*, 500(3):3382–3393, January 2021. doi: 10.1093/mnras/staa3431.

Gopal Hazra, Aline A. Vidotto, Stephen Carolan, Carolina Villarreal D'Angelo, and Ward Manchester. The impact of coronal mass ejections and flares on the atmosphere of the hot Jupiter HD189733b. *MNRAS*, 509(4):5858–5871, February 2022. doi: 10.1093/mnras/stab3271.

J. Sanz-Forcada, G. Micela, I. Ribas, A. M. T. Pollock, C. Eiroa, A. Velasco, E. Solano, and D. García-Álvarez. Estimation of the XUV radiation onto close planets and their evaporation. *A&A*, 532:A6, Aug 2011. doi: 10.1051/0004-6361/201116594.

John McCann, Ruth A. Murray-Clay, Kaitlin Kratter, and Mark R. Krumholz. Morphology of Hydrodynamic Winds: A Study of Planetary Winds in Stellar Environments. *ApJ*, 873(1):89, March 2019. doi: 10.3847/1538-4357/ab05b8.

M. Salz, S. Czesla, P. C. Schneider, and J. H. M. M. Schmitt. Simulating the escaping atmospheres of hot gas planets in the solar neighborhood. *A&A*, 586:A75, February 2016. doi: 10.1051/0004-6361/201526109.

R. D. Kavanagh, A. A. Vidotto, D. Ó. Fionnagáin, V. Bourrier, R. Fares, M. Jardine, Ch Helling, C. Moutou, J. Llama, and P. J. Wheatley. MOVES - II. Tuning in to the radio environment of HD189733b. *MNRAS*, 485(4):4529–4538, June 2019. doi: 10.1093/mnras/stz655.

Winds of Stars and Exoplanets
Proceedings IAU Symposium No. 370, 2023
A. A. Vidotto, L. Fossati & J. S. Vink, eds.
doi:10.1017/S1743921322003556

Physics of the atmospheric escape driven by EUV photoionization heating: Classification of the hydrodynamic escape in close-in planets

Hiroto Mitani[1], **Riouhei Nakatani**[2] **and Naoki Yoshida**[1,3,4]

[1]Department of Physics, School of Science, The University of Tokyo, 7-3-1 Hongo, Bunkyo, Tokyo 113-0033
email: hiroto.mitani@phys.s.u-tokyo.ac.jp

[2]RIKEN Cluster for Pioneering Research, 2-1 Hirosawa, Wako, Saitama 351-0198, Japan

[3]Kavli Institute for the Physics and Mathematics of the Universe (WPI), UT Institutes for Advanced Study, The University of Tokyo, Kashiwa, Chiba 277-8583, Japan

[4]Research Center for the Early Universe, School of Science, The University of Tokyo, 7-3-1 Hongo, Bunkyo, Tokyo 113-0033

Abstract. The intense extreme ultraviolet radiation heats the upper atmosphere of close-in exoplanets and drives the atmospheric escape. The escaping process determines the planetary evolution of close-in planets. The mass loss rate depends on the UV flux at the planet. We introduce the relevant physical quantities which describe the dominant physics in the atmosphere. We find that the equilibrium temperature and the characteristic temperature determine whether the system becomes energy-limited or recombination-limited. We classify the observed close-in planets using the physical conditions. We also find that many of the Lyman-α absorptions detected planets receive intenser flux than the critical flux which can be determined from physical conditions. Our classification method can quantitatively reveal whether the EUV is not strong enough to drive the outflow or the Lyman-α absorption is not detected for some reason (e.g. stellar wind confinement). We also discuss the thermo-chemical structure of hydrodynamic simulations with the relevant physics.

Keywords. hydrodynamics – methods: numerical – planets and satellites: atmospheres – planets and satellites: physical evolution

1. Introduction

The close-in exoplanets lose their atmosphere due to the intense radiation from the host star. Understanding the mass loss process is necessary for the evolution of close-in planets. The process can be an origin of the statistical properties of observed exoplanets like sub-Neptune desert (Fulton et al. 2017; Owen & Wu 2017). In a typical hot Jupiter, extreme ultraviolet (EUV; > 13.6 eV) photons photoionize the hydrogen atoms and the photoelectrons thermalize into gas. The EUV heating process drives the atmospheric escape. There are some theoretical models of the atmospheric escape (Murray-Clay et al. 2009; Tripathi et al. 2015). Radiation hydrodynamics simulations have revealed the thermo-chemical structure of the upper atmosphere and the importance of such escaping processes in planetary evolution.

Previous theoretical studies calculate the mass loss rate due to the EUV heating. Some planetary evolution models have assumed the mass loss rate is proportional to the EUV flux but radiation hydrodynamic simulations have revealed the assumption is not correct every time. In the high UV case, the system becomes recombination-limited. In the recombination-limited regime, the photoionization rate and the recombination rate are balanced and the UV energy is lost to the radiative recombination energy. The mass loss rate is not proportional to the UV flux and is proportional to the square root of the UV flux.

Transit observations have revealed the existence of such escaping outflow (Vidal-Madjar et al. 2003; Ehrenreich et al. 2015). Recent observations have found that the close-in planets without Lyman-α absorption despite the intense UV irradiation (Rockcliffe et al. 2021). The physical conditions which determine the thermo-chemical structure are still unknown though it is important to understand the observed signatures.

Understanding physics in the upper atmosphere is necessary for the planetary evolution theory. We perform the radiation hydrodynamic simulations and construct a theoretical model to understand the escaping outflow in the observed exoplanets using the relevant physical quantities.

2. Relevant physical quantities

The important physical processes in the atmospheric escape are the photoheating, the planetary gravity, and the gas expansion. We can define the relevant temperatures and timescales in the upper atmosphere with photoionization heating. The characteristic temperature is given as (Begelman et al. 1983):

$$kT_{\rm ch} = \frac{\Gamma}{c_p/\mu m_{\rm H}} \frac{R_p}{c_{\rm ch}} \tag{2.1}$$

where Γ is the photoheating rate, R_p is the planetary radius, c_p is the specific heat at constant pressure, μ is the mean molecular weight, and m_H is the mass of the hydrogen atom. The right-hand side represents the deposited EUV energy in the characteristic sound crossing time. The characteristic sound speed $c_{\rm ch}$ is given as:

$$c_{\rm ch} = \left(\frac{\Gamma R_p}{c_p}\right)^{1/3} \tag{2.2}$$

The typical value of the sound speed in the upper atmosphere is $\sim 10\,{\rm km/s}$. The characteristic temperature represents how rapid the photoheating is compared to the gas expansion and does not mean the typical temperature of the system. In the case of EUV photoionization heating of hydrogen atoms, the photoheating rate is given as:

$$\Gamma = \frac{1-x}{m} F_0 \delta \langle\sigma\rangle\langle\Delta E\rangle \tag{2.3}$$

$$= 1.2\times10^{7}\,{\rm erg\,g^{-1}\,s^{-1}} \left(\frac{1-x}{0.5}\right)\left(\frac{m}{1.4m_{\rm H}}\right)\left(\frac{\Phi}{10^{41}\,{\rm s^{-1}}}\right) \tag{2.4}$$

$$\times \left(\frac{r}{1{\rm au}}\right)^{-2}\left(\frac{\delta\langle\sigma\rangle\langle\Delta E\rangle}{10^{-18}\,{\rm cm^2}\times 1\,{\rm eV}}\right) \tag{2.5}$$

where x is the ionization degree, m is the gas mass per hydrogen nucleus, Φ is the EUV photon flux, δ is the attenuation factor, $\langle\sigma\rangle$ is the average cross-section of the photoionization and $\langle\Delta E\rangle$ is the average deposited energy per photoionization. The rate depends on the spectral shape. At higher energies, the cross section is smaller but the deposited energy increases, so these effects cancel each other out overall.

The equilibrium temperature T_{eq} can be determined from the balance of the photoheating and the cooling and becomes $\sim 10^4$ K for EUV photoionization heating of hydrogen atoms.

The gravitational temperature is given as:

$$T_g = \frac{GM_p \mu m_{\rm H}}{R_p k} \tag{2.6}$$

The temperature indicates the strength of the planetary gravity. We can also define the gravitational radius$R_g = GM_p/c_{\rm eq}^2$ using the sound speed at the equilibrium temperature $c_{\rm eq}$.

The ratio between $T_{\rm ch}$ and $T_{\rm eq}$ is

$$\frac{T_{\rm ch}}{T_{\rm eq}} = \left(\frac{R_p \Gamma}{c_{\rm eq}^3 c_p}\right)^{2/3} = \left(\frac{R_p}{R_g}\right)^{2/3} \left(\frac{\Gamma G M_p}{c_p c_{\rm eq}}\right)^{2/3} \tag{2.7}$$

From the second factor, we can define the critical flux as:

$$F_{\rm cr} = \frac{m c_p c_{\rm eq}^5}{G M_p \sigma \Delta E} \tag{2.8}$$

$$= 5.8 \times 10^{12}\,{\rm cm^{-2}\,s^{-1}} \left(\frac{m}{1.4 m_{\rm H}}\right) \left(\frac{c_{\rm eq}}{10\,{\rm km\,s^{-1}}}\right)^5 \tag{2.9}$$

$$\times \left(\frac{M_p}{M_J}\right)^{-1} \left(\frac{\sigma_0}{5 \times 10^{-18}\,{\rm cm^2}}\right)^{-1} \left(\frac{\Delta E}{1\,{\rm eV}}\right)^{-1} \left(\frac{c_p}{5/2}\right) \tag{2.10}$$

and the ratio can be given as:

$$\frac{T_{\rm ch}}{T_{\rm eq}} \sim \xi^{2/3} \left(\frac{F_0}{F_{\rm cr}}\right)^{2/3} \tag{2.11}$$

where $\xi = R_p/R_g$.

We can also define the photoheating timescale and the gravitational timescale:

$$t_g = \sqrt{\frac{R_p^3}{GM_p}} \tag{2.12}$$

$$t_h = \frac{R_p}{c_{\rm ch}} \tag{2.13}$$

The ratio between the gravitational and photoheating timescales is

$$\frac{t_g}{t_h} = \xi^{5/6} \left(\frac{F_0}{F_{\rm cr}} \epsilon (1 - x)\right)^{1/3} \tag{2.14}$$

which represents the relative strength of the gravity to that of the photoheating. ϵ is an attenuation factor that depends on the column density of the hydrogen. If the ratio is larger than unity $t_g/t_h > 1$, the photoheating is faster and the system loses the atmosphere quickly.

We run radiation hydrodynamics simulations with the EUV photoionization and the recombination of hydrogen atoms to understand thermo-chemical structure of the atmosphere. In the simulations, we assume the Ly-α cooling is the dominant radiative cooling process as in the previous studies. We run simulations for the high EUV case ($T_{\rm ch} > T_{\rm eq}$) and the low EUV case ($T_{\rm ch} < T_{\rm eq}$) for typical hot Jupiter ($M_p = 0.7 M_J$, $R_p = 1.4 R_J$). Figure 1 shows the radial profile of the heating and cooling in our simulations. We find that the adiabatic cooling dominates in the case of low EUV and the Ly-α cooling dominates in the case of high EUV planets. $T_{\rm ch}/T_{\rm eq}$ can be used to distinguish the dominant

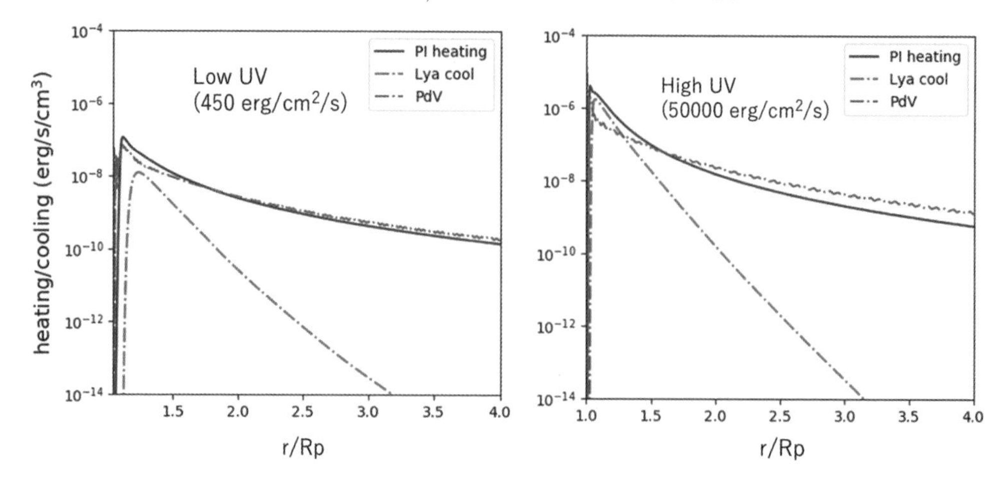

Figure 1. The radial profile of the heating and cooling rate. The ratio between the characteristic temperature and the equilibrium temperature is less than unity ($T_{\rm ch} < T_{\rm eq}$) in the low UV case (left) and larger than unity ($T_{\rm ch} > T_{\rm eq}$) in the high EUV case (right).

cooling process and whether the system is energy-limited or recombination-limited. In the typical hot Jupiters, $T_{\rm eq} = T_{\rm ch}$ when the UV flux is $\sim 10^3$ erg/cm^2/s and this value is consistent with the previous simulations.

3. Classification of the planetary atmosphere

The two physical conditions $t_g = t_h$ and $T_{\rm ch} = T_{\rm eq}$ can be drawn by a straight line on the $\log(F_0/F_{\rm cr}) - \log \xi$ plane. We investigate the conditions in the observed exoplanets using the open dataset (Schneider et al. 2011). We choose close-in planets with the orbital period $P < 10$ day. The EUV flux is intense in the case of a young host star. We estimate the EUV luminosity using the empirical relationships in (Sanz-Forcada et al. 2011)

$$\log L_{\rm EUV} = 29.12 - 1.24 \log \tau \tag{3.1}$$

where τ is the stellar age in Gyr. For stars whose ages were not listed in the catalog, we assumed the solar EUV luminosity. Figure 2. shows the distribution of the observed exoplanets in $\log(F_0/F_{\rm cr}) - \log \xi$ plane. The two conditions above can be drawn by two straight lines in the plane. The atmospheric escape in observed planets can be classified by two lines. We find that the EUV flux is lower than the critical flux in the case of the two of the Ly-α non-detected planets and the mass-loss is weak. K2-25 b receives intense EUV radiation from the host star and the EUV flux is larger than the critical flux. The planet does not show the Ly-α absorption (Lecavelier des Etangs et al. 2012; Rockcliffe et al. 2021). It may mean that unknown factors (e.g. stellar wind confinement (Carolan et al. 2020; Mitani et al. 2022), intense photoionization of the outflow) reduce the absorption signal. The $T_{\rm eq} = T_{\rm ch}$ condition can be used to quantitatively distinguish whether the EUV is not strong enough to drive the outflow. Future hydrodynamic simulations incorporating the effects of stellar winds, flares, and CMEs, and transit observations other than Ly-α are important to understand the structure of the outflow of these planets. There is few planets above $t_g = t_h$ line. The strong mass-loss reduces the planetary radius and planets above the line may move to the left side. In this plane, the young planets are in the upper part. If we assume the planetary radius weakly depends on the mass, ξ becomes smaller as the planet evolves and the old planets move to the left bottom part of the plane.

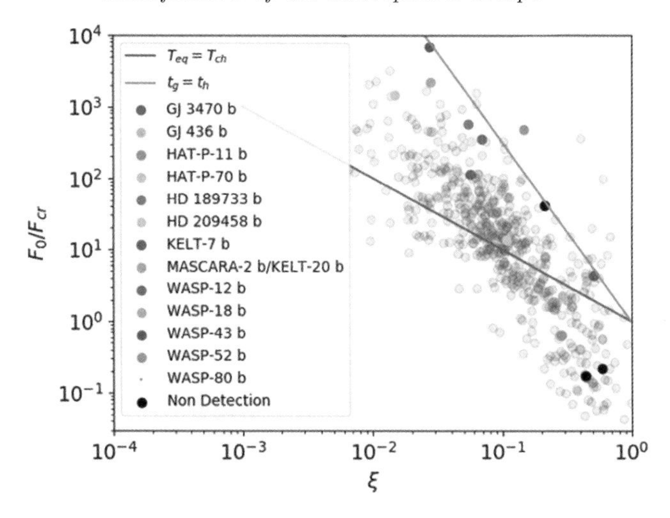

Figure 2. Observed close-in planets in $\log(F_0/F_{\rm cr}) - \log\xi$ plane. Colored points represent H detected planets and the black dots represent the planets with Ly-α non-detection. Blue points represent other discovered close-in planets. Two straight lines show the physical conditions which we derived (Blue; $T_{eq} = T_{ch}$, Orange $t_g = t_h$).

We only consider the planetary gravity but the stellar gravity also plays an important role in gas kinematics around the hill radius. It is known that stellar gravity boosts the mass-loss rate, and there are formulas for planetary atmospheric dissipation rates that take this effect into account (Erkaev et al. 2007).

We also consider the timescale around the hill radius.

$$t_{h\_hill} = \frac{R_{hill}}{(\Gamma R_{hill}/c_p)^{1/3}} \tag{3.2}$$

$$t_{g\_hill} = \sqrt{\frac{a^3}{GM_*}} \tag{3.3}$$

where $R_{hill} = a\sqrt{M_p/3M_*}$ is the hill radius.

We find that the timescales can be used to determine whether the stellar gravity enhances the mass loss rate. When t_{h_hill}/t_{g_hill} is less than unity, the effect of the stellar gravity is small and photoheating energy dominates the system energy. On the bottom part of $\log(F_0/F_{\rm cr}) - \log\xi$ plane, the ratio is less than unity. We find that the host star's gravity boosts the escape when the semi-major axis is less than 0.1 AU.

4. Discussion

FOSSATI: How do you define the equilibrium temperature?

MITANI: We can define the equilibrium temperature from photoheating and recombination cooling balance. The temperature is about 10^4 K in the case of EUV photoionization heating. This definition is different from the normal equilibrium temperature determined from the balance of radiation between the host star and the planet, which is usually used.

VIDOTTO: How do you treat the excess energy goes into heating in your characteristic temperature? In some models, the heating efficiency is used.

MITANI: The photoionization heating rate depends on the UV spectrum. We use the average deposited energy and cross-section to define the characteristic temperature. We do not use the heating efficiency as a parameter.

References

Begelman, M. C., McKee, C. F., & Shields, G. A. 1983, ApJ, 271, 70. doi:10.1086/161178
Carolan, S., Vidotto, A. A., Plavchan, P., et al. 2020, MNRAS, 498, L53
Ehrenreich, D., Bourrier, V., Wheatley, P. J., et al. 2015, Nature, 522, 459
Erkaev, N. V., Kulikov, Y. N., Lammer, H., et al. 2007, A&A, 472, 329
Fulton, B. J., Petigura, E. A., Howard, A. W., et al. 2017, AJ, 154, 109
Lecavelier des Etangs, A., Bourrier, V., Wheatley, P. J., et al. 2012, A&A, 543, L4
Mitani, H., Nakatani, R., and Yoshida, N. 2022, MNRAS, 512, 855
Murray-Clay, R. A., Chiang, E. I., & Murray, N. 2009, ApJ, 693, 23
Owen, J. E. & Wu, Y. 2017, ApJ, 847, 29
Rockcliffe, K. E., Newton, E. R., Youngblood, A., et al. 2021, AJ, 162, 116
Sanz-Forcada, J., Micela, G., Ribas, I., et al. 2011, A&A, 532, A6
Schneider, J., Dedieu, C., Le Sidaner, P., et al. 2011, A&A, 532, A79
Tripathi, A., Kratter, K. M., Murray-Clay, R. A., et al. 2015, ApJ, 808, 173
Vidal-Madjar, A., Lecavelier des Etangs, A., Désert, J.-M., et al. 2003, Nature, 422, 143

Winds of Stars and Exoplanets
Proceedings IAU Symposium No. 370, 2023
A. A. Vidotto, L. Fossati & J. S. Vink, eds.
doi:10.1017/S1743921322004525

Discrete Absorption Components from 3-D spot models of hot star winds

F. A. Driessen[1,2] **and N. D. Kee**[2]

[1]Institute of Astronomy, KU Leuven, Celestijnenlaan 200D/2401, 3001 Leuven, Belgium

[2]National Solar Observatory, 22 Ohiʻa Ku St., Makawao, HI 96768, USA

Abstract. The winds of hot, massive stars are variable from processes happening on both large and small spatial scales. A particular case of such wind variability is 'discrete-absorption components' (DACs) that manifest themselves as outward moving density features in UV resonance line spectra. Such DACs are believed to be caused by large-scale spiral-shaped density structures in the stellar wind. We consider novel 3-D radiation-hydrodynamic models of rotating hot star winds and study the emergence of co-rotating spiral structures due to a local (pseudo-)magnetic spot on the stellar surface. Subsequently, the hydrodynamic models are used to retrieve DAC spectral signatures in synthetic UV spectra created from a 3-D short-characteristics radiative transfer code.

Keywords. hydrodynamics, line: formation, radiative transfer, stars: early-type, stars: winds, outflows

---

1. Introduction

Line-driven wind outflows from hot, massive OB-stars are mainly caused by stellar photons scattering off metallic ions in the stellar atmosphere (Castor et al. 1975). However, instead of being homogeneous the wind outflow is rather structured on small- and large-spatial scales. A type of large-scale *coherent* wind structure is believed to originate from spiral arms akin to co-rotating interactions regions (CIRs) in the solar wind (Mullan 1986) whereby the spiral is initiated close to the stellar surface (e.g. Massa & Prinja 2015). The origin of such spiral structures in hot star winds remains, however, illusive with non-radial stellar pulsations (Lobel & Blomme 2008), local stellar spots (Cranmer & Owocki 1996; David-Uraz et al. 2017), or potentially both mechanisms together as potential causes. More recently, the idea of short-lived stellar prominences has been put forward as an explanation of spiral structures (Sudnik & Henrichs 2016).

Particularly, CIRs in line-driven winds have gained considerable attention as they may explain observed line-profile variability in UV resonance lines among the sample of OB-stars (Howarth & Prinja 1989). This variability presents as 'discrete absorption components' (DACs), which represent localised absorption features that move from line centre toward line edge. The time needed to complete such migration typically amounts to a day up to a few days, a fact which appears to be intimately correlated with the (projected) stellar rotation. Therefore, there is strong believe that DACs are rotationally modulated (e.g. Kaper et al. 1999, and references therein).

To model line-driven wind CIRs and their DAC signature we here present 3-D radiation-hydrodynamic models of a line-driven wind from a typical O-star in our Galaxy. We take into account the effects of a small-scale sub-surface magnetic field and its influence on the wind outflow. Such small-scale, localised magnetic fields are thought to arise from the spatially restricted sub-surface iron-recombination convection zone of OB-stars (Cantiello & Braithwaite 2011) leading to the formation of bright star spots.

2. Magnetic spot model

We model the local emergence of a hot star spot from the iron-recombination zone by considering at the stellar surface a pure vertical magnetic field of strength B in magneto-hydrostatic equilibrium with its surroundings. Horizontal pressure balance dictates that the sum of gas and magnetic pressure inside the spot must balance the pressure of the ambient medium: $P_{\rm spot} + P_{\rm mag} = P_{\rm a}$. Invoking the ideal gas law, $P = \rho k_{\rm B} T/\mu$, it follows that the density inside the spot $\rho_{\rm s}$ is lower compared to the density outside the spot

$$\rho_{\rm spot} = \rho_{\rm a} - \frac{P_{\rm mag}}{k_{\rm B}T/\mu} = \rho_{\rm a} - \frac{B_{\rm s}^2}{8\pi k_{\rm B}T/\mu}. \tag{2.1}$$

To include the effect of the spot on the radiation force in our model we consider the spot to be located in a gray radiative atmosphere. At a given optical depth τ we may then write the atmospheric temperature stratification as

$$T^4(\tau) = 0.75 T_{\rm eff}^4 (\tau + 2/3), \tag{2.2}$$

and the local optical depth above some height z can be expressed as

$$\tau(z) \equiv \int_z^{+\infty} \kappa(z')\rho(z')\, dz' = \kappa\rho H, \tag{2.3}$$

where the last equality assumes a spatially constant opacity κ inside the stratified atmosphere with barometric scale height $H = k_{\rm B}T/(\mu g)$, and g the local gravity. By combining the above relations, it follows that the spot optical depth

$$\tau_{\rm spot} = \tau_{\rm a} - \frac{\kappa H B_{\rm spot}^2}{8\pi k_{\rm B}T/\mu} = \tau_{\rm a} - \frac{\kappa B_{\rm spot}^2}{8\pi g}. \tag{2.4}$$

At the height where the spot reaches optical depth $\tau_{\rm spot} = 2/3$—often defined as the visible surface—the ambient atmosphere has a higher optical depth, $\tau_{\rm a} = 2/3 + \kappa B_{\rm spot}^2/(8\pi g)$, such that the star spot appears *bright* relative to its surroundings.

The brightness of the spot will perturb the radiation flux to be higher inside the spot $F_{\rm spot}$ relative to the unperturbed stellar flux $F_\star$. If we assume that the spot size is small to not distort the overall atmospheric temperature structure the radiative flux enhancement from the spot can be expressed as

$$\frac{F_{\rm spot}}{F_\star} = \left(\frac{T(\tau_{\rm spot} = 2/3)}{T(\tau_{\rm a} = 2/3)}\right)^4 = 1 + \frac{3}{4}\frac{\kappa B_{\rm spot}^2}{8\pi g} = 1 + 0.5\left(\frac{B_{\rm spot}}{B_{\rm crit}}\right)^2, \tag{2.5}$$

where $B_{\rm crit}^2 \equiv 16\pi g/(3\kappa)$ is a critical magnetic field strength for which the magnetic pressure equals to the photospheric gas pressure at the $\tau = 2/3$ surface. For the case that $B_{\rm crit} = B_{\rm spot}$ the small spot will provide a 50% increased amplification in the radiation force compared to the unperturbed stellar radiation force.

3. Wind dynamics from radiation-hydrodynamic simulations

The above (local) spot flux enhancement can be readily incorporated in a radiation-hydrodynamic simulation of a line-driven wind outflow. To that end we solve the 3-D spherically-symmetric hydrodynamic equations for a proto-typical Galactic O-supergiant star with parameters $L_\star = 9 \times 10^5 L_\odot$, $M_\star = 50 M_\odot$, $R_\star = 20 R_\odot$, and $T_{\rm eff} = 40\,000$ K. The star is moderately rotating ($v_{\rm rot} \approx 100$ km/s) at 20% of its critical rotation speed. Such slow rotation relieves us from considering more involved effects like gravity darkening and taking into account the oblateness of the star (Cranmer & Owocki 1996). Finally, we assume an isothermal wind outflow, which is justifiable with the high densities inside the wind that make radiative cooling very efficient

We adopt the Sobolev approximation for the line-driven wind (Castor et al. 1975) meaning that we suppress the effects of a powerful radiation-triggered instability in the wind (e.g. Driessen et al. 2022, and references therein). Within this Ansatz the *vector* radiation line force is computed by performing at every wind point a quadrature over the solid angle, $d\Omega = \sin\Theta\, d\Theta\, d\Phi$, of the stellar core intensity for a set of rays in direction $\hat{\boldsymbol{n}}$, weighted by the line-of-sight velocity gradient:

$$\boldsymbol{g}_{\rm line}(\boldsymbol{r}) = \frac{(\kappa_{\rm e}\bar{Q})^{1-\alpha}}{(1-\alpha)c^{\alpha+1}} \oint_{\Omega} d\Omega\, \hat{\boldsymbol{n}}\, I_{\star}(\hat{\boldsymbol{n}})\, [\hat{\boldsymbol{n}} \cdot \nabla(\hat{\boldsymbol{n}} \cdot \boldsymbol{v})/\rho(\boldsymbol{r})]^{\alpha}\,, \tag{3.6}$$

which is added as a source term to the hydrodynamic momentum equation. Here $\kappa_{\rm e} = 0.34\,{\rm cm}^2/{\rm g}$ is the free-electron scattering opacity and $I_\star$ is the unattenuated stellar core intensity that is ray-by-ray weighted according to where the radiation emerges on the surface—for instance to account for limb darkening. The quantity α is the CAK power-law index and $\bar{Q}$ sets the ensemble-integrated line strength (see Poniatowski et al. 2022, for more details on these atomic quantities). These CAK line-ensemble distribution parameters are taken to be the standard O-star Galactic values of $\alpha = 0.65$ and $\bar{Q} = 2000$. Note that along a given ray the line-of-sight velocity gradient tensor $\hat{\boldsymbol{n}} \cdot \nabla(\hat{\boldsymbol{n}} \cdot \boldsymbol{v})$ depends exclusively on local quantities and can be readily computed. Initial conditions are set from a smooth CAK wind while the stellar boundary adopts a constant mass density with the radial velocity extrapolated into the boundary, the polar velocity is set under a no-slip condition, and the azimuthal velocity is set to the stellar rotation. Polar boundaries are set by symmetric/asymmetric combinations and in azimuth we apply periodic boundary conditions.

The required angle quadrature is evaluated by distributing radiation rays across the stellar disc. Experimentation has shown that a ray quadrature of $(n_\Theta, n_\Phi) = (6, 6)$ rays is sufficient to model the overall wind dynamics. For the radiation polar angle Θ the ray emerging positions and flux weights are computed with a Gauss–Legendre quadrature. For the radiation azimuthal angle Φ the ray emerging positions and flux weights are taken uniformly distributed across the full 2π projected stellar disc. Furthermore points with the same Θ have their Φ adjusted by $\pm 20°$ to better sample the full stellar disc. In case a ray is emerging from inside the bright spot, we increase the ray flux weight by an amount set through Eq. (2.5) (for our chosen star $B_{\rm crit} \approx 300\,{\rm G}$). The spot has a size of $10°$ and has a Gaussian decay of field strength towards its edge. In analogy with previous line-driven CIR models, we simulate a half hemisphere, such that over the entire star two symmetric spots appear.

In Figures 1a and 1b we display a set of 3-D isodensity contours from the resulting (pseudo-)magnetic spot and wind interaction for both of the magnetic field strengths (150 G and 1000 G). Although such 3-D models lend themselves well to display the overall wind dynamics, they are somewhat harder for the quantitative interpretation due to the visualisation. Nonetheless, when we consider 2-D cuts in the meridional and equatorial plane we find overall wind dynamics and properties akin to previous 2-D line-driven wind models aiming to model line-driven CIRs (see Cranmer & Owocki 1996; David-Uraz et al. 2017, for extensive discussions). Notable differences in the 3-D wind dynamics concern the fact that the additional flow direction weakens compressions such that CIRs appear less strong compared to 2-D models. Moreover, the azimuthal line force component can yield a net spin-up of wind material ahead of the spot with a spin-down in regions trailing the spot (Gayley & Owocki 2000). Overall this effect azimuthally broadens the 3-D CIR compared to its 2-D analogue while also softening the effect of the spot. Finally, an effect

Figure 1. (a) 3-D isocontours showing wind density values 1.5% above the mean wind mass density. The displayed wind dynamics results from a pseudo-magnetic spot of strength 1000 G. *Left*: Line-of-sight structure seen under 45° from the pole. *Right*: Line-of-sight structure under 90° from the pole. Centred is the solid, grey star. (b) Same as in (a), but for a spot of strength 150 G. We note that the weak magnetic field model generates an overall weak CIR such that it is challenging to visualise its structure.

intrinsic to the 3-D models is the fact that all wind material pushed by the spot flux is contained within the cone angle covered by the spot.

It should be noted that these previous 2-D line-driven CIR models only performed ad-hoc flux enhancements of some putative star spot—although David-Uraz et al. (2017) aim to more realistically model possible spot flux enhancements using spectroscopic constraints. This is in contrast with the present models wherein a first effect, albeit crude and simplified, of a (pseudo-)magnetic spot is taken into account. The models displayed in Figure 1a and 1b also serve as a direct input to the 3-D short-characteristics radiative transfer computation discussed in the next section. This allows future comparison of constraints on the flux enhancement and DAC strength to limits on the unobserved, localised patches of stellar magnetic field.

4. Synthetic time series of a proto-typical UV resonance line

Using the wind structure from our radiation-hydrodynamic simulations, Figures 2a and 2b display the resulting synthetic UV resonance line spectrum of C IV obtained from solving the 3-D radiation transfer problem with the method of short-characteristics (see Hennicker et al. 2020, for full details).

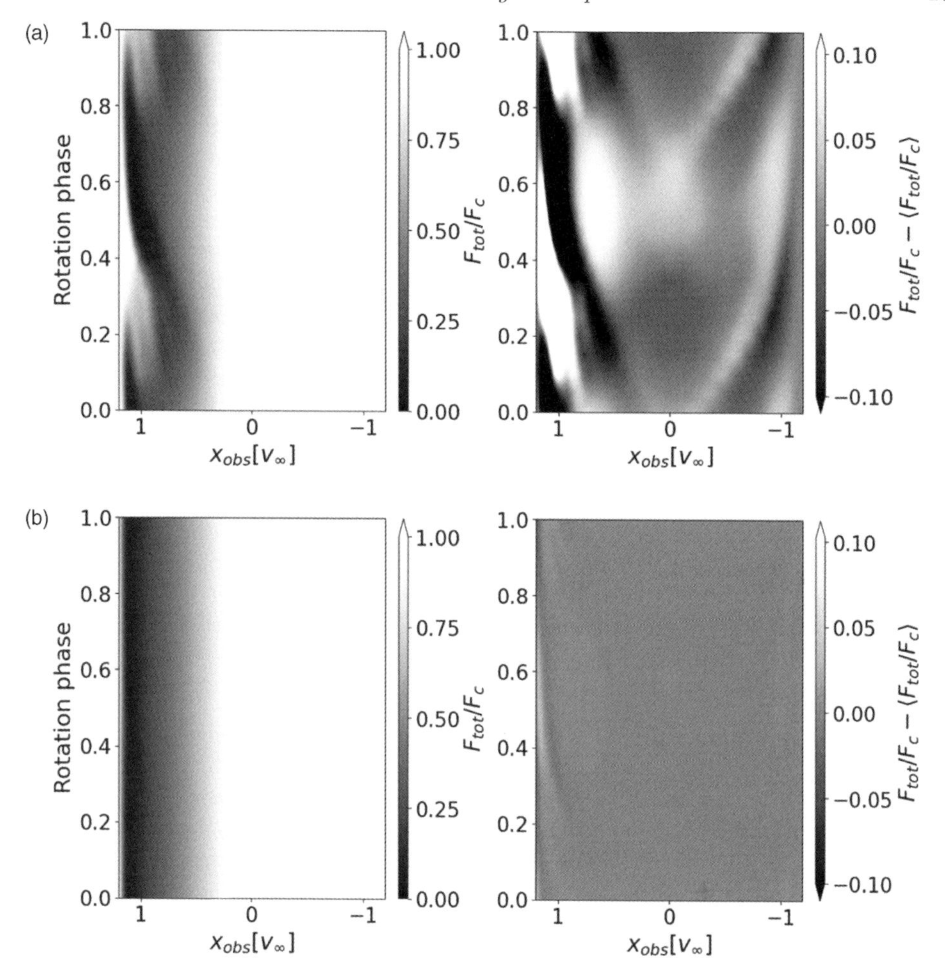

Figure 2. (a) Temporal variation of the C IV UV resonance line for a pseudo-magnetic spot of strength 1000 G. *Left*: Flux with respect to the continuum. *Right*: Temporally averaged normalised spectrum subtracted from the line profile in the left panel with variations limited to ±10% with respect to the mean spectrum. This serves to better show the line-profile variability. (b) Same as in (a), but for a spot of strength 150 G.

The time series of the 1000 G spot model indeed *quantitatively* reproduce the expected DAC shape—we stress that DAC signatures are star-dependent in their shape and duration. Indeed, excess absorption occurs in the blue-edge of the line profile starting from velocities $\approx 0.5v_\infty$. Since line-driven winds already attain supersonic speeds about 1% above the stellar surface and $v \approx 0.5v_\infty$ around $1R_\star$ above the surface, this shows that the CIR creating the DAC should form close to the star (Massa & Prinja 2015). The rather 'broad' absorption signal then gradually moves outward becoming narrow near the blue edge of the line at terminal wind speeds. This can be understood from the fact that near the star a collection of gas parcels, that have the correct optical depth to absorb, also possess a large range of velocity dispersion in the dense inner star region. This leads to a broad initial absorption near $v \approx 0.5v_\infty$. On the contrary, gas parcels that absorb at the blue edge populate the outer edges of the winding spiral arm and are generally all moving at or near the terminal wind speed. This means these gas parcels posses much less velocity dispersion, hence the DAC absorption excess narrows farther away from the star at high wind speeds.

This general behaviour is in contrast to the model with a spot field strength of 150 G. Here a DAC absorption signature is only modestly visible in the right-hand panel of Figure 2b that subtracts the mean spectrum from all spectra. Intuitively one may expect this is due to the lack of density enhancement in the spiral arm (see Figure 1b), but in fact the generation of DACs in line-driven CIRs is mainly dependent on kinks in the *velocity field* (Cranmer & Owocki 1996). The kinks in the velocity field are, however, also weak due to the rather modest flux enhancement the spot generates. As such no significant DAC feature can form. Caution is warranted since the flux enhancement predicted by Eq. (2.5) depends on free parameters, notably, the assumed mass-absorption coefficient κ in the atmosphere. Since a 150 G magnetic field strength is expected to arise from a dynamo-generated spot (Cantiello & Braithwaite 2011), we have performed an additional spot model (not shown here) employing a less conservative value for the opacity, $\kappa = \kappa_{\rm OPAL} \approx 1$, near the iron-recombination zone (Cantiello et al. 2009, their Fig. 1). Adopting this opacity lowers the critical field strength to $B_{\rm crit} \approx 170$ G and yields larger spot flux enhancement. Nonetheless, our simulations predict only a marginally denser, compressed CIR with a modest amount of extra absorption in the blue edge of the DAC compared to Figure 1b. A potential reason for this behaviour at low spot field strengths is that for the assumed spots in this work of 10° in size, the Gaussian decay applied to the spot field strength 'shuts off' too fast the flux enhancement over the geometrical extent the spot can contribute to. Future investigation of characteristic spot properties is thus called for.

5. Conclusion

We find that a bright star spot with a strong magnetic field strength, typically associated with the inferred large-scale field strengths of OB-stars, produces a strong enough absorption signature and expected DAC shape. On the other hand, a spot with a magnetic field strength predicted to arise from a dynamo operating in the sub-surface convection zone (Jermyn & Cantiello 2020) produces almost no absorption features in the DAC. *A word of caution is in place*, however, in the interpretation of these results given the model uncertainties involved. To assess whether magnetic spots can generate line-driven CIRs and explain DACs, full radiation-*magneto* hydrodynamic models are necessary. These models would need improved input physics on the spot size, spot magnetic field strength (strong vs. weak), the local field geometry (simple loops vs. complex topology), the lifetime of the spot (short vs. long compared to stellar rotation), and/or the rotation properties of the spot (co-rotating vs. differentially rotating). This may then solve some of the puzzling outcomes found for the weak field strength spot model here. Future theoretical and observational efforts are thus further required to assess the link between small-scale surface magnetic fields in OB-stars and DAC signatures.

References

Cantiello, M., Langer, N., Brott, I., et. al (2009), A&A, 499, 279
Cantiello, M., & Braithwaite, J. (2011), A&A, 534, A140
Castor, J. I., Abbott, D. C., Klein, R. I. (1975), ApJ, 195,157
Cranmer, S. R., & Owocki, S. P. (1996), ApJ, 462, 469
David-Uraz, A., Owocki, S. P., Wade, G. A., et al. (2017), MNRAS, 470, 3672
Driessen, F. A., Sundqvist, J. O., & Dagore, A. (2022), A&A, 663, A40
Gayley, K. G., & Owocki, S. P. (2000), ApJ, 537, 461
Hennicker, L., Puls, J., Kee, N. D., Sundqvist, J. O. (2020), A&A, 633, A16
Howarth, I. D., & Prinja, R. K. (1989), ApJ, 69, 527
Jermyn, A. S., Cantiello, M. (2020), ApJ, 900, 113
Kaper, L., Henrichs, H. F., Nichols, J. S., et al. (1999), A&A, 344, 231

Lobel, A., & Blomme, R. (2008), ApJ, 678, 408
Massa, D., & Prinja, R. K. (2015), ApJ, 809, 12
Mullan, D. J. (1986), A&A, 165,157
Poniatowski, L. G., Kee, N. D., Sundqvist, J. O., et al. (2022), A&A, in press
Sudnik, N. P., & Henrichs, H. F. (2016), A&A, 594, A16

Winds of Stars and Exoplanets
Proceedings IAU Symposium No. 370, 2023
A. A. Vidotto, L. Fossati & J. S. Vink, eds.
doi:10.1017/S1743921322004550

Hydrodynamic disk solutions for Be stars using HDUST

C. Arcos[1], M. Curé[1], I. Araya[2], A. Rubio[3] and A. Carciofi[3]

[1]Instituto de Física y Astronomía, Universidad de Valparaíso, Chile

[2]Vicerrectoría de Investigación, Universidad Mayor, Chile

[3]Instituto de astronomia, geofísica e ciências atmosféricas, Universidade de São Paulo, Brazil

Abstract. In this work, we implemented a hydrodynamical solution for fast rotating stars, which leaves high values of mass-loss rates and low terminal velocities of the wind. This 1D density distribution adopts a viscosity mimicking parameter which simulates a quasi-Keplerian motion. Then, it is converted to a volumetric density considering vertical hydrostatic equilibrium using a power-law scale height, as usual in viscous decretion disk models. We calculate the theoretical hydrogen emission lines and the spectral energy distribution utilizing the radiative transfer code `HDUST`. Our disk-wind structures are in agreement with viscous decretions disk models.

Keywords. stars: emission-line, Be, stars: winds, outflows, hydrodynamics

---

1. Introduction

Classical Be stars (CBes) are defined as fast rotating (Zorec et al. 2016; Solar et al. 2022) main sequence stars that form an equatorial gas disk rotating in a quasi-Keplerian motion (Rivinius et al. 2013). These disks are built from mass ejected from the stellar surface that acquires sufficient velocity and angular momentum to orbit the star and are referred to as decretion disks. Once ejected from the star, material in a CBe disk is generally governed by gravity and viscous forces. The Viscous Decretion Disk (VDD) model (Lee et al. 1991; Bjorkman & Wood 1997; Okazaki 2001; Bjorkman & Carciofi 2005) is currently the best theoretical framework describing the evolution of these disks once formed. Viscosity acts to shuffle the angular momentum of the circumstellar material. A small fraction (∼1%) of the mass ejected by the star acquires sufficient angular momentum to move outwards, orbiting at increasing radii. In contrast, most of the ejected mass falls back onto the star (Haubois et al. 2012). The strength of viscosity is parameterized as in the model from Shakura (1973) and dictates the timescales over which Be star decretion disks evolve and dissipate.

The disk feeding process is still not well understood. It is probably a combination of some mechanisms such as non-radial pulsation, stellar winds, and fast rotation, since these features are usually observed in these stars. Recent works (Kee et al. 2018b,a) indicate that, in addition to viscosity, radiative ablation may play a role in the mass budget of CBe disks by systematically removing (employing radiative acceleration) some material from the inner part of CBe disks, thus forming a disk-wind type of structure both above and below the disk.

Under the hypothesis that these structures are fed like radiation-driven winds and the rotation of the central star plays an important role in shaping the disk, in this work, we study the interplay of a particular hydrodynamic solution called Ω-slow (Curé 2004; Araya et al. 2017) to test a physical model for Be stars outflows.

2. VDD parametric model

The parametric model for the volumetric density distribution depends on the radial distance from the star, r, and the height above the equatorial plane, z, and has the form

$$\rho(r,z) = \rho_0 \left(\frac{r}{R_\star}\right)^{-m} \exp\left(\frac{-z^2}{2H^2}\right), \tag{2.1}$$

where ρ_0 is the disk base density, $R_\star$ is the stellar radius and H is the height scale given by

$$H = H_0 \left(\frac{r}{R_\star}\right)^{\beta}. \tag{2.2}$$

For isothermal disks β=3/2 and the constant H_0 is defined by

$$H_0 = \left(\frac{2R_\star^3 k T_0}{GM_\star \mu_0 m_H}\right)^{1/2}, \tag{2.3}$$

where $M_\star$ is the stellar mass, G is the gravitational constant, m_H is the mass of a hydrogen atom, k is the Boltzmann constant, μ_0 is the mean molecular weight of the gas and T_0 is an isothermal temperature used only to fix the vertical structure of the disk initially. In this work we set μ_0=0.6 and T_0=0.72 T_{eff}. This steady-state and isothermal power-law approximation has been extensively used in the literature, achieving good agreement with observations (Carciofi & Bjorkman 2006; Carciofi et al. 2009; Jones et al. 2008; Klement et al. 2015; Silaj et al. 2016; Arcos et al. 2017).

3. Ω-slow solution for Be stars

Curé (2004) found a new hydrodynamical solution from the standard radiation-driven winds theory (Castor et al. 1975; Friend & Abbott 1986). This is called Ω-slow solution and appears for fast rotating stars, $\Omega = v_{rot}/v_{crit} \gtrsim 0.75$. This solution is characterized by a slow wind with high values of mass-loss rates. However, it assumes angular momentum conservation and its terminal velocity is too fast to explain the Be disks. Currently, the disk-like structure observed in Be stars is modeled using a quasi-Keplerian motion.

Okazaki (2001) used viscosity and the model of a radiative force due to an ensemble of optically-thin lines from Chen & Marlborough (1994). In their model, the radiation force is always an increasing function of the radial coordinate and does not decay at large distances as it should happen. Therefore, based on the results of the Ω-slow solution, we implemented a radiative force that increases near the stellar surface and after particular distance decay in terms of the radial coordinate. We followed the procedure made by de Araujo (1995) implementing a mimicking viscosity parameter via

$$V_\phi = \Omega \sqrt{\frac{GM_\star(1-\Gamma)}{R_\star}} \left(\frac{R_\star}{r}\right)^{\gamma_{vis}}, \tag{3.1}$$

where V_ϕ is the rotational velocity component of the wind in the equatorial plane, Γ considers the mean opacity of the total flux and γ_{vis} is the viscosity mimicking parameter. For $\gamma_{vis} = 0.5$ a Keplerian outflowing wind is obtained.

In Figure 1 we compare two wind density distributions (dotted blue and dot-dashed red lines) obtained under our procedure described above. Both solutions were generated considering a high rotational speed with Ω=0.99 and fixing the line-force parameters δ=0 and k=0.2 for a γ_{vis}=0.51 (quasi-Keplerian motion). The stellar parameters are for a B2V star, with T_{eff}=21000 K, log g = 4.00 and $R_\star$=4.5 $R_\odot$. We set the base density to ρ_0=5×10^{-11} g/cm^3 which is a typical value for a B2Ve star (see Arcos et al. 2017; Vieira et al. 2017). The wind parameters obtained by these models are summarized in

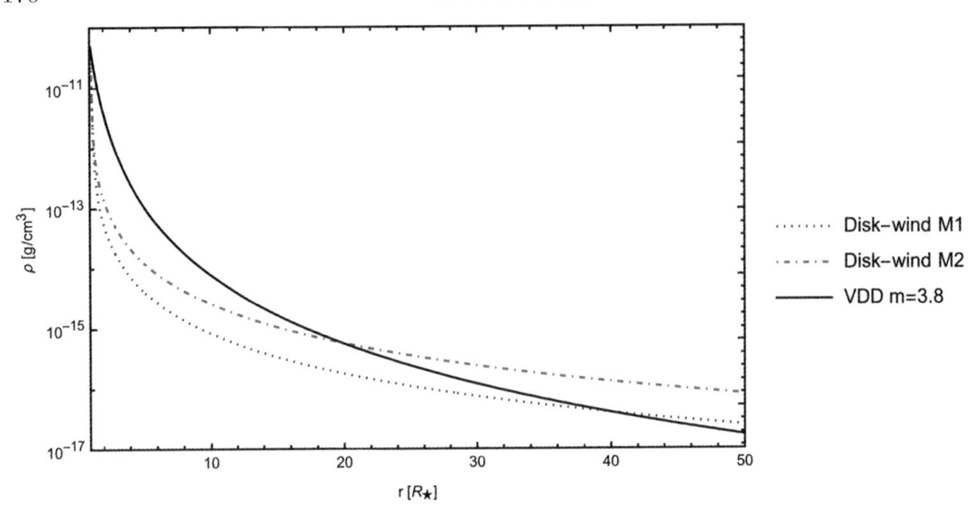

Figure 1. Equatorial density distributions for the models used in this work. Line force parameters for the wind are displayed in Table 1. Both hydrodynamical solutions are compared to a VDD model with an exponent power-law m=3.8 represented by the solid black line.

Table 1. Wind density description. Models were generated with δ=0, Ω=0.99 and a base density ρ_0=5×10^{-11} g/cm^3. The stellar parameters are those for a typical B2V star, with $T_{\rm eff}$=21000 K, log g = 4.00 and $R_\star$=4.5 $R_\odot$.

Model	α	k	γ_{vis}	$\dot{M}$ ($10^{-8}\,M_\odot$/yr)	v_∞ (km/s)
M1	0.50	0.2	0.51	1.17	101
M2	0.62	0.2	0.51	4.98	131

Table 1 where a denser, slower solution is observed compared to a standard fast solution. The best representation of the equatorial density structure from our hydrodynamical models is for a VDD parametric model with exponent m=3.8 (solid black line overplotted in Fig. 1). The figure shows that the density of the hydrodynamical solutions decay much faster near the stellar photosphere ($\sim$ one order of magnitude at 5 $R_\star$) than the standard VDD model. This would imply less flux contribution (emission-line intensity) to the emitting region enclosed by Hα, usually considered as the size of the disk in the optical range. In addition, both hydrodynamical solutions overpass the VDD model at some point, e.g., model M1 does it at $\sim$20 $R_\star$ and M2 at $\sim$40 $R_\star$, then for pole-on views we will expect that the emission-lines will be more intense than the ones obtained with VDD models.

4. Disk-wind structure

To perform the radiative transfer calculations using our hydrodynamical wind solutions, we used of the 3D Monte Carlo radiative transfer code, `HDUST` (Carciofi & Bjorkman 2006, 2008). The code computes the spectral energy distribution (SED), polarization and Hydrogen lines for different lines of sight. In addition, it allows several input options for the density distribution and motion of the disk. Also, it considers the oblateness of the rotating star and, thus, the gravity darkening effects.

Our first step was creating a conversion code to appropriately situate our 1D hydrodynamical solutions to be read by `HDUST`. The adopted grid structure is divided in 50 radial cells (N_r), 42 co-latitudinal cells (N_μ) and 1 azimuthal cells (N_ϕ, axisymmetry). Thus, the 1D density input is distributed into a 3D geometry in the equatorial plane by setting vertical hydrostatic equilibrium with a power-law height scale (see eq. 2.2).

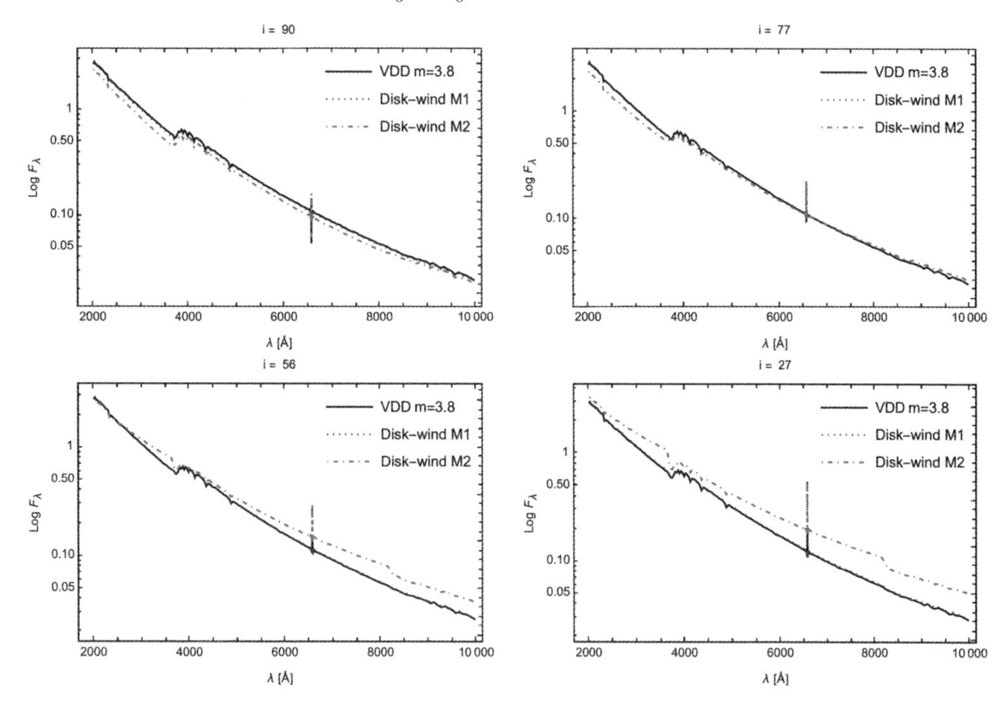

Figure 2. SED computed by HDUST code for the disk-wind and VDD models.

The second step was to use this hydrodynamical density distribution for the wind as input in the HDUST code. We set a disk size of 50 $R_\star$ and a parametric rigid rotator star without oblateness (because our hydrodynamical solution was created with a spherical geometry). The stellar spectrum was defined by a Kurucz (1994) model and the limb darkening from Claret (2000) models. We calculated the SED and the theoretical Hα emission line with this setup. Fig. 2 shows the SED for our wind models computed with HDUST. We present the result for four different inclination angles, starting at edge-on view (i=90°) up to pole-on view (i=27°). The output flux is dimensionless (we are using it only for comparison proposes) and is in logarithm scale to appreciate the differences. The VDD model from Fig. 1 is over-plotted to our wind solutions and is represented with a solid black line. From the figure, we can see that the shape of the density distribution affects directly the flux continuum, having the VDD and model M1 similar slopes (but different fluxes, see Fig. 1) compared to the M2 model which is denser and thus, more optically thick, absorbing more photons at the edge-on view.

In the case of the Hα emission line, the results are displayed in Fig. 3 for the two wind models and the VDD comparison. The sub-figures at the top have a different vertical scale than those at the bottom. At the equatorial view, the VDD model (solid black line) has the highest intensity, as we expected from the density distribution (Fig. 1). Because the density from the model M2 decays faster than the VDD, but from ~20 $R_\star$, it compensates by increasing its intensity, then a similar line-emission is observed but with a different shape. We note models have different densities, but same velocities, therefore since the slower components of the Hα line come from a denser region, they are more intense. The opposite behavior is observed close to the pole when the wind model is almost twice in intensity compared with VDD (see Fig. 3 at i=27°). From the disk temperature distribution, all models show similar behavior to a 'U' shape profile, as shown in Carciofi & Bjorkman (2006), but with different locations for the minimum temperature in the disk, being closer to the star with high slopes of density decay. Model

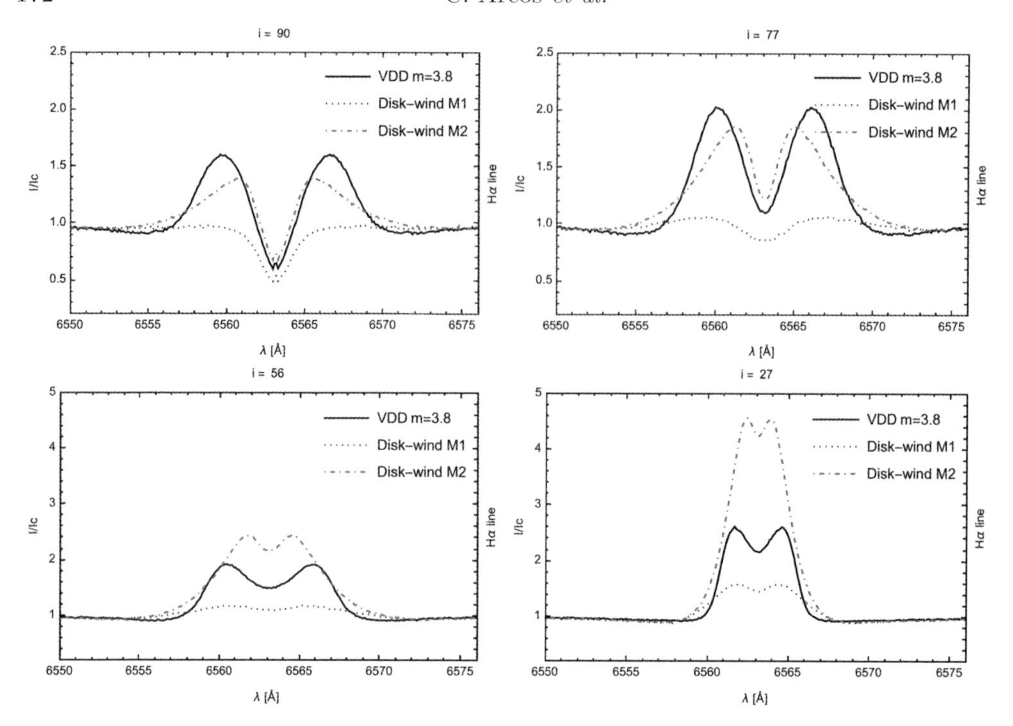

Figure 3. Hα theoretical emission line computed with HDUST for the different wind models and the VDD comparison. The sub-figures at the top have a different vertical scale than those at the bottom. This theoretical line profiles were generated with the density distributions shown in Fig. 1.

M1 (dotted blue line) does not have enough density to form a suitable emission line (as we expected with the other models). Therefore, these density distributions better describe dissipating or forming disk phases. We note that the emission line generated with this solution is all the time under the VDD model, as we expected from Fig. 1.

5. Summary and discussion

Massive stars possess strong stellar winds driven by radiation, usually with high mass-loss rates and terminal velocities. CBes are massive stars on the main sequence, which enhance the mass-loss rates at the stellar equator, forming and dissipating a quasi-keplerian rotating gas disk under some unknown process. In order to understand the physical process these stars go through, we test the Ω-slow hydrodynamical solution for stars rotating faster than 75% of their critical rotation. This solution leads to high values of mass-loss rates and slow terminal velocities. In addition, we adopt a viscosity parameter for the azimuthal velocity to reach a quasi-Keplerian gas motion. By using the radiative transport code HDUST and forcing vertical hydrostatic equilibrium, our preliminary results successfully reproduce a similar theoretical line-emission shape for a typical B2Ve star. These preliminary results were made under a spherical symmetry; thus, gravity darkening effects are not considered here. Also, we note that our density distribution decays too fast near the stellar surface, with differences close to one order of magnitude compared to the VDD model; as a result, a low intensity in the Hα emission line. More tests must be done, with different values of Ω covering other spectral types. Once having several models with disk-wind structures, the next step is to compare with spectroscopic and photometric observations of several Be stars.

Acknowledgments

Authors thank project ANID-FAPESP N°133541. CA and IA thanks FONDECYT 11190945 and 11190147, respectively. MC, CA & IA are grateful to projects FONDECYT 1190485. This work has been possible thanks to the use of AWS-UChile-NLHPC credits. Powered@NLHPC: This research was partially supported by the super-computing infrastructure of the NLHPC (ECM-02). This project has also received funding from the European Union's Framework Programme for Research and Innovation Horizon 2020 (2014–2020) under the Marie Skłodowska-Curie grant Agreement No 823734. ACC acknowledges support from CNPq (grant 311446/2019-1) and FAPESP (grants 2018/04055-8 and 2019/13354-1). ACR acknowledges support from CAPES (grant 88887.464563/2019-00) and DAAD (reference number 91819449).

References

Araya, I., Jones, C. E., Curé, M., et al. 2017, ApJ, 846, 2
Arcos, C., Jones, C. E., Sigut, T. A. A., Kanaan, S., & Curé, M. 2017, ApJ, 842, 48
Bjorkman, J. E. & Carciofi, A. C. 2005, in Astronomical Society of the Pacific Conference Series, Vol. 343, Astronomical Polarimetry: Current Status and Future Directions, ed. A. Adamson, C. Aspin, C. Davis, & T. Fujiyoshi, 270
Bjorkman, J. E. & Wood, K. 1997, in American Astronomical Society Meeting Abstracts, Vol. 191, American Astronomical Society Meeting Abstracts, 12.04
Carciofi, A. C. & Bjorkman, J. E. 2006, ApJ, 639, 1081
Carciofi, A. C. & Bjorkman, J. E. 2008, ApJ, 684, 1374
Carciofi, A. C., Okazaki, A. T., Le Bouquin, J. B., et al. 2009, A&A, 504, 915
Castor, J. I., Abbott, D. C., & Klein, R. I. 1975, ApJ, 195, 157
Chen, H. & Marlborough, J. M. 1994, ApJ, 427, 1005
Claret, A. 2000, A&A, 363, 1081
Curé, M. 2004, ApJ, 614, 929
de Araujo, F. X. 1995, A&A, 298, 179
Friend, D. B. & Abbott, D. C. 1986, ApJ, 311, 701
Haubois, X., Carciofi, A. C., Rivinius, T., Okazaki, A. T., & Bjorkman, J. E. 2012, ApJ, 756, 156
Jones, C. E., Tycner, C., Sigut, T. A. A., Benson, J. A., & Hutter, D. J. 2008, ApJ, 687, 598
Kee, N. D., Owocki, S., & Kuiper, R. 2018a, MNRAS, 474, 847
Kee, N. D., Owocki, S., & Kuiper, R. 2018b, MNRAS, 479, 4633
Klement, R., Carciofi, A. C., Rivinius, T., et al. 2015, A&A, 584, A85
Kurucz, R. 1994, Solar abundance model atmospheres for 0, 19
Lee, U., Osaki, Y., & Saio, H. 1991, MNRAS, 250, 432
Okazaki, A. T. 2001, PASJ, 53, 119
Rivinius, T., Carciofi, A. C., & Martayan, C. 2013, AAPR, 21, 69
Shakura, N. I. 1973, SOVAST, 16, 756
Silaj, J., Jones, C. E., Carciofi, A. C., et al. 2016, ApJ, 826, 81
Solar, M., Arcos, C., Curé, M., Levenhagen, R. S., & Araya, I. 2022, MNRAS, 511, 4404
Vieira, R. G., Carciofi, A. C., Bjorkman, J. E., et al. 2017, MNRAS, 464, 3071
Zorec, J., Frémat, Y., Domiciano de Souza, A., et al. 2016, A&A, 595, A132

Winds of Stars and Exoplanets
Proceedings IAU Symposium No. 370, 2023
A. A. Vidotto, L. Fossati & J. S. Vink, eds.
doi:10.1017/S1743921322003519

Role of Longitudinal Waves in Alfvén-wave-driven Solar/Stellar Wind

Kimihiko Shimizu[1], Munehito Shoda[2] and Takeru K. Suzuki[1,3]

[1]School of Arts & Sciences, The University of Tokyo, 3-8-1 Komaba, Meguro, Tokyo, 153-8902, Japan

[2]Department of Earth and Planetary Science, The University of Tokyo, 7-3-1, Hongo, Bunkyo, Tokyo, 113-0033, Japan

[3]Department of Astronomy, The University of Tokyo, 7-3-1, Hongo, Bunkyo, Tokyo, 113-0033, Japan

Abstract. We study the role the the p-mode-like vertical oscillation on the photosphere in driving solar winds in the framework of Alfvén-wave-driven winds. By performing one-dimensional magnetohydrodynamical numerical simulations from the photosphere to the interplanetary space, we discover that the mass-loss rate is raised up to ≈ 4 times as the amplitude of longitudinal perturbations at the photosphere increases. When the longitudinal fluctuation is added, transverse waves are generated by the mode conversion from longitudinal waves in the chromosphere, which increases Alfvénic Poynting flux in the corona. As a result, the coronal heating is enhanced to yield higher coronal density by the chromospheric evaporation, leading to the increase of the mass-loss rate. Our findings clearly show the importance of the p-mode oscillation in the photosphere and the mode conversion in the chromosphere in determining the basic properties of the wind from the sun and solar-type stars.

Keywords. solar wind, stars: winds, outflows, MHD, turbulence, waves

---

1. Introduction

The sun and low-mass main sequence stars posses a surface convection zone beneath the photosphere. Various types of waves are excited and are propagating upward to the atmosphere (Lighthill 1952; Stepien 1988). Alfvén waves have attracted a great attention as a reliable player in heating and driving the solar wind and stellar winds from solar-type stars because they can efficiently transport the convective kinetic energy to the corona and the wind region owing to the incompressible nature (Belcher 1971; Suzuki & Inutsuka 2005, 2006; Cranmer et al. 2007; Verdini & Velli 2007; Matsumoto & Suzuki 2012; Shoda et al. 2019; Sakaue & Shibata 2021; Matsumoto 2021; Vidotto 2021).

On the other hand, acoustic waves are not regarded to play a substantial role in heating and accelerating the solar and stellar winds because those excited by p-mode oscillations easily steepen to form shocks before reaching the corona (Stein & Schwartz 1972; Cranmer et al. 2007). Recently, however, Morton et al. (2019) reported the tight relation between the p-modes on the photosphere and Alfvénic oscillations in the corona from the comparison of their power spectra and pointed out the possibility of the generation of transverse waves by the mode conversion from longitudinal waves excited by p-modes (Cally & Hansen 2011).

Inspired by this observational finding, we studied the effect of vertical oscillations at the photosphere on Alfvén wave-driven solar winds by magnetohydrodynamical (MHD)

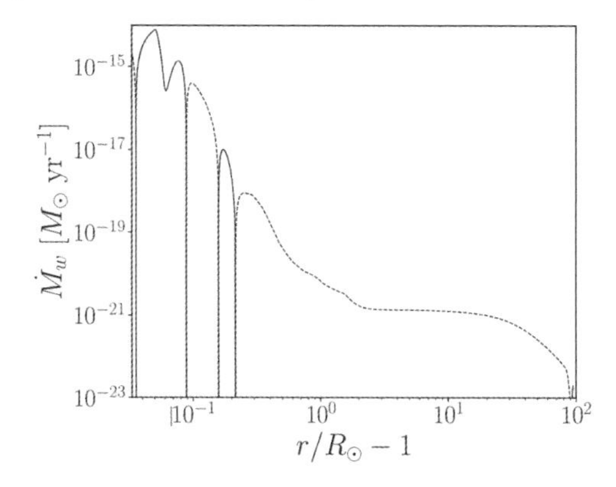

Figure 1. Mass-loss rate (equation 3.1) of the case with the only longitudinal fluctuation, $\langle \delta v_\parallel \rangle = 0.6$ km s^{-1}, on the photosphere.The solid (dashed) line corresponds to positive (negative) $\dot{M}_w$.

simulations (Shimizu et al. 2022). In this talk, we introduce the results of this paper and explain the importance of the longitudinal fluctuation in driving the wind from the sun and solar-type stars from a theoretical point of view.

2. Simulation

We set up one-dimensional (1D) super-radially open magnetic flux tubes along the r direction rooted at the photosphere of a star with the solar mass, $M_\odot$, and radius, $R_\odot$. The radial magnetic field strength is set to be $B_r = 1300$ G on the photosphere, which gives the equipartition between the gas pressure and the magnetic pressure. In the outer region after the super-radial expansion has completed, the radial field decreases with $B_r = 1.3$ G $(r/R_\odot)^{-2}$. We solve compressible MHD equations with radiative cooling and thermal conduction in the flux tube from the photosphere to $\approx 100 R_\odot$. We input three dimensional velocity fluctuation on the photosphere; the two transverse components, $\delta v_{\perp,1}$ and $\delta v_{\perp,2}$, generate Alfvén waves and the longitudinal (radial) component, $\delta v_\parallel$, excites acoustic waves. In order to take into account cascading Alfvénic turbulence, which plays a role in the dissipation of Alfvén waves (Goldreich & Sridhar 1995; Matthaeus et al. 1999), we employ phenomenological dissipation terms in the transverse components of the momentum equation and the induction equation (Shoda et al. 2018). See Shimizu et al. (2022) for the detailed setup of the numerical simulation.

3. Results

Before investigating roles of acoustic waves in Alfvén wave-driven winds, let us first examine whether the solar wind is driven solely by acoustic waves in the absence of Alvfén waves. To do so, we run a simulation with the only longitudinal perturbation, $\langle \delta v_\parallel \rangle = 0.6$ km s^{-1}, switching off the transverse components, $\delta v_{\perp,1} = \delta v_{\perp,2} = 0$, where $\langle \cdots \rangle$ stands for the root-mean-squared average over time. Figure 1 presents mass-loss rate,

$$\dot{M}_w = 4\pi \rho f v_r r^2, \tag{3.1}$$

after time $t = 3500$ minutes from the start of the simulation, where f is the filling factor of open magnetic flux regions. One can recognize negative $\dot{M}_w$ (dashed line) in the most part of the outer region, which indicates that the gas does not stream out but falls down to the surface; the acoustic wave cannot drive the solar wind only by itself.

Next, we study the effect of longitudinal perturbations in the presence of Alfvén waves. We vary the amplitude of the longitudinal perturbation at the photosphere in a wide range of 0 km s$^{-1} \leq \langle \delta v_\parallel \rangle \leq 3.0$ km s^{-1} for a fix amplitude of the transverse components,

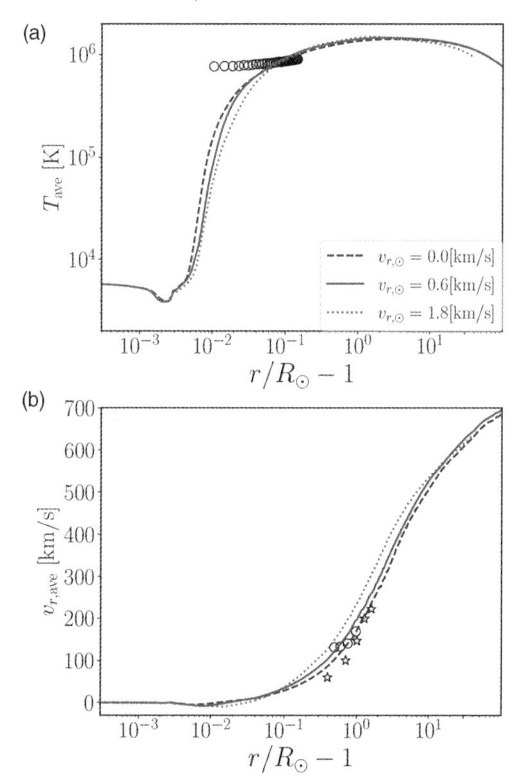

Figure 2. Temperature (top) and radial velocity (bottom) of the cases with the longitudinal amplitude of $\langle \delta v_\parallel \rangle = 0$ km s^{-1} (blue dashed), 0.6 km s^{-1} (green solid) and 1.8 km s^{-1} (red dotted) for the fixed transverse amplitude, $\langle \delta v_\perp \rangle = 0.6$ km s^{-1}. The circles in the top panel represent electron temperature observed by CDS/SOHO (Fludra et al. 1999). The stars (Zangrilli et al. 2002) and the circles (Teriaca et al. 2003) in the bottom panel indicate proton outflow velocities observed in polar regions by SOHO.

$\langle \delta v_{\perp,1} \rangle = \langle \delta v_{\perp,2} \rangle = 0.6$ km s^{-1}. Figure 2 compares the temperatures, T, (top) and radial velocities, v_r, (bottom) of three cases with different $\langle \delta v_\parallel \rangle = 0$ (blue dashed), 0.6 (green solid), and 1.8 km s^{-1} (red dotted). These three cases show quite similar profiles of T and v_r. In contrast, Figure 3 shows that the density in the coronal region is higher for larger $\langle \delta v_\parallel \rangle$; the case with $\langle \delta v_\parallel \rangle = 1.8$ km s^{-1} yields ≈ 4 times denser corona than the case with $\langle \delta v_\parallel \rangle = 0$ km s^{-1}.

The difference in the coronal density directly leads to the change of the mass-loss rate of these cases. Figure 4 shows that $\dot{M}_w$ obtained from the numerical simulations (blue stars) increases with $\langle \delta v_\parallel \rangle$ except for the range of $\langle \delta v_\parallel \rangle \geq 2.7$ km s^{-1}. In addition, these numerical data are roughly reproduced by an analytic relation of $\dot{M}_w \approx L_{\rm A,cb}/v_{\rm g,\odot}^2$ (red circles; Cranmer & Saar 2011), where $L_{\rm A,cb}$ is the Alfvénic Poynting flux at the coronal base (see Shimizu et al. 2022, for the specific expression) and $v_{\rm g,\odot} = \sqrt{2GM_\odot/R_\odot}$ is the escape velocity. Since $v_{\rm esc,\odot}$ is constant in our setup, the dependence of $L_{\rm A,cb}$ on $\langle \delta v_\parallel \rangle$ indicates that the Alfvénic Poynting flux that reaches the corona increases by adding the photospheric longitudinal fluctuation even though the injected Alfvénic Poynting flux at the photosphere is the same. While it is hard to understand this dependence because the longitudinal fluctuation, which generates acoustic-mode waves, does not directly contribute to Alfvénic Poynting flux, the connection between $\langle \delta v_\parallel \rangle$ and $L_{\rm A,cb}$

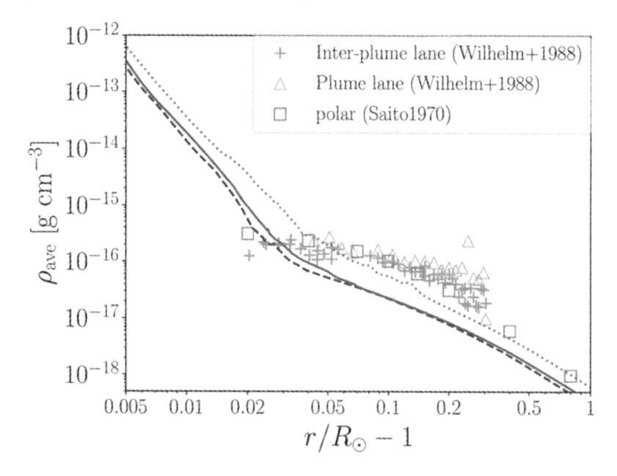

Figure 3. Density of the same three cases as presented in Figure 2. The linetypes are also the same as those in Figure 2. The crosses and triangles are observed density (Wilhelm et al. 1998) in interplumes and plumes, respectively. The squares show the density averaged over multiple solar eclipses during solar minimum phases (Saito et al. 1970).

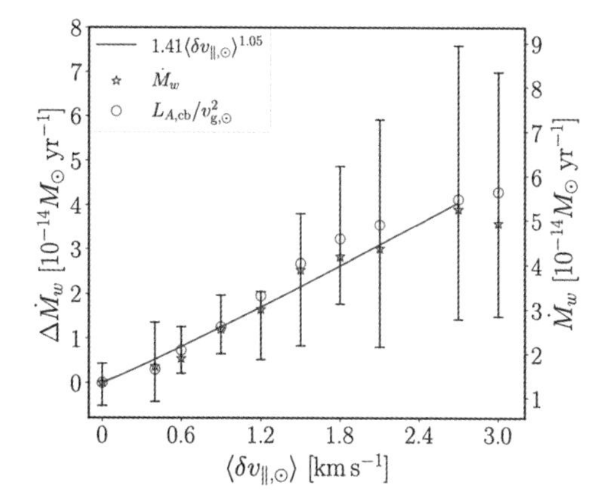

Figure 4. Mass-loss rate against input longitudinal-wave amplitudes on the photosphere. The right axis shows $\dot{M}_w$ (equation 3.1). The left axis indicates the excess, $\Delta\dot{M}_w$, of $\dot{M}_w$ from the value obtained for $\langle\delta v_\parallel\rangle = 0$. The blue star symbols are the values obtained from our numerical simulations. The red circles present the theoretical relation introduced by Cranmer & Saar (2011, see text). The blue solid line is the power-law fitting relation to $\Delta\dot{M}_w$ for $\langle\delta v_\parallel\rangle \leq 2.7$ km s^{-1}.

implies that something is happening between longitudinal and transverse waves in the chromosphere.

To inspect the propagation and interaction of different modes of waves, we show the variation and dissipation of Alfvénic Poynting flux, $L_{\rm A}$, of the three cases with $\langle\delta v_\parallel\rangle = 0$, 0.6, and 1.8 km s^{-1} from the photosphere to the low corona in Figure 5. The top panel compares the vertical profiles of $L_{\rm A}$ of the three cases. $L_{\rm A}$ of the two cases with zero and small $\langle\delta v_\parallel\rangle$ monotonically decreases by the dissipation of Alfvén waves. In contrast, $L_{\rm A}$ of the case with the large $\langle\delta v_\parallel\rangle$ increases with height near the photosphere.

The middle and bottom panels of Figure 5 respectively presents the contributions of the turbulent cascade, $\Delta L_{\rm A,turb}$, and the mode conversion to longitudinal waves, $\Delta L_{\rm A,mc}$, in the dissipation of Alfvén waves, where both $\Delta L_{\rm A,turb}$ and $\Delta L_{\rm A,mc}$ are normalized by the Alfvénic Poynting flux at the photosphere, $L_{\rm A,\odot}$. In the case of no longitudinal fluctuation (blue dashed) both turbulent cascade and mode conversion play a significant role in the decrease of Alfvénic Poynting flux. When the longitudinal perturbation with $\langle\delta v_\parallel\rangle = 0.6$ km s^{-1} is input from the photosphere (green solid), the relative contribution of the mode

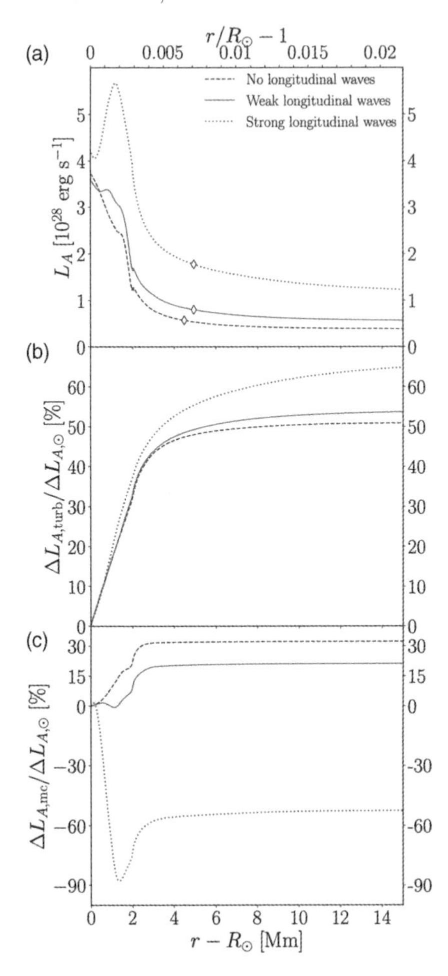

Figure 5. Comparison of the Alfvénic Poynting flux (top) and the relative contributions to the energy loss by the turbulent cascade (middle) and the mode conversion (bottom) of the three cases with different $\langle \delta v_\parallel \rangle$, where the linetypes are the same as in Figure 2.

conversion is reduced and the turbulent cascade is more important in the dissipation of Alfvén waves. The mode conversion from transverse to longitudinal waves is suppressed because acoustic waves are pre-existing in the chromosphere by the longitudinal injection from the photosphere. The tendency of the suppressed mode conversion is quite drastic in the case with large $\langle \delta v_\parallel \rangle = 1.8$ km s^{-1} (red dotted), leading to a negative value of $\Delta L_{\mathrm{A,mc}}$, which indicates that the conversion from longitudinal waves to transverse waves is occurring in the chromosphere (Schunker & Cally 2006; Cally & Goossens 2008). As a result, the Alfvénic Poynting flux increases there as shown in the top panel. Consequently, the Alfvén Poynting flux that reaches the corona is also larger for larger longitudinal fluctuation at the photosphere, giving larger mass-loss rate as discussed in Figure 4.

Acknowledgment

Numerical simulations in this work were partly carried out on Cray XC50 at Center for Computational Astrophysics, National Astronomical Observatory of Japan. M.S. is supported by a Grant-in-Aid for Japan Society for the Promotion of Science (JSPS) Fellows and by the NINS program for cross-disciplinary study (grant Nos. 01321802 and

01311904) on Turbulence, Transport, and Heating Dynamics in Laboratory and Solar/ Astrophysical Plasmas: "SoLaBo-X." T.K.S. is supported in part by Grants-in-Aid for Scientific Research from the MEXT/JSPS of Japan, 17H01105, 21H00033, and 22H01263 and by Program for Promoting Research on the Supercomputer Fugaku by the RIKEN Center for Computational Science (Toward a unified view of the universe: from large-scale structures to planets, grant 20351188-PI J. Makino) from the MEXT of Japan.

References

Belcher, J. W. 1971, ApJ, 168, 509

Cally, P. S. & Goossens, M. 2008, Sol.Phys., 251, 251

Cally, P. S. & Hansen, S. C. 2011, ApJ, 738, 119

Cranmer, S. R. & Saar, S. H. 2011, ApJ, 741, 54

Cranmer, S. R., van Ballegooijen, A. A., & Edgar, R. J. 2007, ApJ Suppl., 171, 520

Fludra, A., Del Zanna, G., & Bromage, B. J. I. 1999, SSRv, 87, 185

Goldreich, P. & Sridhar, S. 1995, ApJ, 438, 763

Lighthill, M. J. 1952, Proceedings of the Royal Society of London Series A, 211, 564

Matsumoto, T. 2021, MNRAS, 500, 4779

Matsumoto, T. & Suzuki, T. K. 2012, ApJ, 749, 8

Matthaeus, W. H., Zank, G. P., Oughton, S., Mullan, D. J., & Dmitruk, P. 1999, ApJ Letters, 523, L93

Morton, R. J., Weberg, M. J., & McLaughlin, J. A. 2019, Nature Astronomy, 3, 223

Saito, K., Makita, M., Nishi, K., & Hata, S. 1970, Annals of the Tokyo Astronomical Observatory, 12, 51

Sakaue, T. & Shibata, K. 2021, ApJ Letters, 906, L13

Schunker, H. & Cally, P. S. 2006, MNRAS, 372, 551

Shimizu, K., Shoda, M., & Suzuki, T. K. 2022, ApJ, 931, 37

Shoda, M., Suzuki, T. K., Asgari-Targhi, M., & Yokoyama, T. 2019, ApJ Letters, 880, L2

Shoda, M., Yokoyama, T., & Suzuki, T. K. 2018, ApJ, 853, 190

Stein, R. F. & Schwartz, R. A. 1972, ApJ, 177, 807

Stepien, K. 1988, ApJ, 335, 892

Suzuki, T. K. & Inutsuka, S.-i. 2005, ApJ Letters, 632, L49

Suzuki, T. K. & Inutsuka, S.-I. 2006, Journal of Geophysical Research (Space Physics), 111, A06101

Teriaca, L., Poletto, G., Romoli, M., & Biesecker, D. A. 2003, ApJ, 588, 566

Verdini, A. & Velli, M. 2007, ApJ, 662, 669

Vidotto, A. A. 2021, Living Reviews in Solar Physics, 18, 3

Wilhelm, K., Marsch, E., Dwivedi, B. N., et al. 1998, ApJ, 500, 1023

Zangrilli, L., Poletto, G., Nicolosi, P., Noci, G., & Romoli, M. 2002, ApJ, 574, 477

Winds of Stars and Exoplanets
Proceedings IAU Symposium No. 370, 2023
A. A. Vidotto, L. Fossati & J. S. Vink, eds.
doi:10.1017/S1743921322004537

ISOSCELES: Grid of stellar atmosphere and hydrodynamic models of massive stars. The first results

Ignacio Araya[1], Michel Curé[2], Natalia Machuca[2] and Catalina Arcos[2]

[1]Vicerrectoría de Investigación, Universidad Mayor, Chile

[2]Instituto de Física y Astronomía, Universidad de Valparaíso, Chile

Abstract. In this work we seek to derive simultaneously the stellar and wind parameters of massive stars, mainly A and B type supergiant stars. Our stellar properties encompass the effective temperature, the surface gravity, the micro-turbulence velocity and, silicon abundance. For wind properties we consider the line–force parameters (α, k and δ) obtained from the standard line-driven wind theory. To model the data we use the radiative transport code FASTWIND considering the hydrodynamic solutions derived with the stationary code HYDWIND. Then, ISOSCELES, a grid of stellar atmosphere and hydrodynamic models of massive stars is created. Together with the observed spectra and a semi-automatic tool the physical properties from these stars are determined through spectral line fittings. This quantitative spectroscopic analysis provide an estimation about the line–force parameters. In addition, we confirm that the hydrodynamic solutions, called δ-slow solutions, describe quite reliable the radiation line-driven winds of B supergiant stars.

Keywords. stars: early-type, stars: mass loss, stars: winds, outflows, hydrodynamics, techniques: spectroscopic

---

1. Introduction

ISOSCELES, GrId of Stellar AtmOSphere and HydrodynamiC ModELs for MassivE Stars, is the first grid of synthetic data for massive stars that involves both, the m-CAK hydrodynamics (instead of the generally used β-law) and the NLTE radiative transport.

ISOSCELES covers the complete parameter space of O-, B- and A-type stars. To produce the grid of synthetic line profiles, we first computed a grid of hydrodynamic wind solutions with our stationary code HYDWIND (Curé 2004). These hydrodynamic wind solutions, based on the CAK theory and its improvements (Castor et al. 1975; Friend & Abbott 1986; Curé 2004; Curé et al. 2011) are used as input in the NLTE radiative transport code FASTWIND (Puls et al. 2005).

In addition, these calculations consider the hydrodynamic solution δ-slow. This type of solution is obtained when the value of the ionization-related parameter δ takes higher values ($\delta \gtrsim 0.3$) than the ones provided by the standard m-CAK solution (Curé et al. 2011). Although these high values of δ are not calculated self-consistently (Gormaz-Matamala et al. 2019), values around 0.3 are similar to the value ($\delta = 1/3$) obtained by Puls et al. (2000) for a wind with neutral hydrogen as a trace element.

Each HYDWIND model is described by six parameters: $T_{\rm eff}$, logg, R_*, α, k, and δ. All these models consider, for the optical depth, the boundary condition $\tau = 2/3$, at the stellar surface. Figure 1 shows the location of the different synthetic models in the

Table 1. Range of values considered in HYDWIND and FASTWIND grids.

$T_{\rm eff}$:	9 000 to 45 000 [K]
log g:	0.6 to 4.5 [dex]
α:	0.45, 0.47, 0.51, 0.53, 0.55, 0.57, 0.61, 0.65
k:	0.05 to 0.60 (step size 0.05)
δ:	0.00, 0.04, 0.10, 0.14, 0.2,0.24, 0.3, 0.31, 0.32,0.33, 0.34, 0.35
$\log \epsilon_{\rm Si}$:	7.21, 7.36, 7.51, 7.66, 7.81
$v_{\rm micro}$:	1.0, 5.0, 10.0, 15.0, 20.0, 25.0 [km/s]

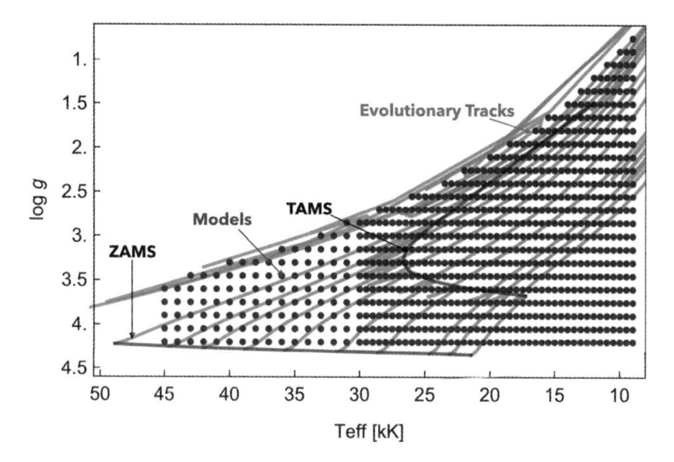

Figure 1. Location of ($T_{\rm eff}$ – log g) pairs (dots) considered in the grid of models. The red lines represent the evolutionary tracks from 7 $M_\odot$ to 60 $M_\odot$ without rotation (Ekström et al. 2008), while the black lines correspond to the zero age main-sequence (ZAMS) and the terminal age main-sequence (TAMS).

$T_{\rm eff}$ – logg plane. The range for the values considered in HYDWIND and FASTWIND grids are listed in Table 1. It is worth noting that not all combinations of these parameters converge to a physical stationary hydrodynamic solution. In addition, the values of δ are necessary to obtain both fast and δ-slow solutions.

In the case of the FASTWIND grid, we calculated a total of 573 433 models. From these models, line profiles of H, He, and Si elements were calculated in the optical and infrared range. Also, it is important to notice, that all our models are calculated without stellar rotation.

2. Method

We develop a methodology (script in Python) to carry out a semi-automatic analysis of an observed spectrum. The observational data was pre-processed using the *IACOB-broad* tool (Simón-Díaz & Herrero 2014) to derive the projected rotational and macroturbulent speeds. Then, these values were used in our search code to perform the spectral fitting with the purpose to obtain stellar and wind parameters. Figure 2 shows the structure of the search code.

First, the code reads an input file that contains the information of the observational data, the convolution parameters and (optional) the type of solution we are searching for, fast or δ-slow. In a second step, it uses multiprocessing tools to search through the grid. The line profiles are rotationally convolved and interpolated with the observed line. Then a χ^2 test is performed. Finally, the code collect all these results and sort them from lower to higher χ^2 values, selecting and returning the one that most resembles to the observational data.

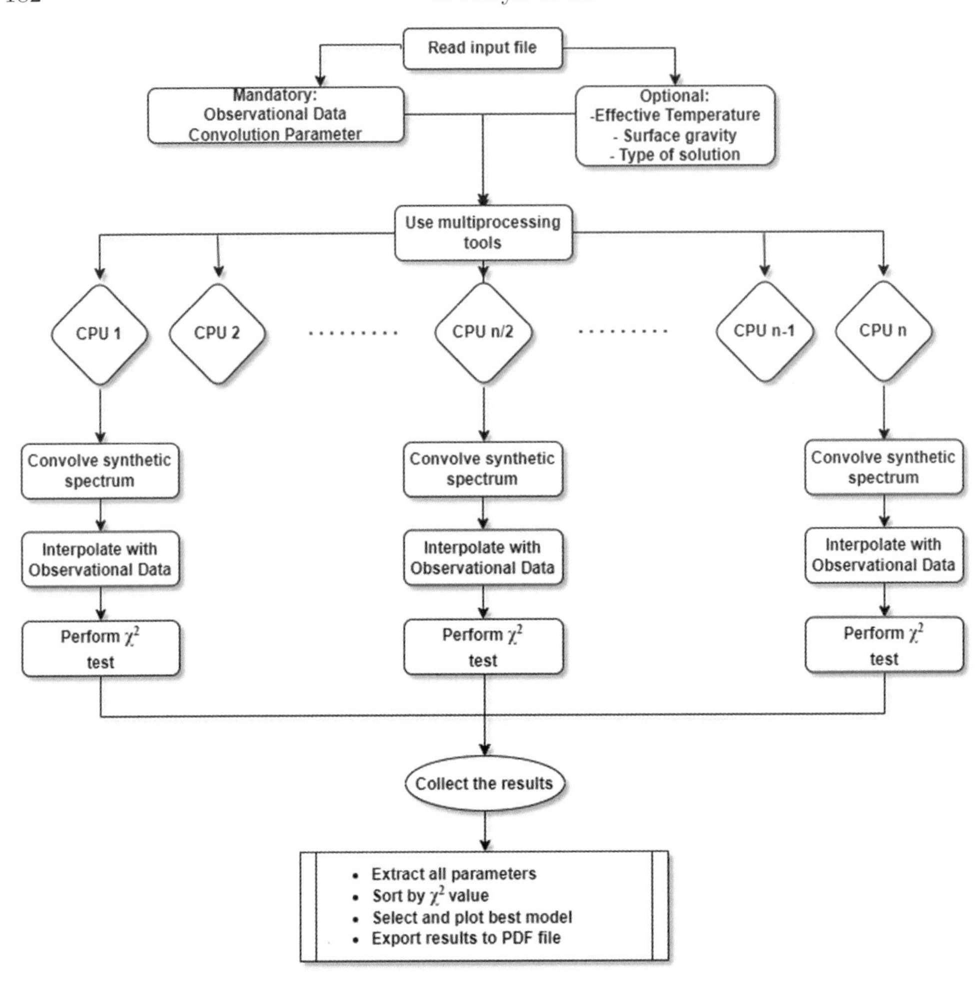

Figure 2. Diagram with our methodology to derive the stellar and wind parameters from observations.

3. First Results

By way of example, we present the results for the star HD 99953 considering six line profiles (see Figure 3). We derived stellar parameters similar to the ones obtained by Haucke et al. (2018). Regarding to the wind parameters, they found $\dot{M} = 0.13 \times 10^{-6}\, M_{\odot}/\mathrm{year}$ and $v_\infty = 500\,\mathrm{km/s}$, with $\beta = 2.0$, for the velocity profile. We obtained line force parameters that corresponds to a δ-slow solution, with the following wind parameters: $\dot{M} = 0.24 \times 10^{-6}\, M_{\odot}/\mathrm{year}$ and $v_\infty = 254\,\mathrm{km/s}$.

From Figure 4, we can observe the difference between the velocity profile (β-law) used by Haucke et al. (2018) and our hydrodynamic profile (δ-slow solution).

4. Conclusions

According to our experience in spectral line fittings, we predict that the models with values of $\beta \gtrsim 1.5$ can be properly reproduced with the hydrodynamic solution δ-slow. Currently, in our methodology, all the lines have the same weight, but we are testing to assign different weight to those lines that could have more impact on stellar parameters (e.g., silicon transitions in the effective temperature).

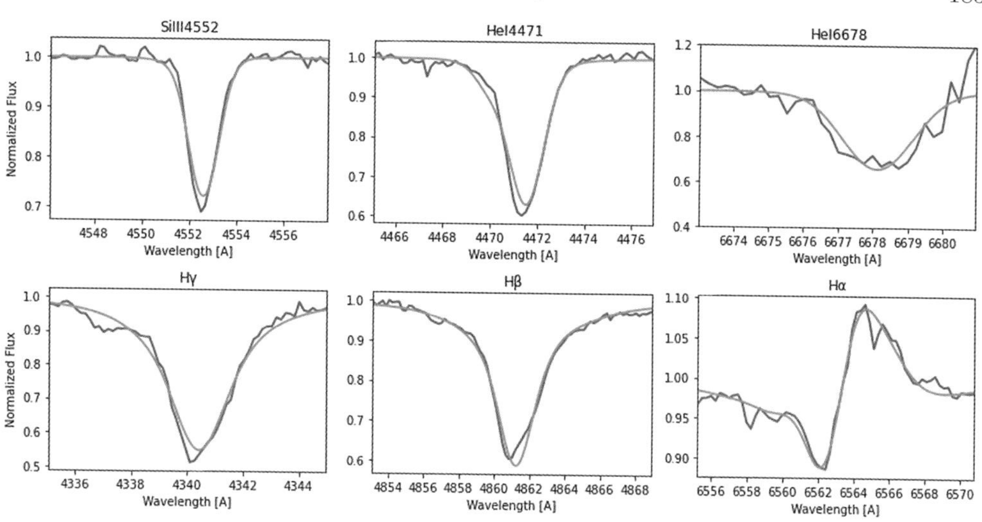

Figure 3. Spectral line fitting for HD 99953. Blue solid line shows the spectra retrieved from Haucke et al. (2018). Orange solid line corresponds to a model with the following parameters: $T_{\rm eff} = 18500$ K, $\log g = 2.4$, $\alpha = 0.45$, $k = 0.15$, $\delta = 0.34$, $v_{\rm micro} = 15$ km/s and $\log \epsilon_{\rm Si} = 7.81$.

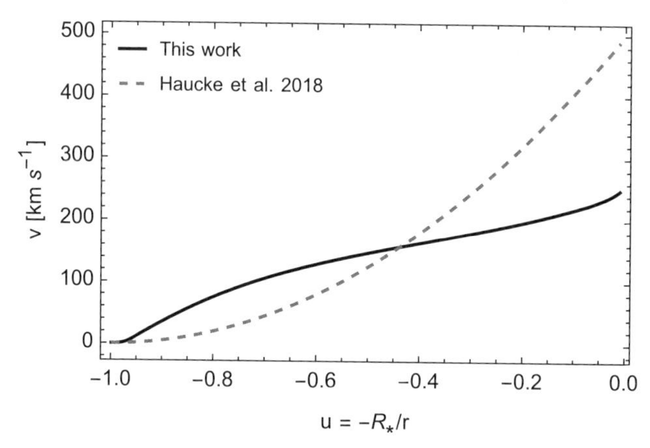

Figure 4. Comparison between the velocities profiles used to fit the spectral lines.

Further studies with more stars and analysis of more spectral lines will give us more precise estimates of the stellar and wind parameters, leading us to calibrate the WLR accurately.

As future work, we expect to analyze the χ^2 distributions and then use it to compute the uncertainties for each derived parameter, also, we are working in a statistical study of the theoretical line profiles shapes in contrast with stellar and wind parameters from ISOSCELES in order to study the influence of the wind in the observed line profiles.

Acknowledgements

M.C., C.A., & I.A. are grateful with the support from FONDECYT project 1190485. I.A. & C.A. also thanks the support from FONDECYT projects 11190147 and 11190945, respectively. This project receive funding from the European Union's Framework Programme for Research and Innovation Horizon 2020 (2014–2020) under the Marie Skłodowska-Curie grant Agreement No. 823734.

References

Castor, J. I., Abbott, D. C., & Klein, R. I. 1975, ApJ, 195, 157
Curé, M. 2004, ApJ, 614, 929
Curé, M., Cidale, L., & Granada, A. 2011, ApJ, 737, 18
Ekström et al., 2008, A&A, 489, 685
Friend, D. B., & Abbott, D. C. 1986, ApJ, 311, 701
Gormaz-Matamala, A. C., Curé, M., Cidale, L. S., & Venero, R. O. J. 2019, ApJ, 873, 131
Haucke, M., Cidale, L. S., Venero, R. O. J., Curé, M., Kraus, M., Kanaan, S., & Arcos, C. 2018, A&A, 614, A91
Puls, J., Springmann, U., & Lennon, M. 2000, A&AS, 141, 23
Puls, J., Urbaneja, M. A., Venero, R., Repolust, T., Springmann, U., Jokuthy, A., & Mokiem, M. R. 2005, A&A, 435, 669
Simón-Díaz, S., & Herrero, A. 2014, A&A, 562, A135

Winds of Stars and Exoplanets
Proceedings IAU Symposium No. 370, 2023
A. A. Vidotto, L. Fossati & J. S. Vink, eds.
doi:10.1017/S174392132200357X

Quantification of the environment of cool stars using numerical simulations

J. J. Chebly[1,2], Julián D. Alvarado-Gómez[1] and Katja Poppenhaeger[1,2]

[1]Leibniz Institute for Astrophysics, An der Sternwarte 16, 14482, Potsdam, Germany

[2]Institute of Physics and Astronomy, University of Potsdam, Potsdam-Golm, 14476, Germany

Abstract. Stars interact with their planets through gravitation, radiation, and magnetic fields. Although magnetic activity decreases with time, reducing associated high-energy (e.g., coronal XUV emission, flares), stellar winds persist throughout the entire evolution of the system. Their cumulative effect will be dominant for both the star and for possible orbiting exoplanets, affecting in this way the expected habitability conditions. However, observations of stellar winds in low-mass main sequence stars are limited, which motivates the usage of models as a pathway to explore how these winds look like and how they behave. Here we present the results from a grid of 3D state-of-the-art stellar wind models for cool stars (spectral types F to M). We explore the role played by the different stellar properties (mass, radius, rotation, magnetic field) on the characteristics of the resulting magnetized winds (mass and angular momentum losses, terminal speeds, wind topology) and isolate the most important dependencies between the parameters involved. These results will be used to establish scaling laws that will complement the lack of stellar wind observational constraints.

Keywords. Stars: low-mass, stars: magnetic fields, stars: mass loss, stars: winds, outflows, methods: numerical, magnetic fields (magnetohydrodynamics:) MHD

---

1. Introduction

The winds from cool, low-mass main sequence stars are weak and cannot be observed directly, except in the case of our own Sun. This means that the nature and behavior of these winds are not well understood. More knowledge about stellar winds is needed because they carry a significant amount of angular momentum that affects the rotational evolution of the star (Johnstone et al. 2015). These magnetized winds also affect the atmospheric evolution of planetary atmospheres (Kislyakova et al. 2014).

Several techniques have been proposed to infer the presence of stellar winds in cool stars. Examples are the free emission at radio wavelengths (Gaidos et al. 2000) and searches for X-ray emission induced by charge exchange (Wargelin & Drake 2002) between ionised stellar winds and neutral interstellar hydrogen. Unfortunately, both methods resulted in non-detections, but provided important upper limits on the wind mass-loss rates of a handful of Sun-like stars. One of the indirect techniques is the Ly-α absorption method, which uses high-resolution UV spectra from the Hubble Space Telescope (HST). As winds propagate from the star, they collide with the Inter Stellar Medium (ISM), forming 'astrospheres' analogous to the 'heliosphere' surrounding the Sun (Wood 2004). Signs of charge exchange occur when the ISM is neutral or at least partially neutral, leading to astrospheric absorption signature in the Ly-α line. This method is the most successful for measuring the mass loss rate ($\dot{M}$) of winds from solar-like stars, with nearly 20 measurements to date (Wood et al. 2021). The new measurements showed, that the

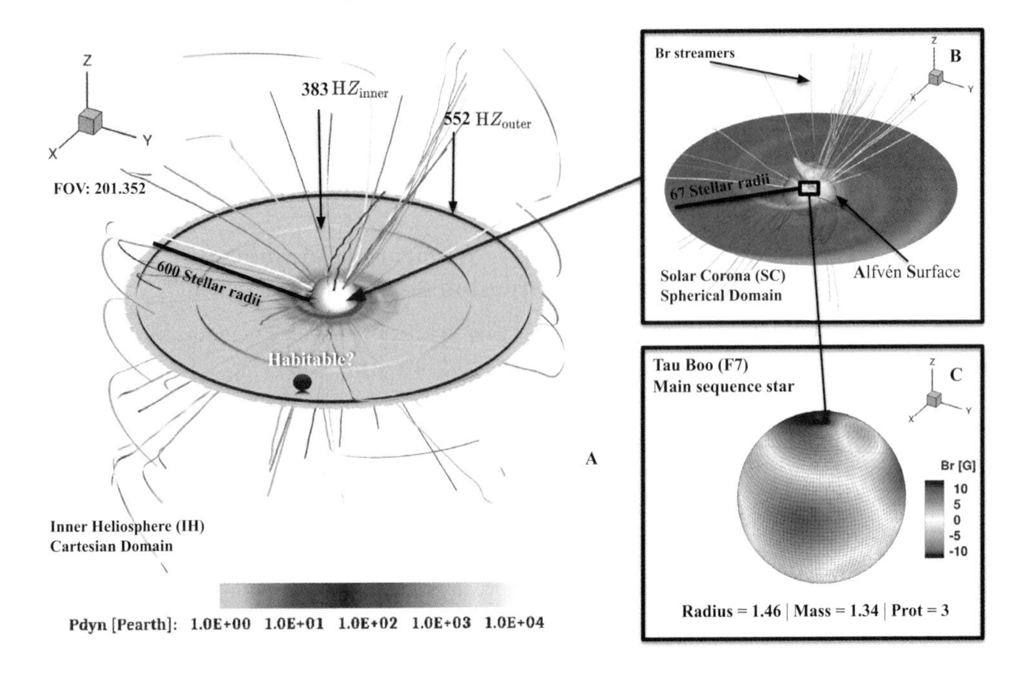

Figure 1. Illustration of a 3D numerical magnetohydrodynamics simulation using the Alfvén wave-solar model (AWSoM). Plots A and B show a steady-state solution of the stellar wind simulation of τ Boo (F7) and some stellar wind properties such as the dynamic pressure and the Alfvén surface ($S_{\rm A}$). In C, we show the magnetogram reconstructed from observed Zeeman Doppler Imaging maps used as the inner boundary of Solar Corona domain. The Inner Heliosphere (IH) is shown in plot A along with the limits of the habitable zone around τ Boo.

coronal activity and spectral type are not enough to determine wind properties. However, this method has not provided measurements for larger numbers of stars because it is only applicable to stars within 10-15 pc. Beyond this distance, ISM becomes fully ionized. Also, Ly-α lines observed from cool stars are always heavily contaminated by interstellar absorption.

The lack of observational data and associated limitations motivate the use of numerical simulations as a way to improve our understanding of stellar winds. Models based on the Alfvén waves are more popular when simulating the stellar wind from stars other than the Sun (Suzuki 2006). These waves are considered to be a likely key mechanism for solar wind heating and acceleration (Van der Holst et al. 2014).

In this study, we use one of the most realistic physics-based solar models to quantify the environment of low-mass main-sequence stars with an outer convective envelope, i.e., late-F to M type. Our goal is to provide a method for estimating stellar wind quantities as other related parameters become available. This will also help to further constrain the habitable zone (HZs) of exoplanets around these types of stars.

2. Stellar Wind Model and Selected Sample

We use the Alfvén Wave Solar Model (AWSoM) to calculate different steady-state solutions for a grid of models. This model reproduces successfully the solar wind in a realistic manner (Oran et al. 2013, Sachdeva et al. 2019). Assuming that stellar winds from solar analogues are driven by the same process as the solar wind, we extend the model to stars other than the sun. AWSoM is part of the Space Weather Modelling

Framework (SWMF), which integrates numerical models from the solar corona to the upper atmosphere into a high-performance coupled model (Tóth et al. 2012).

In our simulation, we couple between two modules: the solar corona (SC), a spherical domain, and the inner heliosphere (IH), a Cartesian domain, as shown in Fig. 1, in the case of F, G, and K stars. Using the IH module was necessary for these stars in order to accommodate their large HZs. For M-dwarfs, the coupling was not necessary since the HZs is closer in. The SC domain in our simulations ranges from 1.05 $R_\star$ up to 67 $R_\star$ for F, G, and K stars. For the M-dwarfs we require a sufficiently large SC domain (250 $R_\star$) so that the resulting $S_{\rm A}$ would be completely contained within it. Once the wind solution in SC reaches a steady state (converges), SC provides the plasma variables at the inner boundary of IH. Domain-overlap (from 62 $R_\star$ to 67 $R_\star$) is used in the coupling procedure between the two domains for F, G, and K type stars. The Inner Heliosphere component uses an ideal MHD approach and covers the range from 62 $R_\star$ up to 600 $\rm R_\star$.

For our stellar wind models, the inner boundary conditions of AWSoM within the SC component (e.g. plasma density, temperature, Alfvén wave pointing flux, and the Alfvén wave correlation length), are kept identical to the values used to simulate the solar wind (Van der Holst et al. 2014). In contrast, the magnetic field and stellar properties are modified for each star. The strength and geometry of the magnetic field is taken from from observed, reconstructed Zeeman Doppler Imaging (ZDI) maps. This is a tomographic imaging technique (Donati & Brown 1997; Kochukhov & Piskunov 2002) that allows us to reconstruct the large-scale magnetic field (intensity and orientation) at the surface of the star from a series of circular polarization spectra (Donati et al. 2006). Therefore, the simulations are more realistic compared to models that assume simplified/idealized field geometries. We only retrieve the radial component of the magnetic field strength from the ZDI maps since solar wind models neglect the contribution from the other components.

We have also restricted the parameter space to stars with an outer convective envelope. The sample considered in this study consists of 21 stars (3F, 4G, 6K, 5M) located within 10 pc (taken from See et al. 2019). Each star has its corresponding stellar rotation period ($P_{\rm rot}$), mass ($m_\star$), and radius ($r_\star$). The fastest star in our data set corresponds to an M6 star with $P_{\rm rot}$ = 0.71 d, and the slowest is a K3 star with a $P_{\rm rot}$ = 42.2 d. As for the magnetic field strength, it ranges from 5G to 2kG.

3. Numerical estimate of stellar wind properties

By definition, the Alfvén surface ($S_{\rm A}$) corresponds to the boundary between magnetically coupled outflows (MA < 1) that do not carry angular momentum away from the star and the escaping stellar wind (MA > 1). In numerical models, the $S_{\rm A}$ is used to extract the mass loss rate ($\dot{\rm M}$), and the angular momentum loss rate ($\dot{\rm J}$) (e.g. Cohen et al. 2010; Garraffo et al. 2015a, Alvarado-Gómez et al. 2016).

The angular momentum loss rate is obtained by integrating over the $S_{\rm A}$. Knowing the wind speed and density, the $\dot{\rm M}$ of a star can be calculated by integrating over a closed area enclosing the star beyond the $S_{\rm A}$. Note that the Alfvén surface does not have a regular shape when visualized in 3D. This is due to several stellar parameters, such as the magnetic field distribution on the stellar surface, the wind velocity, and the density.

$$(\dot{\rm M}) = \rho(\mathbf{u} \cdot {\rm dA}) \tag{3.1}$$

$$(\dot{\rm J}) = \Omega\rho{\rm R}^2 \sin^2\theta(\mathbf{u} \cdot {\rm dA}) \tag{3.2}$$

Here ρ represents the wind density, $\mathbf{u}$ is the wind speed vector and dA is the vector surface element. Ω is the angular frequency of the star and θ is the angle between the lever arm and the rotation axis. Ω changes with the different stellar rotation $\Omega = 2\pi/{\rm P_{rot}}$.

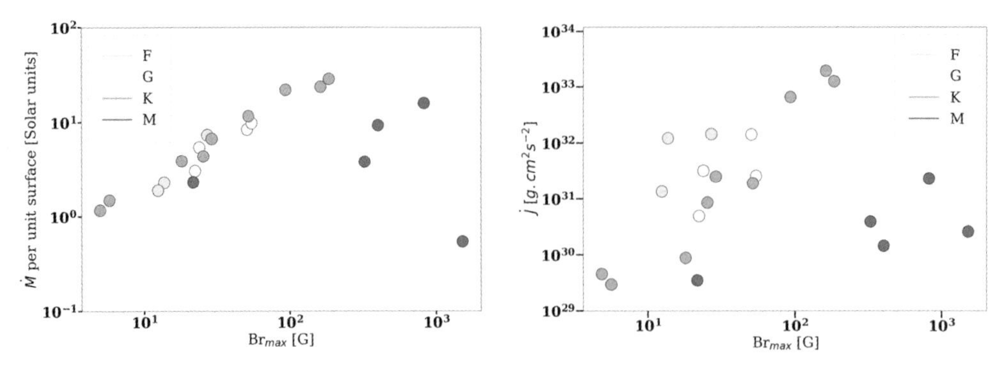

Figure 2. Numerical estimate of $\dot{\mathrm{M}}$ and $\dot{\mathrm{J}}$ in function of the radial magnetic field (B_{R}). The colors correspond to the different spectral types in our sample. The results shows that the loss rates are strongly dependent on the B_{R}.

3.1. $\dot{\mathrm{M}}$ *and* $\dot{\mathrm{J}}$ *as a function of* B_{R}

In Fig. 2, we plot the the $\dot{\mathrm{M}}$ and $\dot{\mathrm{J}}$, as estimated using the method described previously, against B_{Rmax} for our sample of stars. As expected, a stronger B_{R} produces stronger winds. This means either faster or denser winds. This interplay dictates $\dot{\mathrm{M}}$ and $\dot{\mathrm{J}}$ (cf. 3.1, 3.2), which will increase with increasing B strength. We also note that we observe a strong dependence of $\dot{\mathrm{M}}$ and $\dot{\mathrm{J}}$ in stellar winds as a function of B_{Rmax}, regardless of spectral type.

In addition, we see on the left plot in figure 2 the presence of an outlier in M-dwarfs for the strongest field $B_{\mathrm{Rmax}} = 1.5$ kG. One thing that might be causing this, is the way the simulation is being performed. We already pointed out that we had to push further out the SC boundary condition for the M-dwarfs. By doing this we reached a low grid resolution near the outer boundary which might have cause the $\dot{\mathrm{M}}$ to have a small value. Further testing will be performed to identify the possible cause of this outlier.

3.2. $\dot{\mathrm{M}}$ *as function of Rossby number*

The evolution of the magnetic field complexity was proved to be an important asset for explaining the bimodal distribution seen in the spin-down of stars that follows the Skumanich rotation evolution in young Open Clusters (OCs) (Garraffo et al. 2018). This complexity is a function of the Rossby number (R_{o}) which is the ratio between the P_{rot} and the convective turnover time. The Rossby number is considered an important parameter for studying the effects of rotation on the dynamo action in stars with convective outer layers, being related to the dynamo number $D = R_{\mathrm{o}}^{-2}$ (Charbonnneau 2020).

We used the Rossby number as a magnetic field geometry parameter and explored how it changes with the $\dot{\mathrm{M}}$ loss rate of each star in our sample. We adapted empirical convective turnover times defined in Cranmer & Saar (2011), which depends on the effective temperature (Eq. 3.3).

$$\tau_c = 314.24 \exp\left(\frac{-T_{\mathrm{eff}}}{1952.5K} - \left(\frac{T_{\mathrm{eff}}}{6250K}\right)^{18}\right) + 0.002 \tag{3.3}$$

Our $\dot{\mathrm{M}}$ versus R_{o} (Fig. 3) shows a good representation of the trend for the normalized X-ray luminosity as a function of R_{o} in Reiners (2012) and Wright et al. (2018). The magnetic activity increases with decreasing R_{o} ($R_{\mathrm{o}} \leq 0.1$). At fast rotation ($R_{\mathrm{o}} = 0.1$), the dynamo reaches a saturation level that cannot be exceeded even if the star rotates much faster. Moreover, the simulated $\dot{\mathrm{M}}$ appears to have a maximum for stars with R_{o} < 0.1. This raises the question of whether stellar winds also reach a saturated regime.

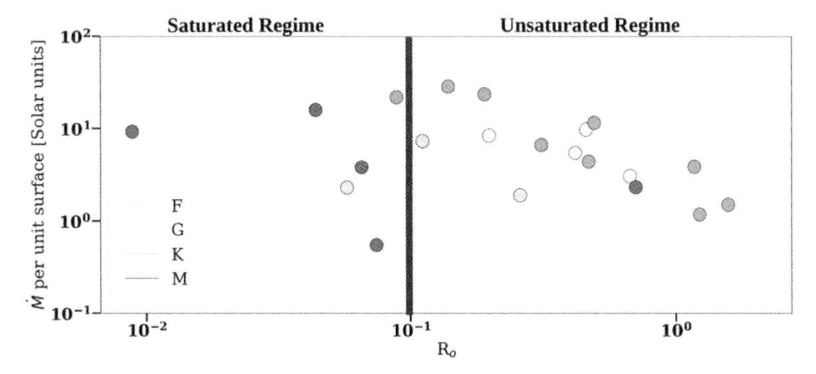

Figure 3. 3D MHD Ṁ estimation in function of the Rossby number (R_o). The colors correspond to the different spectral types in our sample and Solar value is represented by a star symbol. This plot highlights the importance of the field geometry in the magnetic field activity on the star surface.

We also point out that stars with different R_o can still have similar Ṁ. This is expected since the field complexity changes with rotation rate (Vidotto et al. 2014a; Garraffo et al. 2015b; Réville et al. 2015), which in turn influence the amount of mass loss.

4. Discussion and Conclusion

Understanding the behaviour and structure of cool stars' stellar winds and the habitability of its orbiting exoplanets requires detailed information on the strength and topology of the surface magnetic field. Analysis of high-resolution optical and NIR stellar intensity and polarisation spectra is currently the only approach that allows such information to be obtained in a systematic and direct manner. The measurement of quantities on the stellar surface is much simpler and does not suffer from problems with the ISM, which will always be a problem for the astrospheric and for the charge-exchange technique.

Moreover, the number of cool stars with ZDI observations is growing ($\sim$ 200, Marsden et al. 2014), so in principle estimates/models are possible for a larger number of stars that would allow to extract statistical properties about their winds. Quantifying these winds is becoming more crucial as the number of discovered exoplanets is increasing (currently 5178 confirmed exoplanets) and we enter a new era of detailed atmospheric characterization with JWST and E-ELTs. On the other hand, we are reaching the maximum number of systems that can be probed with the astrospheric technique due to the limitations mentioned earlier.

Three-dimensional MHD models that use ZDI maps as inner boundary conditions for the magnetic field strength and geometry, will help us obtain a more realistic estimate of the stellar wind parameters (e.g. dynamic pressure, terminal velocity, sub/super-Alfvénic conditions). Hence, we will be filling the missing observations and making progress in constraining the habitable conditions around low-mass main sequence stars.

References

Alvarado-Gómez, J. D., Hussain, G. A. J., Cohen, O., et al. 2016, A&A, 594, A95
Charbonneau, P. 2020, Living Reviews in Solar Physics, 17, 4
Cohen, O., Drake, J. J., Kashyap, V. L., Sokolov, I. V., & Gombosi, T. I. 2010, ApJ Letters, 723, L64
Cranmer, S. R., & Saar, S. H. 2011, ApJ, 741, 54
Donati, J. F., & Brown, S. F. 1997, A&A, 326, 1135
Donati, J. F., Howarth, I. D., Jardine, M. M., et al. 2006, MNRAS, 370, 629

Gaidos, E. J., Güdel, M., & Blake, G. A. 2000, GeoRL, 27, 501
Garraffo, C., Drake, J. J., & Cohen, O. 2015a, ApJ, 813, 40
—. 2015b, ApJ Letters, 807, L6
Garraffo, C., Drake, J. J., Dotter, A., et al. 2018, ApJ, 862, 90
Johnstone, C. P., Güdel, M., Brott, I., & Lüftinger, T. 2015, A&A, 577, A28
Kislyakova, K. G., Holmström, M., Lammer, H., Odert, P., & Khodachenko, M. L. 2014, Science, 346, 981
Kochukhov, O., & Piskunov, N. 2002, A&A, 388, 868
Marsden, S., Petit, P., Jeffers, S., et al. 2014, in Magnetic Fields throughout Stellar Evolution, ed. P. Petit, M. Jardine, & H. C. Spruit, Vol. 302, 138–141
Oran, R., van der Holst, B., Landi, E., et al. 2013, ApJ, 778, 176
Reiners, A. 2012, Living Reviews in Solar Physics, 9, 1
Réville, V., Brun, A. S., Strugarek, A., et al. 2015, ApJ, 814, 99
Sachdeva, N., van der Holst, B., Manchester, W. B., et al. 2019, ApJ, 887, 83
See, V., Matt, S. P., Folsom, C. P., et al. 2019, ApJ, 876, 118
Suzuki, T. K. 2006, ApJ Letters, 640, L75
Tóth, G., van der Holst, B., Sokolov, I. V., et al. 2012, Journal of Computational Physics, 231, 870
Van der Holst, B., Sokolov, I. V., Meng, X., et al. 2014, ApJ, 782, 81
Vidotto, A. A., Gregory, S. G., Jardine, M., et al. 2014, MNRAS, 441, 2361
Wargelin, B. J., & Drake, J. J. 2002, ApJ, 578, 503
Wood, B. E. 2004, Living Reviews in Solar Physics, 1, 2
Wood, B. E., Müller, H.-R., Redfield, S., et al. 2021, ApJ, 915, 37
Wright, N. J., Newton, E. R., Williams, P. K. G., Drake, J. J., & Yadav, R. K. 2018, MNRAS, 479, 2351

Winds of Stars and Exoplanets
Proceedings IAU Symposium No. 370, 2023
A. A. Vidotto, L. Fossati & J. S. Vink, eds.
doi:10.1017/S174392132200446X

Hydrodynamic solutions of radiation driven wind from hot stars

M. Curé[1], I. Araya[2], C. Arcos[1], N. Machuca[1] and A. Rodriguez[1]

[1]Instituto de Física y Astronomía, Universidad de Valparaíso, Chile

[2]Vicerrectoría de Investigación, Universidad Mayor, Chile

Abstract. We show the application of the δ- and Ω-slow hydrodynamical solutions to describe the velocity profiles of massive stars. In particular, these solutions can help to unravel some of the problems within the winds of massive stars such as the approximation of the β-law for the velocity profile of B supergiant stars and the slow outflow wind observed in Be stars.

Keywords. stars: early-type, stars: winds, outflows, stars: mass loss, stars: atmospheres, hydrodynamics

1. Context

The standard or *fast* solution from Castor et al. (1975) and later improved by Friend & Abbott (1986, hereafter m-CAK) reproduce quiet well the observed winds of massive stars, however some B-types on the main sequence and on the supergiant phases usually present a much slower velocity profile. In the following we present other two hydrodynamical solutions founded by our massive star group, which are useful to describe in particular the wind of B stars.

2. The δ-slow solution

The δ-slow regime (see Curé et al. 2011) is present when $\delta \gtrsim 0.28$ and is characterized by a slower terminal velocity and depending on the value of k, larger or smaller mass loss rate values are obtained compared to the values of the fast solution (Araya et al. 2021). By using the m-CAK hydrodynamic (instead of the generally used β-law) and the Non-LTE radiative transport code FASTWIND, we created a grid of models for fast and δ-slow regimes, setting the line-force parameters as free ones in a suitable range of values (`ISOSCELES` grid, Araya et al. 2022 in preparation) to study the wind of massive stars. The Fig. 1 shows the simultaneous fit of six spectral lines of the star HD99953.

3. The Ω-slow solution

For $\Omega = v_{rot}/v_{crit} \gtrsim 0.75$, the fast solution ceases to exist, and a much slower and denser solution is present, we called this solution Ω-slow (Curé 2004). Following de Araujo (1995), we implemented a mimicking viscosity parameter via

$$V_\phi = \Omega \sqrt{GM(1-\Gamma)/R}\,(R/r)^{\gamma_{vis}}, \tag{3.1}$$

being V_ϕ the rotational velocity component of the wind in the equatorial plane and γ_{vis} a viscosity mimicking parameter. When $\gamma_{vis} = 1$ angular momentum is conserved and $\gamma_{vis} = 0.5$, describes a Quasi-Keplerian outflowing wind. Fig. 2 shows the velocity profile for a Be star with $\gamma_{vis} = 0.52$, and stellar parameters: $T_{\text{eff}} = 21000K$, $\log g = 4.0$,

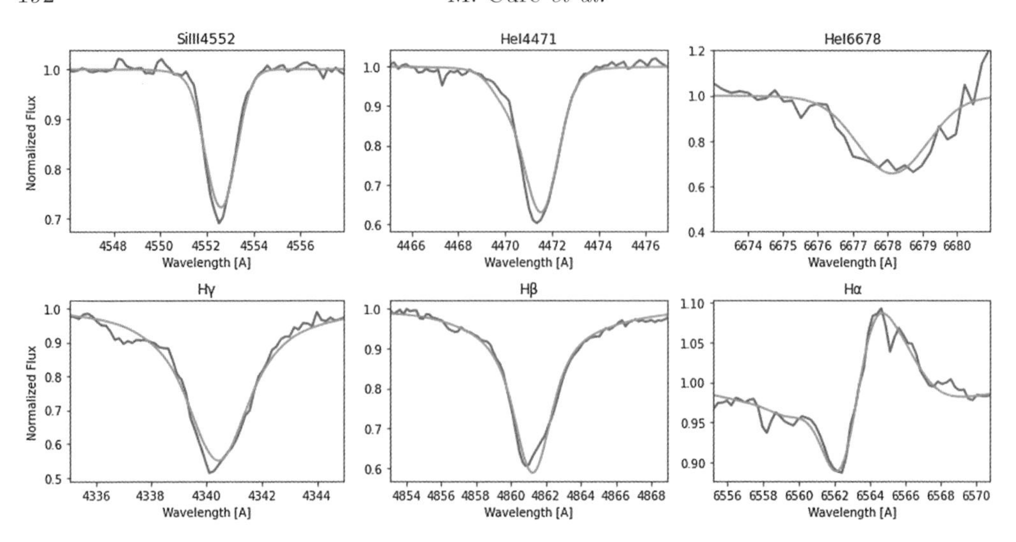

Figure 1. HD99953 (B2 Ia): Best fit for six stellar lines. Observed line profiles are in solid blue line and the synthetic best model in solid orange line. Stellar and wind parameters are the following: $T_{\rm eff} = 18500\,K$, $\log g = 2.4$, $R = 45.6\,R_{\odot}$, $V_{\infty} = 254\,$km/s and $\dot{M} = 2.43 \cdot 10^{-7} M_{\odot}/yr$.

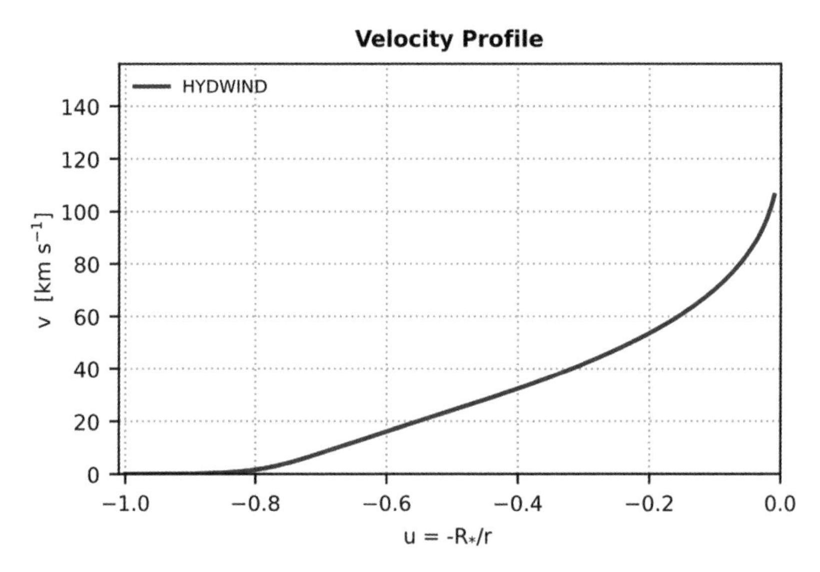

Figure 2. Velocity profile of a Ω-slow solution as a function of the inverse radial coordinate $u = -R/r$, with a γ_{vis}=0.52 which describe a Quasi-Keplerian outflowing decretion disk wind.

$R = 4.5\,R_{\odot}$ and $\Omega = 0.99$. The terminal velocity is 100 km/s, much lower than the case when angular momentum is conserved.

4. Conclusions

We have shown that both δ- and Ω-slow solutions can explain some of the current 'problems' that faces the massive star community. The δ-slow solution describes properly the winds of B stars and the Ω-slow solution with the inclusion of viscosity explains the viscous decretion disks of Classical Be stars.

Acknowledgments

MC, CA & IA are grateful to projects FONDECYT 1190485 and FAPESP NNº133541. CA thanks to FONDECYT 11190945. IA thank to FONDECYT N.11190147. This work has been possible thanks to the use of AWS-UChile-NLHPC credits. Powered@NLHPC: This research was partially supported by the supercomputing infrastructure of the NLHPC (ECM-02). This project has also received funding from the European Union's Framework Programme for Research and Innovation Horizon 2020 (2014–2020) under the Marie Skłodowska-Curie grant Agreement No 823734.

References

Araya, I., Christen, A., Curé, M., et al. 2021, MNRAS, 504, 2550
Castor, J. I., Abbott, D. C., & Klein, R. I. 1975, ApJ, 195, 157
Curé, M. 2004, ApJ, 614, 929
Curé, M., Cidale, L., & Granada, A. 2011, ApJ, 737, 18
de Araujo, F. X. 1995, A&A, 298, 179
Friend, D. B. & Abbott, D. C. 1986, ApJ, 311, 701

Winds of Stars and Exoplanets
Proceedings IAU Symposium No. 370, 2023
A. A. Vidotto, L. Fossati & J. S. Vink, eds.
doi:10.1017/S1743921322003544

Gap opening by planets in discs with magnetised winds

Vardan Elbakyan, Yinhao Wu, Sergei Nayakshin and Giovanni Rosotti

School of Physics and Astronomy, University of Leicester, Leicester LE1 7RH, UK

Abstract. Planets open deep gaps in protoplanetary discs when their mass exceeds a gap opening mass, $M_{\rm gap}$. We use one- and two-dimensional simulations to study planet gap opening in discs with angular momentum transport powered by MHD disc winds. We parameterise the efficiency of the MHD disc wind angular momentum transport through a dimensionless parameter $\alpha_{\rm dw}$, which is an analogue to the turbulent viscosity $\alpha_{\rm v}$. We find that magnetised winds are much less efficient in counteracting planet tidal torques than turbulence is. For discs with astrophysically realistic values of $\alpha_{\rm dw}$, $M_{\rm gap}$ is always determined by the residual disc turbulence, and is a factor of a few to ten smaller than usually obtained for viscous discs. We introduce a gap opening criterion applicable for any values of $\alpha_{\rm v}$ and $\alpha_{\rm dw}$ that may be useful for planet formation population synthesis.

Keywords. planetary systems: formation, planetary systems: protoplanetary disks, planets and satellites: formation

---

1. Introduction

Gap opening by planets embedded in protoplanetary discs has a significant impact on a wide range of phenomena in star and planet formation. The low levels of turbulence observed by ALMA are at odds with the stirred-up vertical distribution of micron sized dust in some discs, and they are also insufficient to account for the significant mass accretion rates onto the stars (cf. the review and references in Miotello et al. 2022). A number of authors have recently argued that a weak net flux of vertical magnetic field present in molecular cloud cores may enhance MHD-driven disc winds that remove angular momentum of gas at a faster rate than disc turbulence (e.g., Lesur 2021). The goal of our paper is to evaluate the gap opening planet mass in discs where mass and angular momentum transfer are dominated by magnetised winds rather than disc turbulence.

2. Analytical estimates

To derive the conditions for a gap opening in a disc with MHD disc winds, we shall assume for simplicity that $\alpha_{\rm v}$ is negligibly small. At $R > a$, where a is the radial distance of the planet, the angular momentum gain by the gas due to the disc-planet interaction results in an outward flow of gas at velocity

$$v_{\rm p} = \frac{2\Lambda_{\rm p}}{v_K}, \tag{2.1}$$

where $\Lambda_{\rm p}$ is the specific torque from the planet on the surrounding gas and can be approximated by (Lin & Papaloizou 1986)

$$\Lambda_{\rm p} = f \,{\rm sign}\,(R-a)\,\frac{q^2}{2}\frac{a^4}{\Delta R^4}v_K^2, \tag{2.2}$$

where v_K is the Keplerian velocity at the radial distance of the planet a, q is the planet-to-star mass ratio, $\Delta R = R - a$, and $0.1 \leq f < 1$ is a constant of order unity that depends on the detail of torque deposition and wave energy dissipation in the disc. In this paper we take $f = 0.15$.

This outward flow of gas, away from the planet, results in the creation of the gap. Due to 3D effects in realistic discs, the torque Λ_p from the planet saturates at distance $|\Delta R| \sim \kappa R_H$ away from the planet, where κ is a constant of order unity.

The MHD disc winds results in an inward flow of gas at the velocity (Tabone et al. 2022)

$$v_{dw} = -\frac{3}{2}\alpha_{dw}\left(\frac{H}{R}\right)^2 v_K \, . \tag{2.3}$$

where α_{dw} is a dimensionless parameter, defined by analogy with the α_v parameter. MHD winds will close a gap opened by the planet when $v_{dw} + v_p \leq 0$. Solving the equation $v_{dw} + v_p = 0$, we get the critical q to open a gap in such a disc:

$$q_{dw} = \left[\frac{\kappa^4}{2f3^{1/3}}\right]^{3/2} \alpha_{dw}^{3/2} h^3 \, , \tag{2.4}$$

where $h = H/R$. The result depends strongly on the parameter κ, which is not possible to constrain from first principles. We find that Eq. (2.4) fits well the gap opening mass determined numerically for $\kappa = 1.65$. With this value for κ,

$$q_{dw} \approx 7 \times 10^{-5} \left[\frac{\alpha_{dw}}{0.01}\right]^{3/2} \left[\frac{h}{0.1}\right]^3 \tag{2.5}$$

We define a modified parameter for gap opening that combines the previously derived (Crida et al. 2006) criteria for gap opening in a turbulent disc with the condition we derived above for a disc with MHD disc wind parameter α_{dw}:

$$\mathcal{P}_{dw} = \frac{3H}{4R_H} + \frac{50\alpha_v}{q}\left(\frac{H}{R}\right)^2 + \frac{70\alpha_{dw}^{3/2}}{q}\left(\frac{H}{R}\right)^3 \, . \tag{2.6}$$

3. Conclusions

In this paper, we showed that planets embedded in protoplanetary discs have easier time opening deep gaps in gas and dust if mass and angular momentum transport is dominated by magnetised winds rather than by the standard turbulent viscosity. In brief, this is because transport of material across the gap region of the planet by advection is less efficient in resisting planet tidal torques compared with that by turbulent diffusion of material in the standard discs.

For more details on this study please see Elbakyan et al. (2022).

References

Crida, A., Morbidelli, A., & Masset, F. 2006, Icarus, 181, 587

Elbakyan, V., Wu, Y., Nayakshin, S., & Rosotti, G. 2022, MNRAS, 515, 3113

Lesur, G. R. J. 2021, A&A, 650, A35

Lin, D. N. C. & Papaloizou, J. 1986, ApJ, 309, 846

Miotello, A., Kamp, I., Birnstiel, T., Cleeves, L. I., & Kataoka, A. 2022, arXiv e-prints, arXiv:2203.09818

Tabone, B., Rosotti, G. P., Cridland, A. J., Armitage, P. J., & Lodato, G. 2022, MNRAS, 512, 2290

Winds of Stars and Exoplanets
Proceedings IAU Symposium No. 370, 2023
A. A. Vidotto, L. Fossati & J. S. Vink, eds.
doi:10.1017/S1743921322003490

Solar Wind and Hydrologic Cycle

Xuguang Leng

freelance
Breinigsville, PA 18031, USA
email: xuguangleng@gmail.com

Abstract. The solar hydrologic cycle is the process of comets delivering water and gasses to the planets by collision, and solar wind stripping water and gasses from the planets and delivering them back to the Kuiper Belt. This new theory of solar hydrologic cycle provides that the solar hydrologic cycle is the continuation of planetary formation, and the cause of outer planets becoming gas giants, inner planets staying small rocks.

Keywords. solar wind, Kuiper Belt, comet, planets formation, Jupiter.

There is constant mass movement in the solar system. Mass is moving from Kuiper Belt to the inner solar system through comet collision with the planets, mass is also moving the opposite direction by solar wind stripping water vapor and gasses from the inner planets. The mass movement forms a cycle, like Earth water cycle, so it is called solar hydrologic cycle. The balance of the cycle determines the mass of the planets today, such that the solar hydrologic cycle is the continuation of planetary formation.

Hydrogen and other light elements are the most abundant elements in the solar system, rocky materials are a tiny fraction. Comparing total comet mass today of 2% solar mass (Mendis *et al.* 1986), total asteroid mass of 12^{-10} solar mass (Pitjeva *et al.* 2015) is negligible. The existence of a separate comet hydrogen envelope (Mancuso 2015) indicates that the comet is hydrogen rich. Movement of mass in the solar system is the movement of lighter elements, primarily hydrogen.

In the young solar system, there were a vast number of comets. Solar hydrologic cycle started strong. Comets rained large quantities of water and gasses on the planets.

Solar wind works in the opposite direction, brings hydrogen, water vapor and other gasses, back to the direction of Kuiper Belt, and at farthest, to the heliopause. In that region, the particles and molecules are bumping into each other forming small chunks, losing electric charge in the process. Ebb and wane of the solar wind, and gravity ripples of planets swinging by congregate the material like waves in the pond congregate leaves. The larger chunks are pulled inward by solar gravity to the Kuiper Belt, where they reform into comets by the positive feedback loop of gravity. There can be many cycles of water and gasses traveling between the Kuiper Belt and inner solar system over billions of years. The gasses react with each other on the warm planets with lightning, producing methane, ammonia, and other compounds to be carried by solar wind back to the Kuiper Belt, changing the composition of next generation comets. Today's comets have a small amount of methane (Mumma *et al.* 1996) and ammonia is the result of the solar hydrologic cycle.

Solar wind exerts great influence in the inner solar system, blows away water and gasses from inner planets, including Mars (Barabash *et al.* 2007). This prevents inner planets mass growth from comet collisions. The solar wind, however, has diminished effect on the

Table 1. Relationship of outer planets mass.

Outer Planet	Distance from the Sun [AU]	Mass [Earth mass]	Orbital Mass Gain Constant	Adjusted Orbital Mass Gain Constant
Jupiter	5.2	318	1654	8599
Saturn	9.6	95	912	8755
Uranus	19.2	14.5	278	5345
Neptune	30.0	17.1	513	15390

outer planets due to the distance. When comets collide with outer planets, their mass stays with the planets. Outer planets are deadends for solar hydrologic cycles.

Outer planets grow larger and larger as the solar hydrologic cycle repeats, locking up more and more comet mass, reducing the number of comets and intensity of solar hydrologic cycle, depriving inner planets of an ample supply of water and gasses. With solar wind relatively constant, the inner planets began to dry up. Earth, with a magnetic field and large mass, is able to retain most of the water and heavier gasses. The present day mass and composition of water and air on Earth are the result of billions of years of seesaw action between comets and solar wind with many sharp spikes.

The comet mass does not distribute to outer planets evenly or randomly, but in accordance with the probability of collisions. There are many factors that contribute to the probability. The comet orbit plane relates to the planet orbit plane is an important factor, yet does not favor any particular planet. Whether perihelion of comet orbit is inside the planet's orbit is another factor, yet minor for outer planets. The deciding factor is the planet orbit size, which can be substituted with mean distance from the Sun. The planet mass gain is in inverse proportion with distance from the Sun. Assuming outer planets started with negligible mass, then, mass gain from the comet equals to planet mass. The equation can be expressed as:

$$m_1 r_1 = m_2 r_2 = c \tag{0.1}$$

Where m is the mass of a planet, r is the distance the planet is from the Sun. c is the Orbital Mass Gain Constant related to the total comet mass at Kuiper Belt.

The next weighty factor is the positive feedback loop where larger mass attracts more comet collisions, which in turn produces even larger mass. The positive feedback loop has many random factors, and can not be predicted by simple formulas. Since distance to the Sun drives mass, and mass drives positive feedback loop, the final mass could be driven by the square of the distance to the Sun, Eq. (0.1) becomes:

$$m_1 r_1^2 = m_2 r_2^2 = c_{adj} \tag{0.2}$$

Calculating the constants for outer planets as in Table 1.

Jupiter and Saturn fully align with Eq. (0.2). Uranus fits less well, perhaps because positive feedback loop effect is not as pronounced when the mass is small. Neptune is an exception, because of its proximity to the Kuiper Belt.

References

Barabash S., Fedorov A., Lundin R., Sauvaud J.-A. 2007, *Science*, 315, 501–503
Mancuso S. 2015, *A&A*, 578:L7
Mendis, D.A., Marconi, M.L. 1986, *Earth Moon Planet*, 36, 187–190
Mumma, M.J., et al 1996, *Science*, 272, 1310-1314
Pitjeva, E., Pitjev, N. 2015, *Proc. of the IAU*

Winds of Stars and Exoplanets
Proceedings IAU Symposium No. 370, 2023
A. A. Vidotto, L. Fossati & J. S. Vink, eds.
doi:10.1017/S1743921322003702

Magnetic confinement in the wind of low mass stars

Rose F. P. Waugh and Moira M. Jardine

School of Physics and Astronomy, University of St Andrews, Scotland

Abstract. Magnetic confinement of material is observed on both high and low mass stars. On low mass stars, this confinement can be seen as slingshot prominences, in which condensations are supported several stellar radii above the surface by strong magnetic fields. We present a model for generating cooled field lines in equilibrium with the background corona, which we use to populate a model corona. We find prominence masses on the order of observationally derived values. We find two types of solutions: footpoint heavy "solar-like prominences" and summit heavy "slingshot prominences" which are centrifugally supported. These can form within the open field region i.e. embedded in the wind. We generate Hα spectra from different field structures and show that all display behaviour that is consistent with observations. This implies that the features seen in observations could be supported by a range of conditions, suggesting they would be common across rapidly rotating stars.

Keywords. stars: activity, stars: coronae, stars: individual: AB Dor, stars: low-mass, stars: magnetic field, stars: pre-main-sequence

1. Introduction

Magnetic confinement is observed in both high and low mass stars. One form of this in low mass stars is "slingshot prominences". With masses of 10-100 times solar prominences and confined at distances of a few stellar radii, they are particularly well observed on AB Doradus (Collier Cameron 1999). We present a model for populating model coronae of *rapidly rotating stars* with cooled field lines, in equilibrium with the background corona, as published in Waugh et al. (2022).

2. Modelling the cooled field lines

We assume a 2D background dipolar field who axis can lie either in the equatorial plane or along the rotation axis. This dipole can also be altered to become purely radial at the "source surface" (r_{ss}). The shapes of cooled field lines are found from the conservation of momentum:

$$\mathbf{0} = -\nabla p + (\mathbf{B}.\nabla)\frac{\mathbf{B}}{\mu} - \nabla\left(\frac{B^2}{2\mu}\right) + \rho\mathbf{g}, \tag{2.1}$$

which combines, from left to right: the gradients of gas pressure, magnetic tension and magnetic pressure forces and the gravitational force. We note that $\mathbf{g}$ is the effective gravity, as equations are constructed in the co-rotating frame: $\mathbf{g} = (-\frac{GM_\star}{r^2} + \omega^2 r \sin^2\theta)\hat{r} + (\omega^2 r \sin\theta\cos\theta)\hat{\theta}$.

Equation 2.1 can be solved along the field line to give the distribution of the gas pressure. Within the equatorial plane, the gas pressure is constant with longitude but decreases with radius, reaching a minimum at the co-rotation radius and then increasing beyond this. In meridional planes the gas pressure shows the same behaviour, however, it

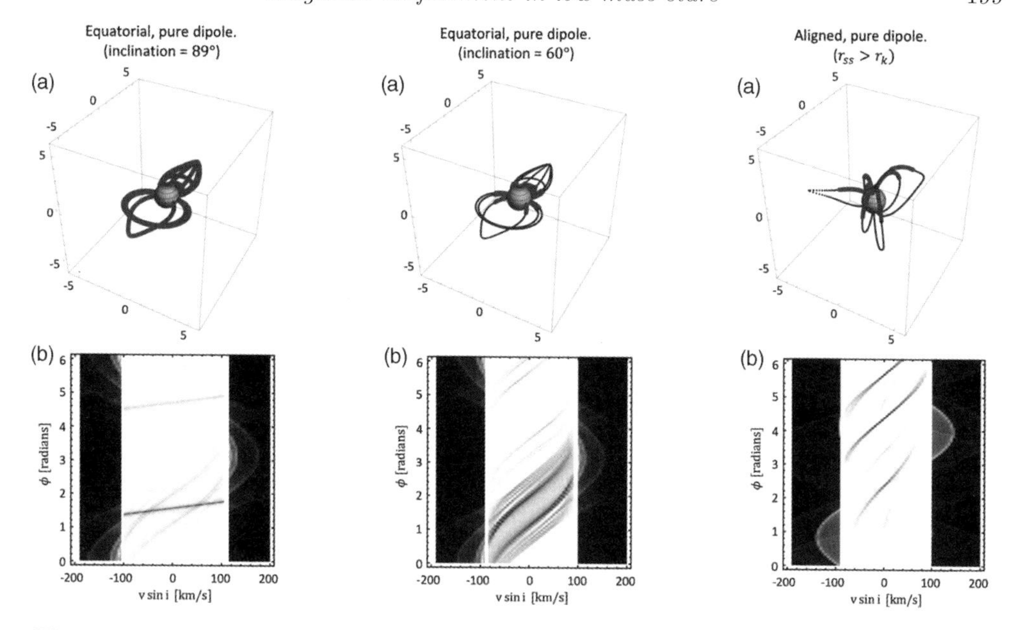

Figure 1. Example magnetic field lines (blue) and visible sections (purple) and below, the associated Hα trails.

is not constant with latitude. The component of the momentum equation perpendicular to the field line, which is a 2nd order ODE, then yields the shape of the cooled field line.

3. Generating synthetic dynamic spectra

Two types of solution are found in this model: loops with summits below co-rotation, with low-density summits and high-density footpoints and very tall loops with high-density summits. The second of these result in dark, slowly transiting features in the Hα spectra, whilst the summits of the low-lying loops do not generate clear absorption features. The dense footpoints can be seen as fast travelling absorption features. Examples of these spectra are shown in Figure 1. All field structures here generate spectra that have features consistent with slingshot prominence observations. Equatorial dipoles differ from aligned dipoles in these spectra by the number of absorption features associated with any given loop: both footpoints are visible for equatorial dipoles whereas only one leg of the loop is visible for an aligned dipole. The inclination of star determines if the summits of the tall, slingshot loops transit the disc and thus are visiblein the spectra, which can be seen by comparing the two most left hand images in Figure 1. This suggests that many field structures could be consistent with the observations for slingshot prominences, adding to the evidence that they could be common features on young stars. Despite this, the stellar inclination is crucial in determining if a prominence will be visible or not.

References

Collier Cameron, Andrew, 1999, ASP Conference Series, 158, 146
Waugh, Rose F. P. and Jardine, Moira M., 2022, MNRAS, 514, 5465

Winds of Stars and Exoplanets
Proceedings IAU Symposium No. 370, 2023
A. A. Vidotto, L. Fossati & J. S. Vink, eds.
doi:10.1017/S1743921323000492

Clumping and X-rays in cool B Supergiants

Matheus Bernini-Peron[1], W. L. F. Marcolino[2] and A. A. C. Sander[1]

[1]Zentrum für Astronomie der Universität Heidelberg, Astronomisches Rechen-Institut
Mönchhofstr. 12-14, 69120, Heidelberg, Germany
email: matheus.bernini@uni-heidelberg.de

[2]Observatório do Valongo, Federal University of Rio de Janeiro,
Ladeira do Pedro Antônio, 20080-090, Rio de Janeiro, Brazil

Abstract. B supergiants (BSGs) are evolved objects on the cool end of the line-driven wind regime. Studying their atmospheres provides important insights on the stellar wind physics of these objects and their evolutionary status. So far important features of their spectra, especially in the UV region, could not be reproduced consistently with atmosphere models. This translates directly into problems of our understanding of their wind properties. Here, we present new insights about the BSGs on the cooler side of the Bi-Stability Jump, corresponding to spectral types later than B1. Using UV and optical data, we analysed a sample of Galactic cool BSGs. Including for the first time X-rays and clumping the wind models, we show that the spectra of cool BSGs cannot be explained without X-rays, despite any clear detection of the target stars.

Keywords. stars: winds, outflows, stars: fundamental parameters, ultraviolet: stars, X-rays: stars

---

1. Introduction

B supergiants (BSGs) are evolved objects with effective temperatures between 12 and 29 kK, located on the cool end of the hot-star wind regime. As such, they mark a key stage for high-mass stellar evolution and are essential to understand the physics of line-driven wind, and phenomena like the Bi-stability Jump (Lamers et al. 1995). Of major concern are in particular the UV P-Cygni profiles in cool BSGs (spectral type later than B1), which could not be reproduced consistently in previous studies (Crowther et al. 2006; Searle et al. 2008). This leads to considerable uncertainties in our understanding of these stars' wind properties.

To address this issue, we analysed a sample of four Galactic cool BSGs with large UV and optical coverage. None of our targets were directly detected in X-rays. In our analysis, we used CMFGEN to verify whether the inclusion of (optically thin) clumping and X-rays in the wind increases the agreement between models and observations. From fitting the optical spectrum we obtain the stellar properties, whilst luminosity and reddening were obtained by reproducing the spectral energy distribution (SED). The wind properties, including the clumping and X-ray parameters, were obtained by reproducing the UV spectrum.

2. Results and Discussion

By obtaining good agreement between the synthetic and observed spectra (in the optical and UV, see Fig. 1), we find stellar properties consistent with evolved objects, namely: the stars are (i) away from the ZAMS and (ii) display chemical alterations relative to the solar abundance (Asplund et al. 2009). Our spectral modeling yielded

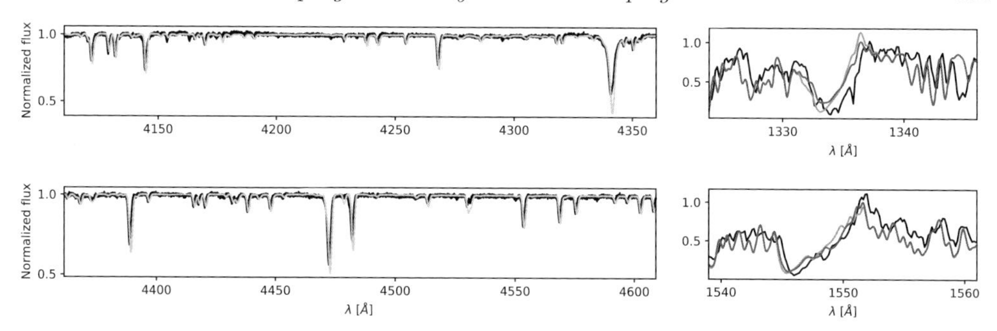

Figure 1. Spectral analysis of HD53138 (B3 Ia): In all panels the observed spectra are depicted as a black line. *Left panels:* Optical regime. *Right Panels:* Reproduction of C II and C IV profiles including clumping and X-rays – green line: model with $f_\infty = 0.5$, $L_x/L = 10^{-7.5}$ and $v_x = 0.8\, v_\infty$ (where v_x is the onset of X-rays); yellow: model with $f_\infty = 0.9$, $L_x/L = 10^{-8.0}$ and $v_x = 0.05\, v_\infty$.

masses discrepant with (single stellar) evolution models including rotation. In the UV, we reproduce key wind spectral profiles. Our results show that both X-rays and clumping need to be included in order to fit C II and C IV simultaneously – and to not overestimate Hα. However, different values than those typically obtained for hotter stars ($\log(L_x/L) = -7$; $f_\infty = 0.1$) are required. Besides $f_\infty > 0.5$, we infer $\log(L_x/L)$ of about -7.5 to -8. Still, we have problems reproducing Hα, Si IV and Al III lines. Motivated by previous work (Prinja & Massa 2010; Petrov et al. 2014), we investigate whether optically thick clumping (with the formalism of Oskinova et al. 2007) could solve or attenuate this problem. Having performed the first study to model cool BSGs atmospheres including clumping and X-rays, we can draw the following conclusions:

- Our results point towards weaker X-ray emission and clumping in BSGs at the cool side of the Bi-Stability Jump compared to their hot counterparts. This is in line with recent theoretical predictions (Driessen et al. 2019) and observational constraints in the infrared (Rubio-Díez et al. 2022).
- While the actual amount of X-ray emission is still unclear for these objects (Berghoefer et al. 1997; Nazé 2009), we show that a certain amount of X-Rays is necessary to reproduce the observed superionization in the UV.
- Compared to the Vink et al. (2000) prediction, the obtained lower amount of clumping would imply a smaller reduction of the empirical mass-loss rates on the cool side than on the hot side of the Bi-Stability Jump.

References

Asplund, M., Grevesse, N., Sauval, A. J., & Scott, P. 2009, *ARA&A*, 47, 481
Berghoefer, T. W., Schmitt, J. H. M. M., Danner, R., & Cassinelli, J. P. 1997, *A&A*, 322, 167
Crowther, P. A., Lennon, D. J., & Walborn, N. R. 2006, *A&A*, 446, 279
Driessen, F. A., Sundqvist, J. O., & Kee, N. D. 2019, *A&A*, 631, A172
Lamers, H. J. G. L. M., Snow, T. P., & Lindholm, D. M. 1995, *ApJ*, 455, 269
Nazé, Y. 2009, *A&A*, 506, 1055
Oskinova, L. M., Hamann, W. R., & Feldmeier, A. 2007, *A&A*, 476, 1331
Petrov, B., Vink, J. S., & Gräfener, G. 2014, *A&A*, 565, A62
Prinja, R. K. & Massa, D. L. 2010, *A&A*, 521, L55
Rubio-Díez, M. M., Sundqvist, J. O., Najarro, F., et al. 2022, *A&A*, 658, A61
Searle, S. C., Prinja, R. K., Massa, D., & Ryans, R. 2008, *A&A*, 481, 777
Vink, J. S., de Koter, A., & Lamers, H. J. G. L. M. 2000, *A&A*, 362, 295

Part 4:
Flow-flow interactions

Winds of Stars and Exoplanets
Proceedings IAU Symposium No. 370, 2023
A. A. Vidotto, L. Fossati & J. S. Vink, eds.
doi:10.1017/S1743921322004501

Interaction between massive star winds and the interstellar medium

Jonathan Mackey

Dublin Institute for Advanced Studies, Astronomy & Astrophysics Section, DIAS Dunsink Observatory, Dublin, D15 XR2R, Ireland

Abstract. Massive stars drive strong winds that impact the surrounding interstellar medium, producing parsec-scale bubbles for isolated stars and superbubbles around young clusters. These bubbles can be observed across the electromagnetic spectrum, both the wind itself and the swept up interstellar gas. Runaway massive stars produce bow shocks that strongly compresses interstellar gas, producing bright infrared, optical and radio nebulae. With the detection of non-thermal radio emission from bow shocks, particle acceleration can now also be investigated. I review research on wind bubbles and bow shocks around massive stars, highlighting recent advances in infrared, radio and X-ray observations, and progress in multidimensional simulations of these nebulae. These advances enable quantitative comparisons between theory and observations and allow to test the importance of some physical processes such as thermal conduction and Kelvin-Helmholtz instability in shaping nebulae and in constraining the energetics of stellar-wind feedback to the interstellar medium.

Keywords. Stars: winds, outflows - ISM: bubbles - Stars: early-type - circumstellar matter - shock waves

---

1. Models for spherical wind bubbles

Massive stars have a strong effect on their surroundings through their intense radiation (especially extreme-UV, ionizing radiation), strong winds, eruptive explosions and supernova explosions at the end of their lives. Quantifying these feedback effects is important for understanding the dynamical and chemical evolution of galaxies, and also the structure and dynamics of the interstellar medium (ISM) in our own Galaxy. On smaller scales, modelling of circumstellar nebulae can also give clues as to the evolutionary history of some nearby massive stars. Comparison of models with observations can give constraints on different physical processes in astrophysical plasmas, such as particle acceleration and thermal conduction.

In one of the first papers studying the effects of stellar winds on the ISM, Mathews (1966) proposed that the optical cavity in the Rosette Nebula around NGC2244 is maintained by dynamical pressure of the strong and high-velocity winds of the massive stars in the central cluster. Dyson & de Vries (1972) built on this work, developing a theory for the dynamics of wind-blown bubbles. This was further generalised by Castor et al. (1975), and again in the classic paper by Weaver et al. (1977). The basic picture is that of a spherical wind expanding from a star or group of stars and displacing the interstellar gas. From the contact discontinuity between the two media, a reverse (or termination) shock propagates backwards towards the star generating a hot bubble of shocked coronal gas with temperature $T \sim 10^6 - 10^8$ K. Similarly a forward shock propagates into the undisturbed ISM as long as the bubble expands supersonically. The ISM is photoionized by extreme-UV radiation from the hot star(s), producing an H II region around the star that usually extends well beyond the wind bubble, but which may be trapped by the shocked ISM (Weaver et al. 1977; Freyer et al. 2003).

Weaver et al. (1977) introduced thermal conduction, which smoothes out the contact discontinuity and produces significant quantities of gas in the temperature range from $10^5 - 10^7$ K, that emits UV and thermal X-ray radiation. This can reproduce the observed [O VI] line emission that is observed, but tends to overpredict thermal X-rays (see, e.g., Toalá et al. 2016). The dynamical model of Weaver has the bubble radius, r, expanding with time, t, as $r \propto t^{3/5}$. This is the solution in the adiabatic limit, i.e., in the limit that the thermal energy in the wind bubble is not lost and can power the pressure-driven expansion of the bubble. In the momentum conserving limit, where the wind energy is efficiently radiated away, a simple dimensional analysis shows that $r \propto t^{1/2}$. If thermal conduction is efficient at transporting thermal energy out of the bubble and into the mixing layer, then the second expansion law holds.

García-Segura et al. (1996a) studied wind-ISM and wind-wind interactions with multi-dimensional simulations, showing that some shocks become effectively isothermal, forming thin and unstable layers that break up into clumps. This multi-dimensional effect is important for the expansion rate and radiative emission of the nebulae, and can explain some of the observational properties of Wolf-Rayet nebulae. This line of research was extended by Freyer et al. (2006) with the inclusion of photoionizing radiative transfer, modelling the H II region and wind-bubble evolution for the full stellar lifetime, expanding into a uniform ISM. They showed that dissipation and instabilities in expanding layers can significantly affect X-ray and UV emission from wind bubbles, potentially resolving the disagreement between large predicted X-ray luminosities, and relatively weak X-ray emission from observed nebulae. Toalá & Arthur (2011) used different stellar evolution sequences and higher-resolution simulations, including calculations with and without thermal conduction, investigating the structure and X-ray emission from massive-star nebulae. They concluded that both thermal conduction and dynamical mixing by hydrodynamic instabilities are affecting the X-ray emission. Geen et al. (2015) modelled the evolution of the CSM around a massive star including effects of both photoionization and stellar wind in 3D simulations with a similar setup. The latest work using as initial conditions wind expanding from a star at rest with respect to a uniform ISM is by Meyer et al. (2020), who used high-resolution 2D simulations to study the CSM of a $60\,M_\odot$ star through various evolutionary phases followed by explosion as a supernova.

These simulations are very useful for elucidating the physical processes at work in wind bubbles, but they assume certain symmetries that are not always present. In particular stellar motion or bulk flows in the ISM lead to distorted wind bubbles (Weaver et al. 1977; Mackey et al. 2015) that produce a bow shock when the star is moving supersonically with respect to the ISM (e.g. van Buren & McCray 1988). Furthermore the turbulent ISM is neither static nor homogeneous, and these density and velocity fluctuations also affect the shape of a wind bubble (see Fig. 1, taken from Geen et al. 2021). The combination of stellar winds, radiative feedback, stellar clustering, and supernovae leads to a very complex multiphase ISM with strong pressure gradients on many scales (Rathjen et al. 2021).

2. Runaway Stars and bow shocks

Not long after the first calculations showing that the interaction between Solar wind (Baranov et al. 1971) or stellar wind (Weaver et al. 1977) and local ISM could produce a bow shock, the bow shock around the closest O star to Earth, ζ Ophiuchi, was detected in nebular emission lines by Gull & Sofia (1979). Bow shocks provide an excellent laboratory for studying the wind-ISM interaction for a number of reasons: *(i)* the star has generally

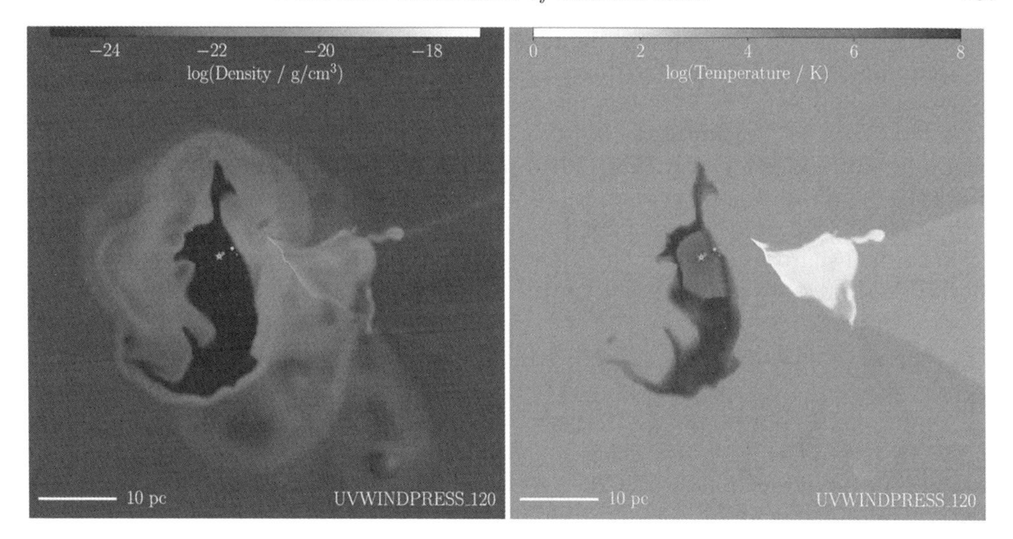

Figure 1. Slice through a 3D simulation of a turbulent molecular cloud, 0.4 Myr after a massive star has formed, showing gas density (left) and temperature (right). The hot, low-density and asymmetric wind bubble surrounds the star. Reproduced from fig. 4 of "The geometry and dynamical role of stellar wind bubbles in photoionized H II regions", Geen et al. (2021), MNRAS, 501, 1352.

moved far from its place of birth and there are no other massive stars in the immediate vicinity that could induce wind-wind interactions; *(ii)* the star is typically moving in the diffuse (warm) phase of the ISM through a less structured medium than in a molecular cloud; and *(iii)* ram pressure provided by stellar motion compresses the bow shock into a structure that is overdense (and more easily observable) and that has a relatively short dynamical timescale.

The structure of a bow shock is well-described in fig. 1 of Comerón & Kaper (1998), again consisting of two shocks separated by a contact discontinuity. Stellar motion at velocity $v_\star$ with respect to the ISM leads to an asymmetric ISM ram pressure of $\rho_0 v_\star^2$ in the reference frame of the star (ρ_0 is the ISM gas density). Pressure balance gives the characteristic size of the bow shock (Baranov et al. 1971), the standoff distance, R_0, as

$$R_0 = \sqrt{\frac{\dot{M} v_\infty}{4\pi\rho_0(v_\star^2 + v_\mathrm{A}^2 + c^2)}}, \qquad (2.1)$$

where $\dot{M}$ and v_∞ are the mass-loss rate and terminal wind velocity of the star, respectively. The Alfvén (v_A) and sound (c) speeds represent the magnetic and thermal pressure contributions to the total pressure, but usually the ram pressure is the dominant term. The dynamical timescale of the bow shock is $t_\mathrm{dyn} = R_0/v_\star$, typically $\sim 10^5$ yr for bow shocks with $R_0 \lesssim 1$ pc and $v_\star \sim 30\,\mathrm{km\,s^{-1}}$, much shorter than the lifetime of a massive star.

This equation provides a measurement of $\dot{M}$ that is independent from stellar-atmosphere spectral-line studies (e.g. Gull & Sofia 1979), if the other quantities on the right-hand side of the equation are well constrained. Gvaramadze et al. (2012) were able to use the properties of the H II region around the runaway O star ζ Oph to constrain ρ_0 and thereby measure $\dot{M} \approx 2.2 \times 10^{-8}\,\mathrm{M_\odot\,yr^{-1}}$, below the estimate derived from optical lines in the atmosphere and the Vink et al. (2000) theoretical estimate, but well above the estimate derived from UV lines (Marcolino et al. 2009). Kobulnicky et al. (2018) use Eq. 2.1 to measure $\dot{M}$ for a sample of bow shocks around 20 O stars by making the assumptions that all outer shocks are adiabatic with a density jump of 4×, that ρ_0 can be accurately measured from far-IR dust emission, and all stars are moving through the ISM with $v_\star = 30\,\mathrm{km\,s^{-1}}$. This was followed up by Kobulnicky et al. (2019) who used *Gaia* DR2 data to measure $v_\star$ for a larger sample of stars and thereby obtaining more accurate results. Henney & Arthur (2019) made a careful analysis of the statistical and systematic uncertainties of measuring $\dot{M}$ from bow-shock observations, also comparing their method with that of Kobulnicky et al. (2018). They find that $\dot{M}$ measurements for individual sources may have large uncertainties (from uncertain dust properties and shocked ISM pressure support), but statistically for a large group of sources the methods are promising.

2.1. *Multiwavelength emission from bow shocks*

The first detection of a significant number of bow shocks was made with the *IRAS* observatory by van Buren & McCray (1988) in the mid-IR (see also van Buren et al. 1995). The broadband IR emission is thermal radiation from interstellar dust grains photo-heated by the intense UV-radiation field of the massive star that drives the bow shock. The *AKARI* observatory obtained higher-resolution IR images of some bow shocks (Ueta et al. 2008), but a major advance came with the *Spitzer* and *WISE* missions. The E-BOSS catalog of Peri et al. (2012, 2015) contains 73 confirmed and candidate bow shocks, and Kobulnicky et al. (2016) compiled a larger catalog of 709 candidate bow shocks selected by mid-IR morphology. Meyer et al. (2014) showed that bow shocks are most luminous in the mid-IR (see also Acreman et al. 2016; Henney & Arthur 2019), explaining why these surveys have been so successful. Asymmetric wind bubbles with subsonic relative motion between star and ISM can also produce bright mid-IR arcs (Mackey et al. 2016).

Despite the first detection of a bow shock being in optical lines, this has proven to be a difficult method for detection because the massive stars are so optically bright that identifying nearby nebular emission is challenging (see Gull & Sofia 1979). A few other detections have been made: HD 165319 appears to show a bow shock in narrow-band Hα imaging (Gvaramadze & Bomans 2008), and the red supergiant IRC -10414 also has a bow shock detected in Hα imaging (Gvaramadze et al. 2014). It seems likely that the bow shocks of Betelgeuse (Mohamed et al. 2012) and ζ Ophiuchi (Green et al. 2022) could be detectable in nebular optical lines if the exceptionally bright stellar emission could be masked.

The first radio detection of a bow shock was by Benaglia et al. (2010), around the runaway O supergiant BD+43 3654, further studied by Benaglia et al. (2021) and Moutzouri et al. (2022) over a wider frequency range. There is significant interest in radio studies at present because of the rapid advances in instrumentation and also the potential to detect both thermal (Bremsstrahlung) and non-thermal (synchrotron) radiation. From its radio spectrum it seems that the bow shock of BD+43 3654 exhibits both. Moutzouri et al. (2022) also detected the Bubble Nebula, NGC 7635, as predicted by Green et al. (2019), and the detections of this and BD+43 3654 are shown in Fig. 3. Some low significance radio emission was found in NVSS survey data for E-BOSS bow shocks by Peri et al. (2015), and some of these are now confirmed radio emitters with ASKAP observations (Van den Eijnden et al. 2022). The bow shock of Vela X-1 was also detected with MeerKAT (van den Eijnden et al. 2022). Given these recent discoveries and the promise of wide-field, broadband surveys with the new interferometric arrays, I anticipate many more radio detections of bow shocks in the coming years.

Thermal bremsstrahlung gives a direct measure of the emission measure, allowing to constrain the gas density in both the pre-shock and shocked ISM in the bow shock.

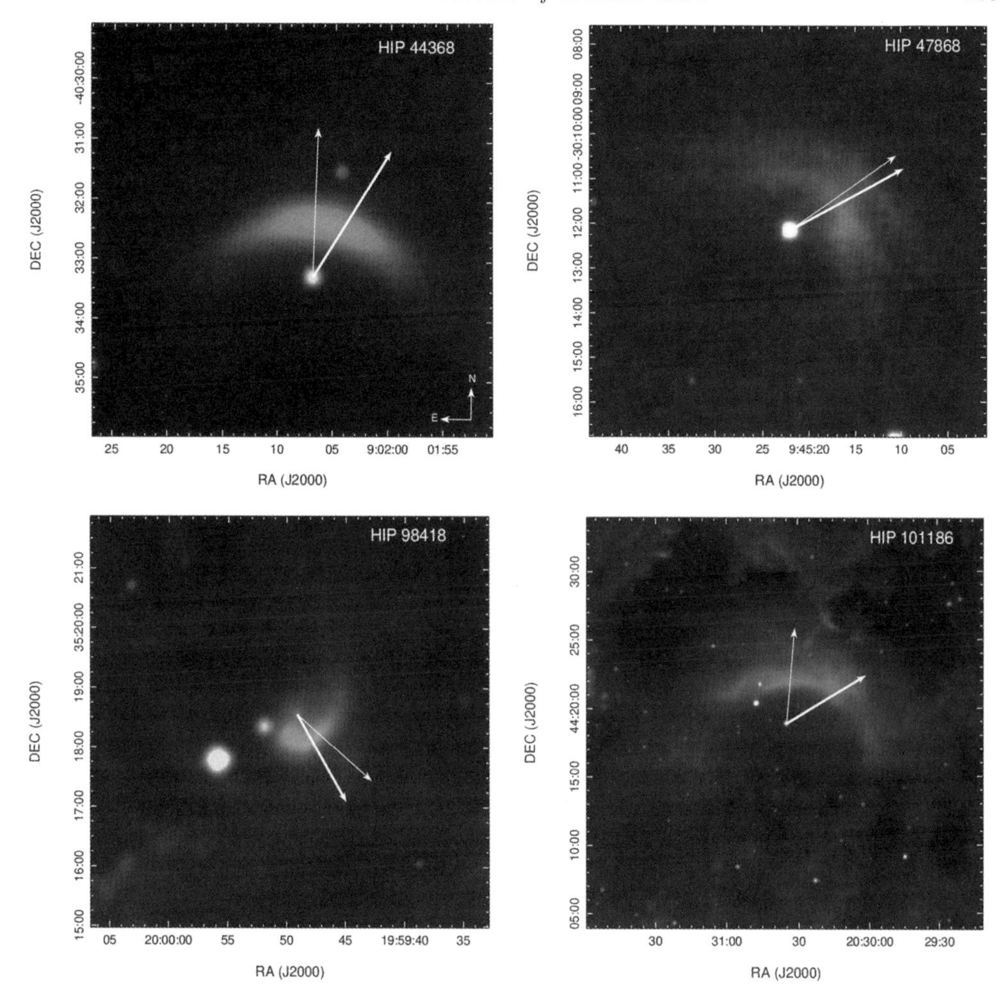

Figure 2. Four bow shocks from the catalog of Peri et al. (2015), imaged by *WISE*. Red: band 4, 22.2 μm. Green: band 3, 12.1 μm. Blue: band 1, 3.4 μm. Credit: Peri, Benaglia & Isequilla, 2015, Astronomy & Astrophysics, 578, A45, fig. 5.

Obtaining gas density from IR emission has significant uncertainties arising from dust composition and grain size distribution (Pavlyuchenkov et al. 2013; Mackey et al. 2016), and nebular line emission can have uncertainties from patchy extinction along the line of sight. Radio has the distinct advantage that is it unaffected by extinction.

With a detection of synchrotron radiation at radio wavelengths one can make model-dependent predictions for synchrotron and inverse-Compton emission at X-ray and γ-ray energies (del Valle & Romero 2012; del Palacio et al. 2018; Moutzouri et al. 2022). Searches for non-thermal X-rays (Toalá et al. 2016, 2017) and γ-rays (Schulz et al. 2014; H.E.S.S. Collaboration et al. 2018) from bow shocks have so far only obtained upper limits, although this is not altogether surprising given the predictions and sensititivies of current instruments. Detection of high-energy radiation may need to wait for the next generation of instruments for bow shocks around single stars (Moutzouri et al. 2022).

3. The wind-ISM boundary layer – theoretical models

The contact discontinuity between shocked stellar wind and interstellar material is very strong for hot stars with fast winds. Typical post-shock temperature of the shocked

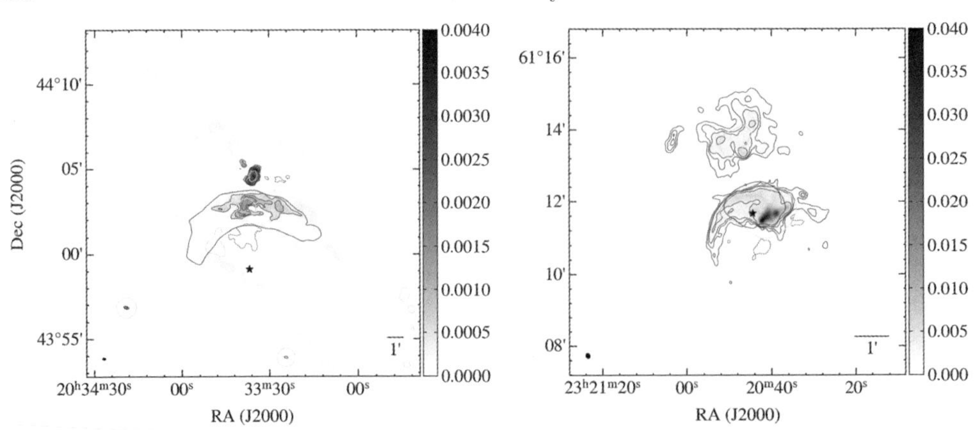

Figure 3. Radio detection of the bow shocks around BD+43 3654 (left) and BD+60 2522 (right), obtained with the VLA at 4-8 GHz (greyscale and grey contours). Blue contour shows *WISE* band 4 emission in mid-IR, and cyan contours show 1.4 GHz emission from the *NVSS* survey. Credit: Moutzouri *et al.*, 2022, Astronomy & Astrophysics, 663, A80, fig. 1.

wind is $T_{\rm PS} \sim 10^8$ K for an adiabatic shock, easily obtainted from the Rankine-Hugoniot jump conditions:

$$T_{\rm PS} \approx 5.6 \times 10^7 \,{\rm K} \left(\frac{v_{\rm sh}}{2000\,{\rm km\,s^{-1}}} \right)^2 , \tag{3.2}$$

where an exact expression depends on the gas composition and ionization state. Given a photoionized ISM temperature of $T_{\rm ISM} \approx 8000$ K, this implies a temperature jump of a factor of 10^4 across the contact discontinuity, and an associated density jump of the same factor in the opposite sense. For comparison this is similar to the density ratio of air and rock. It is easy to see that mixing a small volume of interstellar gas across the contact discontinuity can have a large effect on the density and temperature (and hence radiative emissivity at different energies) of the wind bubble or bow shock.

Electronic thermal conduction (Cowie & McKee 1977) transports heat from the hot and low-density gas phase across a contact discontinuity into a boundary layer of cooler and denser gas. The thickness of this boundary layer is determined by the mean free path of electrons, or the gyroradius in the presence of a magnetic field component parallel to the discontinuity. The heated boundary layer expands, resulting in a region with strong temperature and density gradients rather than a well-defined contact discontinuity. This can be seen in hydrodynamic bow shock simulations by Comerón & Kaper (1998) and Meyer et al. (2014), where the structure of the bow shock is signficantly modified by thermal conduction. Meyer et al. (2017) showed, however, that the inclusion of an interstellar magnetic field dramatically reduces the effects of thermal conduction and, given that the ISM is pervaded by magnetic fields, one may deduce that the hydrodynamic models of wind bubbles (Weaver et al. 1977) and bow shocks (Comerón & Kaper 1998; Meyer et al. 2014) significantly overestimate the effects of thermal conduction.

An alternative physical process that can thicken and distort the wind-ISM boundary layer is often referred to as dynamical or turbulent mixing (e.g. Slavin et al. 1993, applied to supernova remnants). This is generally driven by Kelvin-Helmholtz instability (KHI) because the hot phase has a sound speed up to 100× that of the warm ISM phase, and so the velocity shear at the interface is typically large. It cannot be captured in 1D simulations, but 2D simulations with an unstable wind-ISM interface (due to e.g. Rayleigh-Taylor instability or non-linear thin-shell instability) or with an anisotropic external pressure naturally generate such shear flows (Toalá & Arthur 2011;

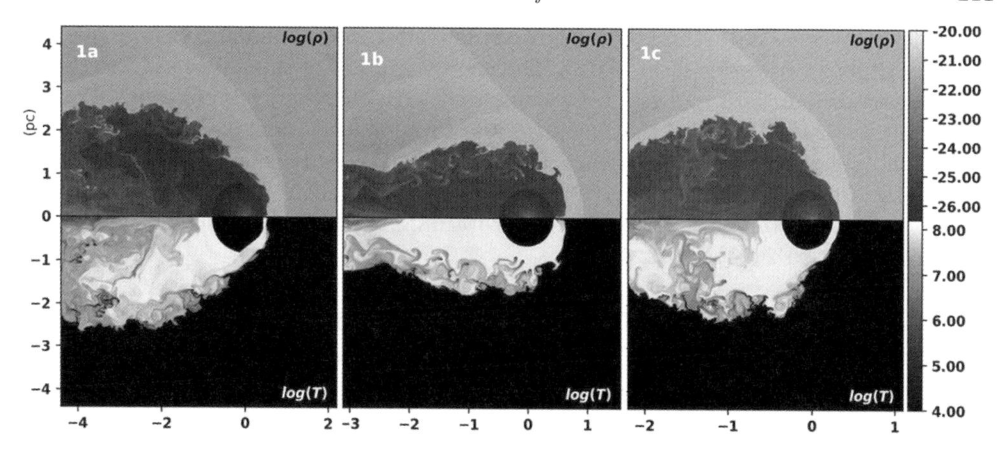

Figure 4. Gas density (g cm^{-3}, upper half-plane) and temperature (K, lower half-plane) on a logarithmic colour scale for high-resolution 2D axisymmetric simulations of bow shock models for the Bubble Nebula, NGC 7635. Credit: Green *et al.*, 2019, Astronomy & Astrophysics, 625, A4, fig. 2.

Mackey et al. 2015; Green et al. 2019). This KHI-induced turbulent mixing is clearly shown in Fig. 4 (taken from Green et al. 2019), where waves initiated near the apex of the bow shock grow to non-linear amplitudes and the resulting vortices strongly mix wind and ISM material in the wake behind the bow shock.

If thermal conduction is not explicitly modelled then the degree of mixing is dependent on numerical resolution through both numerical diffusivity and the degree to which the numerical scheme has the resolution to capture the development of KHI (e.g. Green et al. 2022). Recently, Lancaster et al. (2021a) and Lancaster et al. (2021b) developed a theory of turbulent mixing in wind bubbles expanding into clouds with a fractal density structure, arguing that this should lead to strongly cooled wind bubbles. The degree of energy dissipation is a key constraint for larger scale simulations of the ISM of galaxies, for which it is currently not clear how efficient wind feedback (implemented in a sub-grid model) should be (e.g. Fichtner et al. 2022).

4. The wind-ISM boundary layer – observations

One way to constrain these processes observationally is to study the thermal X-ray or UV emission from the wind-ISM interface. The brightest X-ray emitting wind bubbles are around Wolf-Rayet (WR) stars (Toalá et al. 2015). These are wind-wind interactions where the fast wind of the WR star sweeps up the slow wind (García-Segura et al. 1996b) or mass ejection from binary systems (Jiménez-Hernández et al. 2021). Here there is usually considerable uncertainty about the density structure of the wind in the previous evolutionary phase. UV lines from intermediate temperature gas at the hot-cold interface were also detected by Boroson et al. (1997) within a WR ring nebula and interpreted as a conduction front.

So far the only clear detection of diffuse X-rays from stellar wind of a single main-sequence star is very weak emission from within the bow shock around ζ Oph (Toalá et al. 2016; Green et al. 2022). Diffuse X-rays were also predicted (Mackey et al. 2015) and later detected (Townsley et al. 2018) from the massive star forming region RCW 120, which likely includes a contribution from the O star that is the main driver of the nebula's evolution. A quantitative comparison between theory and observation for this object has not been made.

On larger scales, X-ray emission has been detected from a number of massive star-forming regions (Townsley et al. 2018). The energy content of the hot phase was derived from observational data and has been compared with the energy input from winds by Rosen et al. (2014), finding that most of the input energy cannot be accounted for. This is strong observational evidence for the effectiveness of energy transport across the contact discontinuity by e.g. turbulent mixing, although 'leakage' of energy into the low-density coronal gas surrounding star-forming regions also could not be excluded (Rogers & Pittard 2014).

5. 3D MHD models of bow shocks and astrospheres

We have seen that magnetic fields are important for moderating the effects of thermal conduction at the contact discontinuity of bow shocks, and it is also known that the operation of KHI is markedly different in the presence of a magnetic field (Frank et al. 1996; Keppens et al. 1999). Axisymmetric simulations in 2D do not allow a general magnetic field configuration because only ISM fields that are parallel with (or anti-parallel to) the velocity vector of the star are permitted. There is therefore a need for 3D MHD simulation of bow shocks, and a number of groups have worked on this over the past few years, developing efficient methods that enable 3D simulations with reasonable computational resources.

The computationally cheapest calculations are for stars with slow winds (because the timestep is inversely proportional to the fastest speed on the domain), and so the first 3D hydrodynamic simulations were for isothermal calculations of winds from cool stars (Blondin & Koerwer 1998). These showed very strong dynamical instability that quickly grew to nonlinear amplitudes. More detailed calculations by Mohamed et al. (2012) applied to the red supergiant Betelgeuse explored the stability of the bow shock as a function of space velocity of the star, with a detailed radiative cooling prescription. Recently Meyer et al. (2021) added magnetic fields to these models of bow shocks around cool stars, showing that the field has a strong stabilising influence on the bow shock.

Simulations of bow shocks around O stars are much more challenging because of the larger Mach number of the shocks (affecting stability of the scheme) and the timestep constraint (increasing the computational cost). Scherer et al. (2020) developed a 3D MHD scheme using a spherical coordinate system that naturally has varying spatial resolution and hence is quite efficient. They studied the differences between hydrodynamic and MHD simulations for a bow shock similar to that around the massive stars λ Cep. Baalmann et al. (2021) used this model to study the effects of ISM inhomogeneities on the structure and observable properties of bow shocks.

Using a Cartesian coordinate system with static mesh-refinement, Mackey et al. (2021) introduced 3D MHD simulations of bow shocks with the software PION. This was used by Green et al. (2022), who presented the first 3D MHD simulations dedicated to modelling the bow shock of ζ Oph. This spectacular system (see Fig. 5) provides arguably the cleanest laboratory for testing mixing processes at the wind-ISM boundary. It is a very nearby massive star (136 pc) with well-characterised proper motion, moving through a large photoionized H II region and driving a bright and well-resolved bow shock. Crucially, diffuse X-ray emission has been detected with *Chandra* (Toalá et al. 2016). This source is also far out of the Galactic Plane, which can be valuable for multi-wavelength studies because background emission is less of an issue. Becuase of rapid stellar rotation the radial velocity of the star is actually poorly constrained, and this introduced some uncertainty to the modelling. Nevertheless the authors were able to show that the X-ray emission predicted by the simulated bow shock is somewhat weaker than the observed emission. There are uncertainties, particularly the role of instabilities and effects of limited numerical resolution, but this comparison of 3D MHD simulations with high-resolution observations is a

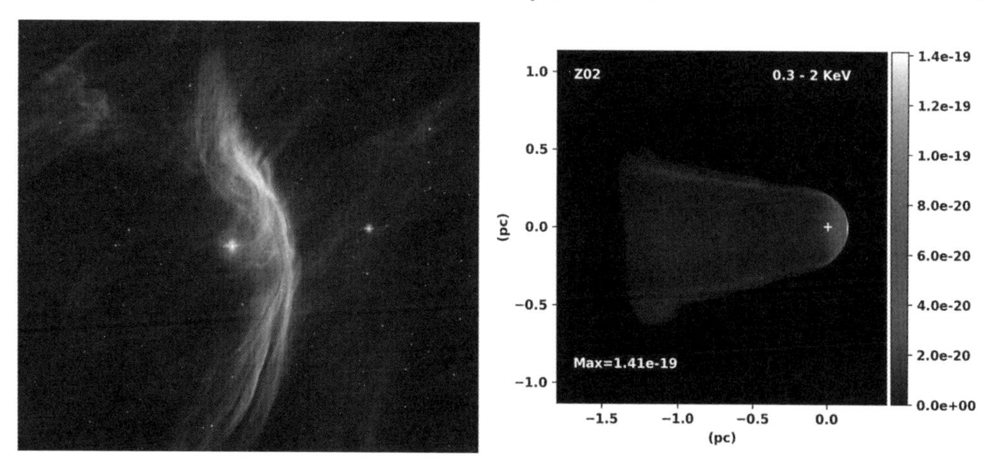

Figure 5. Left: The bow shock of ζ Ophiuchi, seen in infrared (red and green, from *Spitzer Space Telescope*) and X-rays (blue, *Chandra X-ray Observatory*). Image credit: X-ray: NASA/CXC/Dublin Inst. for Advanced Studies/S. Green et al.; Infrared: NASA/JPL/Spitzer. **Right:** synthetic X-ray emission from 3D MHD simulation of the bow shock, from Green et al. (2022) (Astronomy & Astrophysics, 665, A35, fig. 13).

promising avenue for obtaining detailed information on the physical processes operating at shocks and contact discontinuities in the hot and warm ISM phases.

6. Outlook

The interaction of stellar winds with the ISM is important for interpreting observations of circumstellar nebulae, but also has broader application to a number of other areas of astrophysics. The effectiveness of thermal conduction in astrophysical plasmas is so far only experimentally measured in the Solar wind (e.g. Bale et al. 2013), and the degree to which it is inhibited by both large-scale coherent and small-scale turbulent magnetic fields can potentially be investigated at the wind-ISM interface. It is important to constrain mixing of interstellar matter into stellar-wind bubbles and how this affects the efficiency of mechanical feedback from winds to the ISM, with consequences for models of galaxy formation and evolution. The wind-ISM interaction sets the initial conditions into which a supernova blast wave expands, which can have important consequences for the evolution of supernova remnants (Das et al. 2022). Bow shocks and wind bubbles can be a useful laboratory for studying particle acceleration in shocks (Benaglia et al. 2010; H.E.S.S. Collaboration et al. 2022) and associated particle transport and radiation processes (del Valle & Pohl 2018).

This review has focussed on wind bubbles and bow shocks around single stars, but I want to emphasise that the coming decade will see great advances in modelling and observing wind-ISM and wind-wind interactions of binary and higher-order multiple star systems. Unlike low-mass stars, almost all massive stars begin their lives in binaries and the majority will undergo interaction with a companion during their lifetime (Sana et al. 2012). Colliding-wind binaries are one of the few astrophysical systems where time-dependent shock physics can be probed (Pittard & Dougherty 2006; H.E.S.S. Collaboration et al. 2020), and they are bright Galactic sources across the electromagnetic spectrum up to TeV gamma rays.

Progress in our understanding of wind-ISM interaction depends on both new data and more detailed models. In this regard we can anticipate breakthroughs within the next decade. The rapid improvement in large field-of-view radio interferometry driven by *SKA* pathfinder instruments is leading to detection of both thermal and non-thermal radio emission from circumstellar nebulae and bow shocks. This gives important insights into gas density and thermal state, and the population of non-thermal particles. In the next few years the *Cherenkov Telescope Array (CTA)* will come online, providing unprecedented sensitivity to TeV gamma-ray emission, with potential detections of populations of very-high-energy particles accelerated in wind bubbles and bow shocks, and detection of a large population of Galactic binary systems, potentially including the colliding-wind binaries. Looking further ahead, the *ATHENA* mission promises huge sensitivity improvements compared with current X-ray telescopes, leading to much better characterisation of the hot thermal plasma of the shocked stellar-wind, and potentially detection of non-thermal emission via synchrotron or inverse-Compton radiation.

At the same time, rapid improvements in software for astrophysical fluid dynamics, including open-source community projects, mean that high-fidelity simulations of bow shocks and wind bubbles can now be performed. In comparison with new high-resolution datasets coming from observations across the electromagnetic spectrum, these models have significant power to constrain uncertain physical parameters, discriminate between different physical models, and in general give a deeper and clearer understanding of astrophysical gas dynamics.

Acknowledgements

JM acknowledges support from a Royal Society-Science Foundation Ireland University Research Fellowship and an Irish Research Council (IRC) Starting Laureate Award, and the DJEI/DES/SFI/HEA Irish Centre for High-End Computing (ICHEC) for computational facilities and support. It is a pleasure to acknowledge the contributions of members of my research group at DIAS to this review through the papers cited and through many discussions over the past 7 years. I am very grateful to the SOC of IAUS 370 for the invitation to present this review at the IAUGA 2022 in South Korea.

References

Acreman, D. M., Stevens, I. R., & Harries, T. J. 2016, MNRAS, 456, 136
Baalmann, L. R., Scherer, K., Kleimann, J., Fichtner, H., Bomans, D. J., & Weis, K. 2021, A&A, 650, A36
Bale, S. D., Pulupa, M., Salem, C., Chen, C. H. K., & Quataert, E. 2013, ApJL, 769, L22
Baranov, V. B., Krasnobaev, K. V., & Kulikovskii, A. G. 1971, Sov. Phys. Dokl., 15, 791
Benaglia, P., del Palacio, S., Hales, C., & Colazo, M. E. 2021, MNRAS, 503, 2514
Benaglia, P., Romero, G. E., Martí, J., Peri, C. S., & Araudo, A. T. 2010, A&A, 517, L10
Blondin, J. M., & Koerwer, J. F. 1998, NewA, 3, 571
Boroson, B., McCray, R., Oelfke Clark, C., Slavin, J., Mac Low, M.-M., Chu, Y.-H., & van Buren, D. 1997, ApJ, 478, 638
Castor, J., McCray, R., & Weaver, R. 1975, ApJL, 200, L107
Comerón, F., & Kaper, L. 1998, A&A, 338, 273
Cowie, L. L., & McKee, C. F. 1977, ApJ, 211, 135
Das, S., Brose, R., Meyer, D. M. A., Pohl, M., Sushch, I., & Plotko, P. 2022, A&A, 661, A128
del Palacio, S., Bosch-Ramon, V., Müller, A. L., & Romero, G. E. 2018, A&A, 617, A13
del Valle, M. V., & Pohl, M. 2018, ApJ, 864, 19
del Valle, M. V., & Romero, G. E. 2012, A&A, 543, A56
Dyson, J. E., & de Vries, J. 1972, A&A, 20, 223
Fichtner, Y. A., Grassitelli, L., Romano-Díaz, E., & Porciani, C. 2022, MNRAS, 512, 4573
Frank, A., Jones, T. W., Ryu, D., & Gaalaas, J. B. 1996, ApJ, 460, 777
Freyer, T., Hensler, G., & Yorke, H. W. 2003, ApJ, 594, 888
—. 2006, ApJ, 638, 262
García-Segura, G., Langer, N., & Mac Low, M. 1996a, A&A, 316, 133
García-Segura, G., Mac Low, M., & Langer, N. 1996b, A&A, 305, 229

Geen, S., Bieri, R., Rosdahl, J., & de Koter, A. 2021, MNRAS, 501, 1352
Geen, S., Rosdahl, J., Blaizot, J., Devriendt, J., & Slyz, A. 2015, MNRAS, 448, 3248
Green, S., Mackey, J., Haworth, T. J., Gvaramadze, V. V., & Duffy, P. 2019, A&A, 625, A4
Green, S., Mackey, J., Kavanagh, P., Haworth, T. J., Moutzouri, M., & Gvaramadze, V. V. 2022, A&A, 665, A35
Gull, T. R., & Sofia, S. 1979, ApJ, 230, 782
Gvaramadze, V. V., & Bomans, D. J. 2008, A&A, 490, 1071
Gvaramadze, V. V., Langer, N., & Mackey, J. 2012, MNRAS, 427, L50
Gvaramadze, V. V., Menten, K. M., Kniazev, A. Y., Langer, N., Mackey, J., Kraus, A., Meyer, D. M.-A., & Kamiński, T. 2014, MNRAS, 437, 843
Henney, W. J., & Arthur, S. J. 2019, MNRAS, 489, 2142
H.E.S.S. Collaboration et al. 2018, A&A, 612, A12
—. 2020, A&A, 635, A167
—. 2022, A&A, 666, A124
Jiménez-Hernández, P., Arthur, S. J., Toalá, J. A., & Marston, A. P. 2021, MNRAS, 507, 3030
Keppens, R., Tóth, G., Westermann, R. H. J., & Goedbloed, J. P. 1999, Journal of Plasma Physics, 61, 1
Kobulnicky, H. A., Chick, W. T., & Povich, M. S. 2018, ApJ, 856, 74
—. 2019, AJ, 158, 73
Kobulnicky, H. A., et al. 2016, The Astrophysical Journal Supplement Series, 227, 18
Lancaster, L., Ostriker, E. C., Kim, J.-G., & Kim, C.-G. 2021a, ApJ, 914, 89
—. 2021b, ApJ, 914, 90
Mackey, J., Green, S., Moutzouri, M., Haworth, T. J., Kavanagh, R. D., Zargaryan, D., & Celeste, M. 2021, MNRAS, 504, 983
Mackey, J., Gvaramadze, V. V., Mohamed, S., & Langer, N. 2015, A&A, 573, A10
Mackey, J., Haworth, T. J., Gvaramadze, V. V., Mohamed, S., Langer, N., & Harries, T. J. 2016, A&A, 586, A114
Marcolino, W. L. F., Bouret, J.-C., Martins, F., Hillier, D. J., Lanz, T., & Escolano, C. 2009, A&A, 498, 837
Mathews, W. G. 1966, ApJ, 144, 206
Meyer, D. M.-A., Mackey, J., Langer, N., Gvaramadze, V. V., Mignone, A., Izzard, R. G., & Kaper, L. 2014, MNRAS, 444, 2754
Meyer, D. M.-A., Mignone, A., Kuiper, R., Raga, A. C., & Kley, W. 2017, MNRAS, 464, 3229
Meyer, D. M. A., Mignone, A., Petrov, M., Scherer, K., Velázquez, P. F., & Boumis, P. 2021, MNRAS, 506, 5170
Meyer, D. M. A., Petrov, M., & Pohl, M. 2020, MNRAS, 493, 3548
Mohamed, S., Mackey, J., & Langer, N. 2012, A&A, 541, A1
Moutzouri, M., et al. 2022, A&A, 663, A80
Pavlyuchenkov, Y. N., Kirsanova, M. S., & Wiebe, D. S. 2013, Astronomy Reports, 57, 573
Peri, C. S., Benaglia, P., Brookes, D. P., Stevens, I. R., & Isequilla, N. L. 2012, A&A, 538, A108
Peri, C. S., Benaglia, P., & Isequilla, N. L. 2015, A&A, 578, A45
Pittard, J. M., & Dougherty, S. M. 2006, MNRAS, 372, 801
Rathjen, T.-E., et al. 2021, MNRAS, 504, 1039
Rogers, H., & Pittard, J. M. 2014, MNRAS, 441, 964
Rosen, A. L., Lopez, L. A., Krumholz, M. R., & Ramirez-Ruiz, E. 2014, MNRAS, 442, 2701
Sana, H., et al. 2012, Science, 337, 444
Scherer, K., Baalmann, L. R., Fichtner, H., Kleimann, J., Bomans, D. J., Weis, K., Ferreira, S. E. S., & Herbst, K. 2020, MNRAS, 493, 4172
Schulz, A., Ackermann, M., Buehler, R., Mayer, M., & Klepser, S. 2014, A&A, 565, A95
Slavin, J. D., Shull, J. M., & Begelman, M. C. 1993, ApJ, 407, 83
Toalá, J. A., & Arthur, S. J. 2011, ApJ, 737, 100
Toalá, J. A., Guerrero, M. A., Chu, Y.-H., & Gruendl, R. A. 2015, MNRAS, 446, 1083
Toalá, J. A., Oskinova, L. M., González-Galán, A., Guerrero, M. A., Ignace, R., & Pohl, M. 2016, ApJ, 821, 79

Toalá, J. A., Oskinova, L. M., & Ignace, R. 2017, ApJL, 838, L19
Townsley, L. K., Broos, P. S., Garmire, G. P., Anderson, G. E., Feigelson, E. D., Naylor, T., & Povich, M. S. 2018, ApJS, 235, 43
Ueta, T., et al. 2008, PASJ, 60, 407
van Buren, D., & McCray, R. 1988, ApJL, 329, L93
van Buren, D., Noriega-Crespo, A., & Dgani, R. 1995, AJ, 110, 2914
Van den Eijnden, J., Saikia, P., & Mohamed, S. 2022, MNRAS, 512, 5374
van den Eijnden, J., et al. 2022, MNRAS, 510, 515
Vink, J. S., de Koter, A., & Lamers, H. J. G. L. M. 2000, A&A, 362, 295
Weaver, R., McCray, R., Castor, J., Shapiro, P., & Moore, R. 1977, ApJ, 218, 377

Winds of Stars and Exoplanets
Proceedings IAU Symposium No. 370, 2023
A. A. Vidotto, L. Fossati & J. S. Vink, eds.
doi:10.1017/S1743921323000066

Numerical Modeling of Galactic Superwinds with Time-evolving Stellar Feedback

A. Danehkar[1], M. S. Oey[2] and W. J. Gray

[1]Eureka Scientific, 2452 Delmer Street, Suite 100, Oakland, CA 94602-3017, USA

[2]Department of Astronomy, University of Michigan, Ann Arbor, MI 48109, USA
email: danehkar@eurekasci.com

Abstract. Mass-loss and radiation feedback from evolving massive stars produce galactic-scale superwinds, sometimes surrounded by pressure-driven bubbles. Using the time-dependent stellar population typically seen in star-forming regions, we conduct hydrodynamic simulations of a starburst-driven superwind model coupled with radiative efficiency rates to investigate the formation of radiative cooling superwinds and bubbles. Our numerical simulations depict the parameter space where radiative cooling superwinds with or without bubbles occur. Moreover, we employ the physical properties and time-dependent ionization states to predict emission line profiles under the assumption of collisional ionization and non-equilibrium ionization caused by wind thermal feedback in addition to photoionization created by the radiation background. We see the dependence of non-equilibrium ionization structures on the time-evolving ionizing source, leading to a deviation from collisional ionization in radiative cooling wind regions over time.

Keywords. Stars: winds – ISM: bubbles – hydrodynamics – galaxies: starburst

1. Introduction

Galactic-scale superwinds emerging from star-forming galaxies have commonly been seen in several multiwavelength observations (see e.g., Heckman et al. 1990; Rupke et al. 2005; Veilleux et al. 2005), which are sometimes accompanied by large-scale bubbles (e.g., Veilleux et al. 1994; Sakamoto et al. 2006; Tsai et al. 2009). Moreover, some observations of compact starburst regions pointed to unexpected cooling and suppressed superwinds (Oey et al. 2017; Turner et al. 2017; Jaskot et al. 2017), which could not be completely explained by the standard model based on the adiabatic assumption (Chevalier & Clegg 1985; Cantó et al. 2000). However, these phenomena could be related to heat being lost through radiation. In particular, semi-analytical studies of superwind models with radiative cooling found that the wind temperature could deviate from the adiabatic result depending on the stellar mass-loss rate and wind velocity (Silich et al. 2004; Tenorio-Tagle et al. 2005), which has been confirmed by recent hydrodynamic simulations (Gray et al. 2019a; Danehkar et al. 2021).

While mechanical feedback from massive OB stars could create pressure-driven bubbles, some bubbles seem to expand more slowly than predicted (see e.g., Brown et al. 1995; Oey 1996). This could be explained by the time-dependent stellar feedback from evolving massive stars, which undergo stellar evolution, i.e., OB star → RSG/LBV → WR star. Recently, Danehkar et al. (2021, 2022) implemented hydrodynamic simulations and photoionization models of superwinds for different wind parameters using the stellar feedback from a stellar population at a fixed age of 1 Myr. However, taking the entire mass-loss history of massive stars can significantly change the theoretical predictions of pressure-driven bubbles over a timescale larger than 1 Myr (see e.g., Oey & García-Segura 2004; Krause et al. 2013). Moreover, a time-evolving photoionizing source could modify non-equilibrium photoionization predictions, which primarily rely on time-dependent ionization states made by hydrodynamic simulations using the radiation background.

2. Numerical Modeling of Galactic Superwinds

We conducted hydrodynamic simulations of a spherically symmetric superwind model coupled with the radiative efficiency rates using the non-equilibrium chemistry package MAIHEM (Gray et al. 2019b) built on the adaptive mesh hydrodynamics code FLASH (Fryxell et al. 2000), which solves the following fluid equations:

$$\frac{d\rho}{dt} + \frac{1}{r^2}\frac{d}{dr}\left(\rho u r^2\right) = q_m(t), \tag{2.1}$$

$$\frac{d\rho u}{dt} + \rho u \frac{du}{dr} + \frac{dP}{dr} = -\, q_m(t) u, \tag{2.2}$$

$$\frac{d\rho E}{dt} + \frac{1}{r^2}\frac{d}{dr}\left[\rho u r^2 \left(\frac{u^2}{2} + \frac{\gamma}{\gamma-1}\frac{P}{\rho}\right)\right] = \sum_i n_i \Gamma_i(t) - \sum_i n_i n_e \Lambda_i + q_e(t), \tag{2.3}$$

$$\begin{aligned}\frac{1}{n_\mathrm{e}}\frac{dn_i}{dt} =& n_{i+1}\alpha_{i+1} - n_i\alpha_i + n_{i-1}S_{i-1} \\ & - n_i S_i + \frac{1}{n_\mathrm{e}} n_{i-1}\zeta_{i-1}(t) - \frac{1}{n_\mathrm{e}} n_i \zeta_i(t),\end{aligned} \tag{2.4}$$

where r, ρ, u, P, and E are the radius, density, velocity, pressure, and energy per mass of the fluid, respectively, $\gamma = 5/3$ is the specific heat ratio, $q_m(t) = \dot{M}(t)/(\frac{4}{3}\pi R_{\mathrm{sc}}^3)$ and $q_e(t) = \frac{1}{2}\dot{M}(t)V_\infty(t)^2/(\frac{4}{3}\pi R_{\mathrm{sc}}^3)$ are the time-dependent mass and energy injection rates per volume according to the mass-loss rate $\dot{M}(t)$ and wind velocity $V_\infty(t)$ of the time-evolving stellar population produced by Starburst99, respectively, n_i is the ion densities, n_e the electron density, Λ_i the radiative cooling rates for the given temperature T derived from the cooling atomic data (Gnat & Ferland 2012), $\Gamma_i(t) = \int_{\nu_{0,i}}^{\infty}(4\pi J_\nu(t)/\nu)(\nu - \nu_{0,i})\sigma_i(\nu)\mathrm{d}\nu$ and $\zeta_i(t) = \int_{\nu_{0,i}}^{\infty}(4\pi J_\nu(t)/h\nu)\sigma_i(\nu)\mathrm{d}\nu$ are respectively the time-dependent photo-heating and photoionization rates calculated using the photoionization cross-section atomic data $\sigma_i(\nu)$ (Verner & Yakovlev 1995; Verner et al. 1996) and the radiation field $J_\nu(t)$ of the time-evolving ionizing stellar population generated by Starburst99, ν and $\nu_{0,i}$ the frequency and the ionization frequency, respectively, h the Planck constant, α_i the ionic recombination rate including radiative data (Badnell 2006) and dielectronic data (see references in Gray et al. 2015), and S_i is the collisional ionization rates from Voronov (1997).

2.1. *Boundary and Initial Conditions*

To perform hydrodynamic simulations, we assumed the analytic solutions of the fluid model derived by Chevalier & Clegg (1985) and extended by Silich et al. (2004) for radiative cooling. Based on these solutions, we set the time-dependent boundary conditions for the density, velocity, and pressure at the cluster radius $r = R_{\mathrm{sc}}$ as $\rho_{\mathrm{sc}}(t) = \dot{M}(t)/[2\pi R_{\mathrm{sc}}^2 V_\infty(t)]$, $u_{\mathrm{sc}}(t) = \frac{1}{2}V_\infty(t)$, and $P_{\mathrm{sc}}(t) = \dot{M}(t)V_\infty(t)/(\gamma 8\pi R_{\mathrm{sc}}^2)$, respectively, where $V_\infty(t) = V_{\infty,0} f_{\mathrm{v}}(t)$ and $\dot{M}(t) = \dot{M}_0 g_{\dot{\mathrm{m}}}(t)$ are the time-dependent wind velocity and mass-loss rate, respectively, $V_{\infty,0}$ and $\dot{M}_0$ the user-defined wind velocity and mass-loss rate at $t = 0$, $f_{\mathrm{v}}(t)$ is a dimensionless function associated with the time-evolving wind velocity normalized using the initial wind velocity calculated from the mechanical luminosities and mass-loss rates predicted by Starburst99, and $g_{\dot{\mathrm{m}}}(t)$

is a dimensionless function made using the time-evolving mass-loss rates produced by Starburst99 normalized with the initial mass-loss rate. The initial conditions of the density, velocity, and pressure outside the cluster radius at $t=0$ are: $\rho_0=\mu m_{\rm p} n_{\rm amb}$, $u_0=0$, and $P_0=k_{\rm B} n_{\rm amb} T_{\rm amb}$, where $n_{\rm amb}$ and $T_{\rm amb}$ are the number density and temperature of the ambient medium, respectively, μ is the mean atomic weight ($\mu=0.61$ for a fully ionized gas), $m_{\rm p}$ the proton mass, and $k_{\rm B}$ the Boltzmann constant. The ambient temperature $T_{\rm amb}$ is calculated by CLOUDY for a stationary medium with $n_{\rm amb}$.

2.2. *Time-evolving Stellar Feedback*

We used the evolutionary synthesis code Starburst99 (Levesque et al. 2012; Leitherer et al. 2014) to generate the time-evolving radiation field and stellar feedback for stellar population evolution from 1 to 7 Myr. with an initial total stellar mass of $M_\star=2\times10^6\,{\rm M}_\odot$ and an IMF with the Salpeter $\alpha=2.35$ for the stellar masses ranging from 0.5 to 150 ${\rm M}_\odot$, using the rotational Geneva population (Ekström et al. 2012; Georgy et al. 2012) and Pauldrach/Hillier atmosphere (Hillier & Miller 1998; Pauldrach et al. 2001). The time-dependent ionizing luminosity $L_{\rm ion}(t)$ and spectrum $J_\nu(t)$ computed by Starburst99 were employed by the photo-heating efficiencies $\Gamma_i(t)$ in our hydrodynamic simulations and the photoionization rates $\zeta_i(t)$ in our photoionization calculations. The time-dependent mass-loss rate $\dot{M}(t)$ at 0.1 Myr interval calculated by Starburst99 was also used to gradually modify the mass and energy injection rates – $q_m(t)$ and $q_e(t)$ – in our hydrodynamic simulation while it is running.

3. Numerical Results

3.1. *Galactic Superwind Modes*

Danehkar et al. (2021) classified galactic superwinds into different wind modes according to the deviation of the wind temperature ($T_{\rm w}$) from the adiabatic solution ($T_{\rm adi}$). The adiabatic and quasi-adiabatic modes (AW, AB, and AP) are those with mean wind temperatures having $f_T\equiv T_{\rm w}/T_{\rm adi}\geqslant0.75$. The adiabatic wind (AW) and adiabatic bubble (AB) modes are without and with bubbles, respectively, while the bubble expansion is stalled by the ambient pressure in the adiabatic, pressure-confined (AP) mode. The catastrophic cooling (CC) and catastrophic cooling bubble (CB) modes are those with and without bubbles, respectively, but with $f<0.75$, while the cooling, pressure-confined (CP) mode describes radiatively cooling with a stalled bubble. Additionally, the no wind (NW) and momentum-conserving (MC) modes describe suppressed superwinds, which were controlled by high ambient pressures and substantial cooling effects, respectively.

Figure 1 presents different wind modes in the space parameters of the ambient density $n_{\rm amb}$, metallicity $Z/{\rm Z}_\odot$, and age (t) with the time-evolving wind velocity $V_\infty(t)=500\hat{Z}^{0.13}f_{\rm v}(t)$, km s^{-1} and mass-loss rate $\dot{M}(t)=10^{-3}\hat{Z}^{0.72}g_{\dot{\rm m}}(t)\,{\rm M}_\odot\,{\rm yr}^{-1}$, where the stellar cluster has a radius of $R_{\rm sc}=1\,{\rm pc}$ and a total mass of $M_\star=2\times10^5\,{\rm M}_\odot$. We see the formation of radiative cooling in older ages in time-evolving models, while higher metallicities and weaker wind velocities contribute to stronger radiative cooling. However, the formation of a bubble cannot always be suppressed by cooling effects, so we have several superwinds in the CB mode (see also Fig. 4 in Danehkar et al. 2021).

In Figure 2, the temperature and density profiles (solid red lines) of a superwind predicted by our hydrodynamic simulation are plotted in the left panels against the adiabatic solutions (red dashed lines). The profiles are divided into four different regions according to Weaver et al. (1977), namely (1) wind, (2) bubble, (3) shell, and (4) ambient medium (see dotted, dashed, and dash-dotted gray lines). The Strömgren sphere (solid gray line) was also shown, which is predicted by a pure photoionization model.

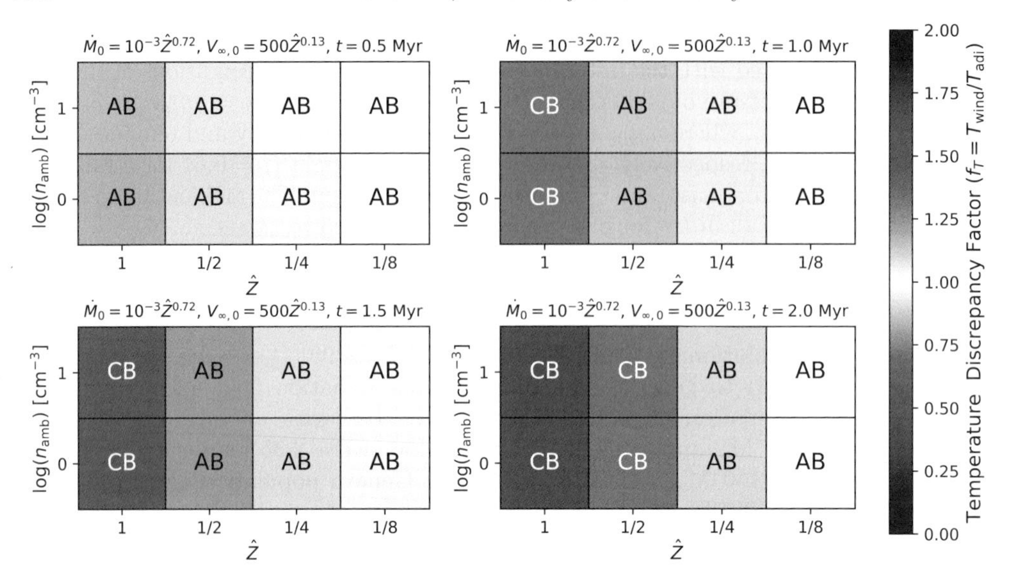

Figure 1. The mean wind temperature $T_{\rm wind}$ produced by MAIHEM with respect to the mean adiabatic solution $T_{\rm adi}$ for the time-dependent wind parameters $V_{\infty}(t) = 500\hat{Z}^{0.13} f_{\rm v}(t)$, km s^{-1} and $\dot{M}(t) = 10^{-3}\hat{Z}^{0.72} g_{\dot{\rm m}}(t)\,{\rm M}_{\odot}\,{\rm yr}^{-1}$, metallicities $\hat{Z} \equiv Z/{\rm Z}_{\odot} = [1/8,\ 1/4,\ 1/2,\ 1]$, ambient densities $n_{\rm amb} = [1,\ 10,\ 10^2,\ 10^3]\,{\rm cm}^{-3}$, and a stellar cluster with $R_{\rm sc} = 1$ pc, $M_{\star} = 2 \times 10^5\,{\rm M}_{\odot}$, and ages $t = [0.5,\ 1,\ 1.5,\ 2]$ Myr. The wind models are identified as adiabatic bubble (AB) and catastrophic cooling bubble (CB), based on criteria defined by Danehkar et al. (2021).

3.2. *Collisional Ionization versus Non-equilibrium Ionization*

To create collisional ionization, Danehkar et al. (2021) employed the density and temperature profiles produced by our hydrodynamic simulations, along with the radiation field, to calculate the emission-line profiles with the photoionization code CLOUDY (Ferland et al. 2017). Our hydrodynamic simulations also generate time-dependent ionization states using Eq. (2.4), which can make non-equilibrium ionization (NEI) in the cooling ($< 10^6$ K) regions where collisional ionization takes longer than radiative cooling. To produce non-equilibrium photoionization, Danehkar et al. (2022) performed zone-by-zone CLOUDY computations by running one individual photoionization model for the given temperature, density, and NEI states of each zone, while each hydrodynamic simulation typically contains 1024 zones. Pure photoionization is applied to the ambient medium.

Figure 2 shows the emissivity profiles of different emission lines predicted by hybrid collisional ionization and photoionization (CPI; top-right panels) and non-equilibrium photoionization calculations (NPI; bottom-right panels) made with the physical properties and time-dependent ionization states produced by our hydrodynamic simulations (for a wider parameter range, see Fig. 3 in Danehkar et al. 2022). It can be seen that the O VI emission line does not have the same emissivity profile in the NPI status as it does in the CPI status, especially at temperatures below 10^6 K in the radiative cooling region. This effect is a consequence of time-dependent photoionization, which occurs when the radiative cooling timescale is much faster than the collisional ionization timescale.

4. Applications in Starburst Regions

Danehkar et al. (2022) found that the O VI $\lambda\lambda$1032,1038 emission-line fluxes predicted by non-equilibrium photoionization demonstrate noticeable enhancements in metal-poor models. As proposed by Gray et al. (2019a,b), the enhanced O VI lines could be linked

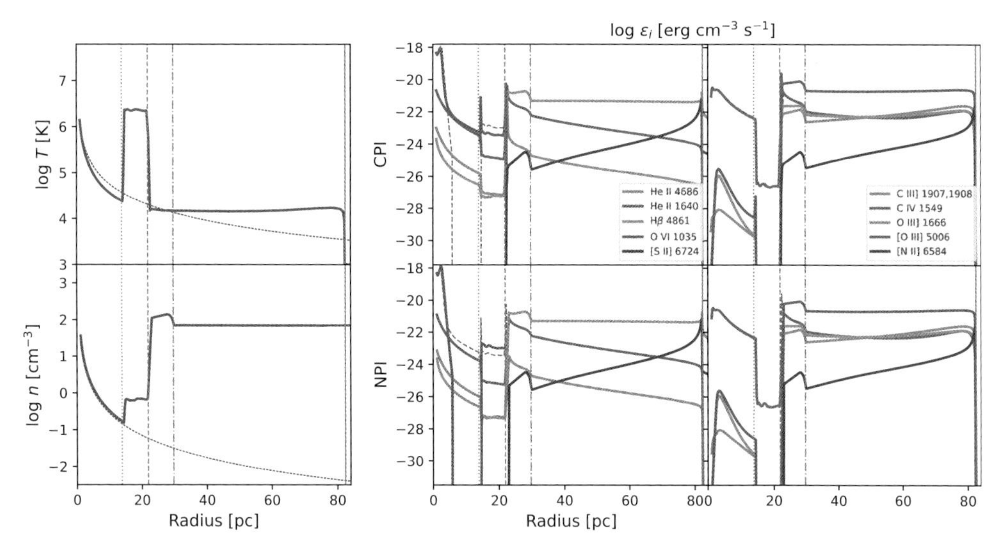

Figure 2. The temperature (T) and density (n) profiles (solid red lines) plotted against the adiabatic solutions (red dashed lines). The logarithmic emissivities $\log \varepsilon_i$ of different emission lines predicted by hybrid collisional ionization and photoionization (CPI) and non-equilibrium photoionization (NPI) calculations. The boundaries of the bubble, shell end, and Strömgren sphere are split using dotted, dashed, dash-dotted, and solid gray lines. The model parameters are: $V_\infty = 418\,\mathrm{km\,s^{-1}}$, $\dot{M} = 0.369 \times 10^{-2}\,\mathrm{M_\odot\,yr^{-1}}$, $t = 1\,\mathrm{Myr}$, $R_{\mathrm{sc}} = 1\,\mathrm{pc}$, $M_\star = 2 \times 10^6\,\mathrm{M_\odot}$, $n_{\mathrm{amb}} = 100\,\mathrm{cm^{-3}}$, $Z/Z_\odot = 0.25$. The O VI lines predicted by NPI are overplotted by dashed green lines in the CPI panel, and vice versa.

to strong radiative cooling in starburst regions. O VI emission with a velocity offset of about 50–100 km s^{-1} in the rest frame was detected toward a hot bubble in the nearby spiral galaxies NGC 4631, providing evidence for gas cooling (Otte et al. 2003). For the intense starburst J1156+5008 at $z = 0.236$ with an O VI absorption blueshifted by 380 km s^{-1}, Hayes et al. (2016) concluded that the O VI -carrying gas must be cooling in situ via the coronal phase. Moreover, Chisholm et al. (2018) proposed the creation of O VI absorption in the high-redshift ($z = 2.92$) galaxy J1226+2152 by either the conductive evaporation of cool gas or a cooling flow between a hot outflow and a cooler photoionized gas.

The nearest ($d = 82$ Mpc), Lyman-break analog Haro 11, also depicts O VI $\lambda\lambda$1032,1038 absorption features with a wind velocity of 200–280 km s^{-1}, which could be attributed to radiative cooling superwinds (Grimes et al. 2007). O VI emission was also identified in Haro 11, for which Grimes et al. (2007) estimated that up to 20% of the supernova feedback was lost to possible radiative cooling. It is one of the nearest analogues to high-redshift galaxies because of its high star formation and relatively low metallicities. Haro 11 includes three main knots (A, B, and C) consisting of several super star clusters with a cluster age distribution up to 40 Myr and a peak age of 3.5 Myr (Adamo et al. 2010). While Knot B could produce energy-driven superwinds due to the presence of visible bubbles, the lack of any bubbles around Knot C may be an indication of momentum conservation or substantial radiative cooling. The ionized gas around Knot C was found to have a metallicity of 0.12 Z$_\odot$, a factor of 3 lower than the ISM in Knot B (James et al. 2013), so the metallicity does not seem to have a major role in the formation of momentum-driven or radiatively cooled superwinds in Knot C. Future high-resolution UV spectroscopic measurements of different regions of the knots in Haro 11 will allow us to determine which of them primarily bears O VI absorbing winds. Our hydrodynamic simulations and non-equilibrium photoionization calculations with time-evolving stellar feedback at ages beyond 1 Myr and wider parameter ranges of superwinds and star clusters will certainly improve our understanding of observed O VI lines and their possible implications for radiative cooling in starburst regions.

References

Adamo A., Östlin G., Zackrisson E., Hayes M., et al., 2010, MNRAS, 407, 870
Badnell N. R., 2006, ApJS, 167, 334
Brown A. G. A., Hartmann D., Burton W. B., 1995, A&A, 300, 903
Cantó J., Raga A. C., Rodríguez L. F., 2000, ApJ, 536, 896
Chevalier R. A., Clegg A. W., 1985, Nature, 317, 44
Chisholm J., Bordoloi R., Rigby J. R., Bayliss M., 2018, MNRAS, 474, 1688
Danehkar A., Oey M. S., Gray W. J., 2021, ApJ, 921, 91
Danehkar A., Oey M. S., Gray W. J., 2022, ApJ, 937, 68
Ekström S. et al., 2012, A&A, 537, A146
Ferland G. J. et al., 2017, RMxAA, 53, 385
Fryxell B. et al., 2000, ApJS, 131, 273
Georgy C., Ekström S., Meynet G., Massey P., Levesque E. M., et al., 2012, A&A, 542, A29
Gnat O., Ferland G. J., 2012, ApJS, 199, 20
Gray W. J., Oey M. S., Silich S., Scannapieco E., 2019a, ApJ, 887, 161
Gray W. J., Scannapieco E., Kasen D., 2015, ApJ, 801, 107
Gray W. J., Scannapieco E., Lehnert M. D., 2019b, ApJ, 875, 110
Grimes J. P. et al., 2007, ApJ, 668, 891
Hayes M., Melinder J., Östlin G., Scarlata C., et al., 2016, ApJ, 828, 49
Heckman T. M., Armus L., Miley G. K., 1990, ApJS, 74, 833
Hillier D. J., Miller D. L., 1998, ApJ, 496, 407
James B. L., Tsamis Y. G., Walsh J. R., et al., 2013, MNRAS, 430, 2097
Jaskot A. E., Oey M. S., Scarlata C., Dowd T., 2017, ApJ, 851, L9
Krause M., Fierlinger K., Diehl R., Burkert A., Voss R., Ziegler U., 2013, A&A, 550, A49
Leitherer C., Ekström S., Meynet G., Schaerer D., et al., 2014, ApJS, 212, 14
Levesque E. M., Leitherer C., Ekstrom S., Meynet G., Schaerer D., 2012, ApJ, 751, 67
Oey M. S., 1996, ApJ, 467, 666
Oey M. S., García-Segura G., 2004, ApJ, 613, 302
Oey M. S., Herrera C. N., Silich S., Reiter M., James B. L., et al., 2017, ApJ, 849, L1
Otte B., Murphy E. M., Howk J. C., Wang Q. D., et al., 2003, ApJ, 591, 821
Pauldrach A. W. A., Hoffmann T. L., Lennon M., 2001, A&A, 375, 161
Rupke D. S., Veilleux S., Sanders D. B., 2005, ApJS, 160, 115
Sakamoto K. et al., 2006, ApJ, 636, 685
Silich S., Tenorio-Tagle G., Rodríguez-González A., 2004, ApJ, 610, 226
Tenorio-Tagle G., Silich S., Rodríguez-González A., Muñoz-Tuñón C., 2005, ApJ, 620, 217
Tsai A.-L. et al., 2009, PASJ, 61, 237
Turner J. L., Consiglio S. M., Beck S. C., Goss W. M., Ho P. T. P., et al., 2017, ApJ, 846, 73
Veilleux S., Cecil G., Bland-Hawthorn J., 2005, ARA&A, 43, 769
Veilleux S., Cecil G., Bland-Hawthorn J., Tully R. B., et al., 1994, ApJ, 433, 48
Verner D. A., Ferland G. J., Korista K. T., Yakovlev D. G., 1996, ApJ, 465, 487
Verner D. A., Yakovlev D. G., 1995, A&AS, 109, 125
Voronov G. S., 1997, Atom. Data Nucl. Data Tabl., 65, 1
Weaver R., McCray R., Castor J., Shapiro P., Moore R., 1977, ApJ, 218, 377

Winds of Stars and Exoplanets
Proceedings IAU Symposium No. 370, 2023
A. A. Vidotto, L. Fossati & J. S. Vink, eds.
doi:10.1017/S1743921322004720

Winds of OB stars: impact of metallicity, rotation and binary interaction

Varsha Ramachandran

Zentrum für Astronomie der Universität Heidelberg, Astronomisches Rechen-Institut, Mönchhofstr. 12-14, 69120 Heidelberg

Abstract. Winds of massive stars are an important ingredient in determining their evolution, final remnant mass, and feedback to the surrounding interstellar medium. We compare empirical results for OB star winds at low metallicity with theoretical predictions. Observations suggest very weak winds at SMC metallicity, but there are exceptions. We identified promising candidates for rotationally enhanced mass-loss rates with two component wind and partially stripped stars hiding among OB stars with slow but dense wind in the SMC. A preliminary analysis of these systems, derived parameters, and their implications are discussed. Finally, we briefly discuss the interaction of OB winds near black holes in X-ray binaries.

Keywords. massive stars, stellar wind, low metallicity, etc.

1. Introduction

Massive stars are hot, luminous objects that lose mass due to stellar winds driven by the scattering of UV photons by metal lines (e.g. Castor et al. 1975; Gräfener & Hamann 2005). Stellar winds have a major role in the evolution of massive stars throughout the course of their lifetimes, as well as in determining their ultimate fate and the masses of compact remnants. Detection of gravitational waves from coalescing black holes further highlighted the need for a better understanding of massive star winds, especially at low metallicity.

The winds of hot stars are characterized by two main parameters: the terminal velocity ν_∞ and the rate of mass-loss $\dot{M}$. We see direct evidence of mass-loss in the UV spectral line profiles of highly ionized species such as C IV, Si IV, and N V. Radiatively-driven winds are by nature metallicity dependent. Moreover, fast rotation and binary interaction can have a strong impact on the stellar wind and the evolution of the massive star. In this study, we highlight empirical studies of OB stars at low metallicity, focusing on their wind properties. Furthermore, we discuss the influence of rapid rotation and binary interaction on the wind properties.

2. Weak winds of massive stars at low metallicity

The spectroscopic analysis using stellar atmosphere models enables us to quantify wind mass-loss rates. UV spectroscopy along with optical observations are particularly suitable for this. Previous theoretical and empirical studies suggest that mass-loss rates depend on Z^α, with α between 0.7 and 0.8 (Vink et al. 2001; Mokiem et al. 2007). However, there is growing evidence for lower mass-loss rates for OB stars (Bouret et al. 2003; Martins et al. 2004) compared to widely used theoretical recipes (Vink et al. 2000, 2001). Lower mass-loss rates are observed for Galactic massive stars with luminosities less than $\log(L/L_\odot) \lesssim 5.2$ (late O and early B-type dwarfs), also known as the "weak-wind problem" (Martins et al. 2004; Puls et al. 2008). However, empirical determination of mass-loss rates of OB stars at SMC metallicity was sparse before. The major reasons include the absence of optical emission lines in the optical spectra due to low metallicity and unavailability of UV spectra.

Figure 1 depicts the mass-loss rate ($\dot{M}$) versus luminosity for SMC OB stars. The samples consist of non-supergiant OB stars in the Wing of the SMC Ramachandran et al. (2019) and in the young massive cluster NGC 346 Rickard et al. (2022). The spectral analyses of these samples were performed using PoWR† model atmospheres. A linear regression to this $\log \dot{M} - \log L$ relation, which accounts for the individual error bars, shows a steeper relation and more than an order of magnitude offset across all luminosity ranges compared to the theoretical predictions (Vink et al. 2001) at SMC metallicity. The results are consistent with earlier spectroscopic investigations of massive stars in the SMC (Bouret et al. 2003; Martins et al. 2004) and other low-metallicity dwarf galaxies like IC 1613 and WLM (Bouret et al. 2015; Lucy 2012). Although the empirical results are close to Björklund et al. (2021), they still overestimate the mass-loss rates by a factor of two or more. It also to be noted that the mass-loss rates of luminous supergiants in the SMC (Bouret et al. 2021) are found to be in agreement with Vink et al. (2001) predictions unlike OB dwarfs and giants.

The weak winds of OB stars at low metallicity should only have a small influence on their evolution. The fact that theoretical evolutionary tracks are typically constructed using conventional mass-loss recipes, which are significantly overestimated compared to observations, may modify our understanding of massive star evolution. This also affects the gravitational wave population synthesis and stellar feedback in galaxy evolution simulations. We have secured HST UV spectroscopy of low metallicity O stars in the Magellanic Bridge which will be further studied in detail to understand the stellar winds of sub-SMC metallicity stars (Ramachandran et al. 2021).

3. Stellar wind in rapidly rotating stars

Rotation modifies the shape of the star, deviating from spherical symmetry and thereby influencing stellar parameters such as effective gravity, effective temperature, and radiative flux. The winds in fast rotating massive stars do not remain isotropic but become increasingly anisotropic as the rotation approaches the critical rotation (Maeder & Meynet 2000). Very rapid rotators are expected to have cooler photospheres closer to their equatorial regions (due to gravity darkening), and that the local mass loss rate attains a minimum at the equator and the strongest winds occur at the poles (Owocki et al. 1996; Maeder & Meynet 2000; Müller & Vink 2014). However if the temperature at the equator reaches below the bistability jump due to gravity darkening, mass-loss rate might be enhanced over equatorial regions (Lamers & Pauldrach 1991). In summary, latitude dependent temperature and gravity by fast rotation are expected to result in asymmetric winds and enhanced mass-loss rates.

Few such systems have been studied in the Galaxy (Bjorkman et al. 1994; Massa 1995; Prinja et al. 1997) and in the LMC (Shepard et al. 2020). However no such fast rotating OB stars with rotationally enhanced wind were reported in the SMC. One exception is the luminous O supergiant AV 83, showing extreme wind mass-loss rate despite slow rotation velocity and no asymmetric wind line profiles (Hillier et al. 2003). Investigating rotationally enhanced mass-loss rate at low metallicity is of foremost importance since most of the OB stars have very weak winds whereas fast rotation is common.

† http://www.astro.physik.uni-potsdam.de/~wrh/PoWR

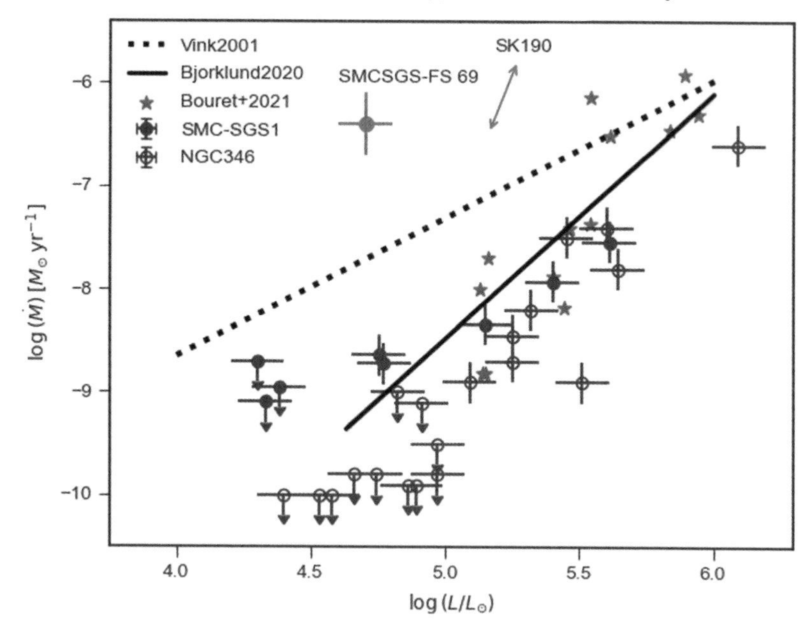

Figure 1. Mass-loss rate as a function of stellar luminosity for OB stars in SMC (Ramachandran et al. 2019; Rickard et al. 2022). For comparison, the theoretical predictions from Vink et al. (2000, 2001) (dashed black line), and Björklund et al. (2021) (solid black line) corresponding to SMC metallicities are marked. Mass-loss rates of the stripped star and rapid rotator are marked in red and giants/supergiants in the SMC are shown in green asterisks for comparison (Bouret et al. 2021).

We found one such good candidate, namely SK 190, an O giant/ supergiant star in the Wing of the SMC. Based on the optical spectral analysis using PoWR, we derived the stellar parameters of the star to be $T_* = 30\,\mathrm{kK}$, $\log g_* = 3.2$, $\log L/L_\odot = 5.3$ and $\nu \sin i = 300$ km $s^{-1} \sim 0.8\nu_{\mathrm{crit}}$. However, the UV lines show evidence for latitude dependent parameters: a hot pole and a cool equator with $\approx 6\,\mathrm{kK}$ difference in temperature, higher surface gravities at the pole by ≈ 0.4 dex and two wind components. Saturate UV P-Cygni profiles C IV and Si IV show a narrow absorption component and a broad emission part, suggesting that the terminal wind velocity increases from equator to poles (Figure 1). In the spectroscopic analysis we had to assume a slow wind of 400 km s^{-1} to account for the narrow absorption and a fast wind of 1600 km s^{-1} to fit the broad emission. In addition high ionization lines such N V and O V present in the spectra can be only reproduced using hotter temperature models than those used to fit optical spectra. Based on models assuming cooler temperature, lower surface gravity and slow wind velocity we estimated a mass-loss rate of $\log \dot{M} = -6.4 M_\odot\,\mathrm{yr}^{-1}$ at the equator whereas hotter, higher surface gravity models corresponding to the pole with faster wind velocities suggested a much higher mass-loss rate $\log \dot{M} = -5.8 M_\odot\,\mathrm{yr}^{-1}$. The mass-loss rate constraints for this star is compared with that of OB stars and supergiants in the SMC in Figure 1. Only a few luminous supergiants in the SMC show such high mass-loss rates. Moreover, stars of similar luminosity as that of Sk 190 show many orders of magnitude lower mass-loss rates.

4. Stellar wind in stripped star

Interactions in OB binaries can often result in the stripping of the primary's envelope (Paczyński 1967), generating hot and compact helium core stars with only a thin layer of hydrogen on top (e.g., Yoon et al. 2010; Claeys et al. 2011). Depending on their

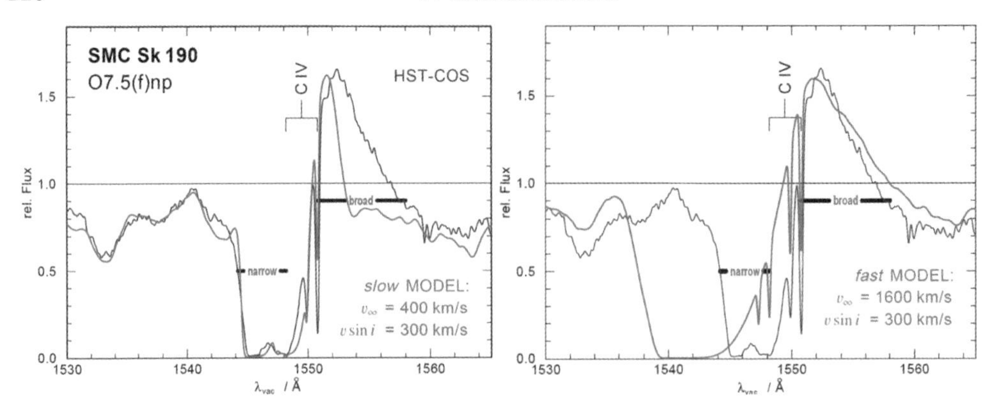

Figure 2. The C IV UV wind profile of Sk 190 (blue) compared with models calculated with hot, fast and dense wind corresponding to the poles ($T_* = 36$ kK, ν_∞=1600 km s^{-1}, $\log \dot{M} = -5.8 M_\odot\,\mathrm{yr}^{-1}$) and cool slow wind at the equator ($T_* = 30$ kK, ν_∞=400 km s^{-1}, $\log \dot{M} = -6.4 M_\odot\,\mathrm{yr}^{-1}$).

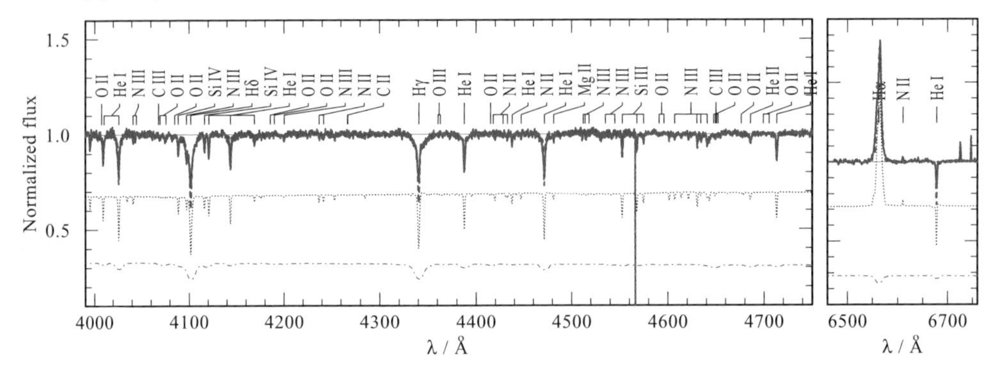

Figure 3. Comparison of observed spectra (blue) with composite PoWR models (red) for SMCSGS-FS 69. The final model is a combination of stripped primary (brown) and fast rotating secondary (green).

initial masses, the stripped envelop primaries would have spectral characteristics ranging from hot subdwarf to Wolf-Rayet (WR) stars (Paczyński 1967; Vanbeveren 1991; Eldridge et al. 2008; Götberg et al. 2017). On the other hand, the secondaries would become rapidly rotating stars (de Mink et al. 2013), showing disk emission features like Be stars (Shao & Li 2014).

Interestingly, stripped stars with masses in between classical WR stars and low-mass subdwarfs are rarely observed. The only known candidate for an intermediate-mass He stars is the so-called qWR star HD 45166 (Groh et al. 2008) with other known stripped stars being in the lower mass range of $< 1.5 M_\odot$ and the appearance of subdwarfs (Wang et al. 2021; Schootemeijer et al. 2018). Some of the recently proposed X-ray quiet BH + Be binaries such as LB1 and HR6819 (Liu et al. 2019; Rivinius et al. 2020) have been later disputed (e.g., Bodensteiner et al. 2020; Shenar et al. 2020) or confirmed (Frost et al. 2022) to be a stripped star + Be star binary. However, both are recently detached systems in which the stripped star appears as a partially stripped cool supergiant. Recent state-of-the-art evolutionary models predict that metallicity has a substantial effect on the course and outcome of mass transfer evolution of massive binaries, leading to a large fraction of such partially stripped donors at low metallicity (Klencki et al. 2022). These systems are predicted to be hidden among OB binaries as main sequence or supergiant stars.

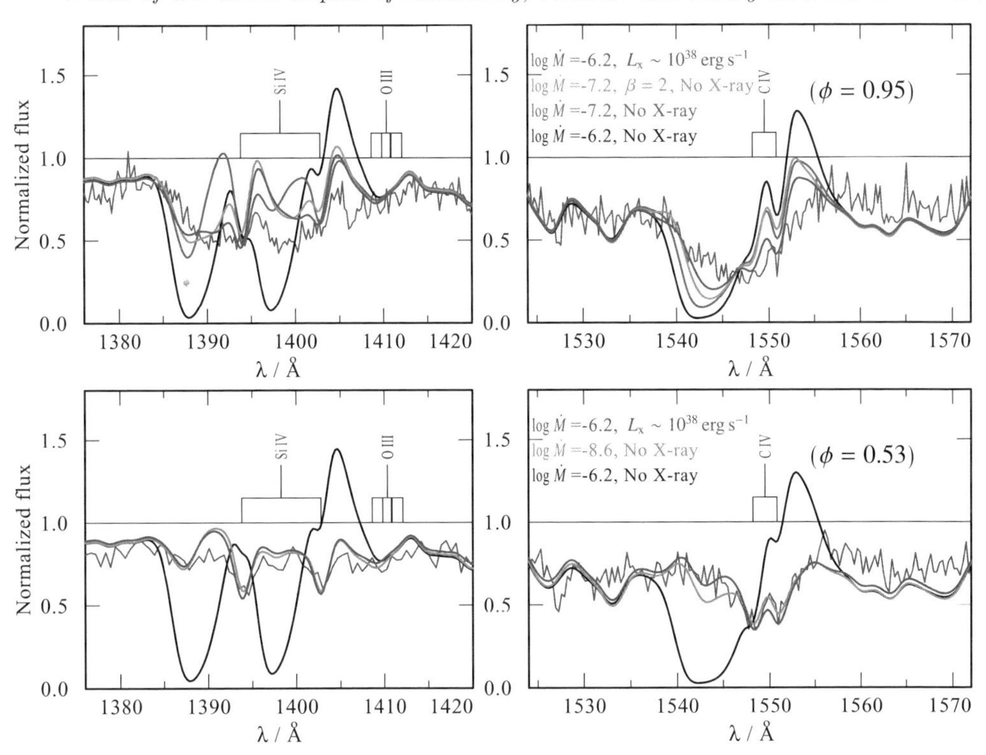

Figure 4. Observed Si IV (left and C IV (right) wind lines (blue solid) of M33 X-7 taken at eclipse (upper) and inferior conjunction (lower) compared to the models with different wind parameters and X-rays.

We identified a partially stripped star in a binary for the first time at low metallicity, SMCSGS-FS 69 (Ramachandran et al. in prep). A careful analysis of the optical spectra of SMCSGS-FS 69 revealed the presence of a narrow-lined and broad-lined star (see Figure 3). The narrow line star has a slowly rotating B supergiant-like spectral features but with very low luminosity ($\log L/L_\odot = 4.7$). The star is likely to be core-He burning, showing very high nitrogen enrichment, enhanced He and CO depletion. The spectroscopic mass of the star is only $\lesssim 3M_\odot$ making it one of the first reports of a partially stripped intermediate-mass star. The mass gainer secondary is found to be a rapidly rotating main sequence B star. As shown in Figure 3 (left), the $H\alpha$ is in strong emission. By reproducing this spectral feature assuming it is formed in the wind of the stripped star, we suggest an upper-limit for the mass-loss rate to be $\log \dot{M} = -6.2 M_\odot\,\mathrm{yr}^{-1}$. However, the $H\alpha$ emission could be a composite of wind from the stripped star and disk emission from the fast rotating secondary, suggesting $\log \dot{M}$ could be lower ($\lesssim -6.6 M_\odot\,\mathrm{yr}^{-1}$). Nevertheless, the mass-loss rate of the stripped star is significantly higher than that of OB stars of the same luminosity (Figure 1). Similar systems with high mass-loss rates may be lurking among ordinary OB stars, where they will be significant contributors to stellar wind feedback.

5. Stellar wind of OB star in black hole X-ray binary

To understand the complex behaviour of High Mass X-ray binaries (HMXBs) with black hole (BH) companions, detailed knowledge of the massive star donors is essential. However, only a few such systems are known so far. To remedy this situation, we performed a multi-wavelength phase-resolved analysis of the extragalactic HMXB M33 X-7.

This eclipsing BH HMXB is reported to contain a very massive O supergiant donor and a massive black hole in a short orbit (Orosz et al. 2007; Pietsch et al. 2006). However, previous spectroscopic analyses were limited to plane-parallel models which are optimized for hot stars with no significant wind.

Using phase-resolved simultaneous *HST*- and *XMM-Newton*-observations, we trace the interaction of the stellar wind with the BH. The UV resonance lines show the Hatchett-McCray effect with a large reduction in absorption strength when the BH is in the foreground due to the strong X-ray ionization (see Figure 4). Our comprehensive spectroscopic investigation of the donor star (X-ray+UV+optical) yields new stellar and wind parameters for the system that differ significantly from previous estimates (see Ramachandran et al. 2022, for more details). In particular, the masses of the components are considerably reduced to $\approx 38 M_{\odot}$ for the O-star donor and $\approx 11.4 M_{\odot}$ for the black hole. The O giant is overfilling its Roche lobe and shows surface He enrichment.

The derived mass-loss rate of the donor is in good agreement with the Vink et al. (2001) prediction assuming a high depth-dependent microclumping. By incorporating observed X-ray luminosities in models corresponding to different orbital phases, we were able to reproduce the spectral variations at three phases with the same stellar and wind parameters. We investigated the wind driving contributions from different ions and the changes in the ionization structure due to X-ray illumination. Towards the black hole, the wind is strongly quenched due to strong X-ray illumination. For this system, the standard wind-fed accretion scenario alone cannot explain the observed X-ray luminosity, pointing towards an additional mass overflow, in line with our acceleration calculations. The classical distinction between wind-fed and Roche-lobe overflow systems becomes meaningless for our system. Our investigations on wind driving and the impact of X-rays in M33 X-7 can be also applied to other high luminosity HMXB systems in general.

References

Björklund, R., Sundqvist, J. O., Puls, J., & Najarro, F. 2021, A&A, 648, A36
Bjorkman, J. E., Ignace, R., Tripp, T. M., & Cassinelli, J. P. 1994, ApJ, 435, 416
Bodensteiner, J., Shenar, T., Mahy, L., et al. 2020, A&A, 641, A43
Bouret, J.-C., Lanz, T., Hillier, D. J., et al. 2003, ApJ, 595, 1182
Bouret, J. C., Lanz, T., Hillier, D. J., et al. 2015, MNRAS, 449, 1545
Bouret, J. C., Martins, F., Hillier, D. J., et al. 2021, A&A, 647, A134
Castor, J. I., Abbott, D. C., & Klein, R. I. 1975, ApJ, 195, 157
Claeys, J. S. W., de Mink, S. E., Pols, O. R., Eldridge, J. J., & Baes, M. 2011, A&A, 528, A131
de Mink, S. E., Langer, N., Izzard, R. G., Sana, H., & de Koter, A. 2013, ApJ, 764, 166
Eldridge, J. J., Izzard, R. G., & Tout, C. A. 2008, MNRAS, 384, 1109
Frost, A. J., Bodensteiner, J., Rivinius, T., et al. 2022, A&A, 659, L3
Götberg, Y., Mink, S. E. d., & Groh, J. H. 2017, Astronomy & Astrophysics, 608, A11, publisher: EDP Sciences
Gräfener, G. & Hamann, W.-R. 2005, A&A, 432, 633
Groh, J. H., Oliveira, A. S., & Steiner, J. E. 2008, A&A, 485, 245
Hillier, D. J., Lanz, T., Heap, S. R., et al. 2003, ApJ, 588, 1039
Klencki, J., Istrate, A., Nelemans, G., & Pols, O. 2022, A&A, 662, A56
Lamers, H. J. G. & Pauldrach, A. W. A. 1991, A&A, 244, L5
Liu, J., Zhang, H., Howard, A. W., et al. 2019, Nature, 575, 618
Lucy, L. B. 2012, A&A, 543, A18
Maeder, A. & Meynet, G. 2000, ARA&A, 38, 143
Martins, F., Schaerer, D., Hillier, D. J., & Heydari-Malayeri, M. 2004, A&A, 420, 1087
Massa, D. 1995, ApJ, 438, 376
Mokiem, M. R., de Koter, A., Vink, J. S., et al. 2007, A&A, 473, 603
Müller, P. E. & Vink, J. S. 2014, A&A, 564, A57

Orosz, J. A., McClintock, J. E., Narayan, R., et al. 2007, Nature, 449, 872
Owocki, S. P., Cranmer, S. R., & Gayley, K. G. 1996, ApJ, 472, L115+
Paczyński, B. 1967, Acta Astron., 17, 355
Pietsch, W., Haberl, F., Sasaki, M., et al. 2006, ApJ, 646, 420
Prinja, R. K., Massa, D., Fullerton, A. W., Howarth, I. D., & Pontefract, M. 1997, A&A, 318, 157
Puls, J., Vink, J. S., & Najarro, F. 2008, A&A Rev., 16, 209
Ramachandran, V., Hamann, W. R., Oskinova, L. M., et al. 2019, A&A, 625, A104
Ramachandran, V., Oskinova, L. M., & Hamann, W. R. 2021, A&A, 646, A16
Ramachandran, V., Oskinova, L. M., Hamann, W. R., et al. 2022, arXiv e-prints, arXiv:2208.07773
Rickard, M. J., Hainich, R., Hamann, W. R., et al. 2022, arXiv e-prints, arXiv:2207.09333
Rivinius, T., Baade, D., Hadrava, P., Heida, M., & Klement, R. 2020, A&A, 637, L3
Schootemeijer, A., Götberg, Y., de Mink, S. E., Gies, D., & Zapartas, E. 2018, A&A, 615, A30
Shao, Y. & Li, X.-D. 2014, ApJ, 796, 37
Shenar, T., Gilkis, A., Vink, J. S., Sana, H., & Sand er, A. A. C. 2020, A&A, 634, A79
Shepard, K., Gies, D. R., Lester, K. V., et al. 2020, ApJ, 888, 82
Vanbeveren, D. 1991, A&A, 252, 159
Vink, J. S., de Koter, A., & Lamers, H. J. G. L. M. 2000, A&A, 362, 295
Vink, J. S., de Koter, A., & Lamers, H. J. G. L. M. 2001, A&A, 369, 574
Wang, L., Gies, D. R., Peters, G. J., et al. 2021, AJ, 161, 248
Yoon, S. C., Woosley, S. E., & Langer, N. 2010, ApJ, 725, 940

Winds of Stars and Exoplanets
Proceedings IAU Symposium No. 370, 2023
A. A. Vidotto, L. Fossati & J. S. Vink, eds.
doi:10.1017/S1743921322004148

X-ray view of colliding winds in WR 25

Bharti Arora[1], Jeewan C. Pandey[1] and Michaël De Becker[2]

[1]Aryabhatta Research Institute of Observational Sciences, Nainital-263 002, Uttarakhand, India

[2]Space Sciences, Technologies and Astrophysics Research (STAR) Institute, University of Liège, Quartier Agora, 19c, Allée du 6 Août, B5c, B-4000 Sart Tilman, Belgium

Abstract. The long-term behavior of a colliding wind binary WR 25 is presented using archival X-ray data obtained over a time span of ~16 years. The present analysis reveals phase-locked variations repeating consistently over many consecutive orbits of the source (with binary orbital period ~208 days). A significant deviation of the X-ray flux with respect to the theoretical 1/D trend (D is the binary separation) close to periastron passage has been observed. This may occur due to the shifting of the adiabatic wind collision to the radiative regime in that part of the orbit. Further, no signature of X-ray emission in 10.0-79.0 keV energy range attributable to inverse Compton scattering is detected by *NuSTAR*.

Keywords. Stars:early-type; binaries:colliding-winds; X-rays:stars; stars:individual: WR 25

1. WR 25: An Introduction

WR 25 (HD 93162) is a bright (V = 8.1 mag) WR star located in the Carina Nebula region and is classified as O2.5If*/WN6+OB (Crowther & Walborn 2011). Gamen et al. (2006) studied the radial velocity profile of WR 25 and suggested that it has an eccentric binary orbit (eccentricity = 0.5) with an orbital period of about 207.85 ± 0.02 days. Previous X-ray studies of WR 25 were based on limited X-ray observations and suggested that it's a colliding wind binary (CWB) system (Pandey et al. 2014). Therefore, in order to investigate this system and the associated winds deeply, we have carried out its X-ray study using the observations made by *NuSTAR*, *Suzaku*, *Swift*, and *XMM-Newton* at 226 epochs during 2000–2016. It is one of the rare instances where a massive binary has been explored in the high energies up to 79 keV which was possible because of *NuSTAR* monitoring of WR 25 along with other X-ray observatories.

2. Results from X-ray exploration of WR 25

Below 10 keV, colliding stellar winds of the binary components of WR 25 results in the enhanced X-ray luminosity. The system approaches a brighter X-ray state as the two binary components move close to the periastron passage in all the broad (0.3–10.0 keV), soft (0.3–2.0 keV), and hard (2.0–10.0 keV) energy bands as shown in Figure 1. This is because the wind interaction is maximum at periastron as wind density is largest in that part of the orbit. However, it gradually becomes fainter when the line of sight passes through the denser wind of the WR star in front and/or when the two binary components move away from each other in the orbit.

The X-ray flux obtained from spectral analysis of WR 25 in 0.3-10.0 keV energy range reveals that the wind collision is mostly adiabatic in WR 25 but significant deviation from the adiabatic cooling is seen around the periastron passage. The main indicator of this effect is the deviation from the expected 1/D (D is the binary separation) dependence of

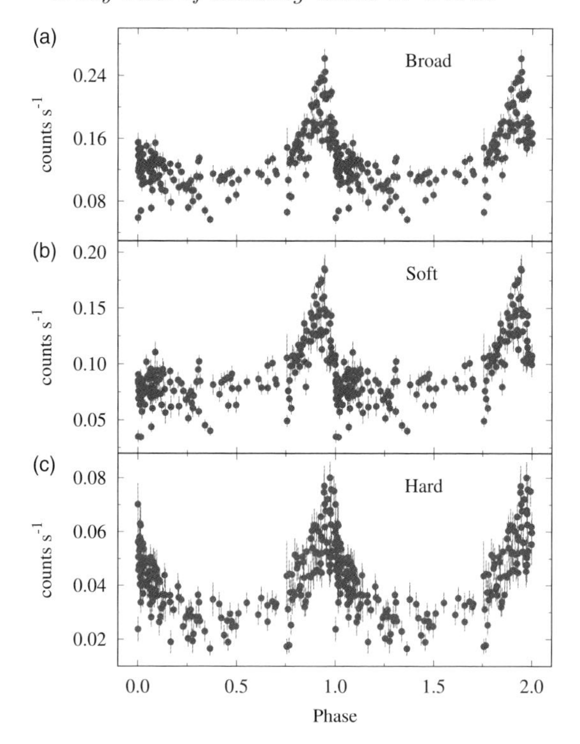

Figure 1. *Swift* observed X-ray light curves folded over the orbital period of WR 25.

the X-ray flux for the long period binaries. The inhibited acceleration of massive stars winds close to the periastron before interacting with each other might be a reason for the brief switch of wind plasma to the radiative regime. The sudden radiative braking of the wind of one component by another may further enhance the velocity drop. However, the temperature of post-shock plasma estimated by the spectral fitting of WR 25 at different orbital phases doesn't seem to support this interspretation. This study has significantly improved the orbital phase coverage of WR 25 as compared to previous studies and hence provided a deeper view of the wind properties (see Arora et al. 2019).

It has also been noticed that some CWBs also act as sources of particle acceleration in their wind collision region through diffusive shock acceleration mechanism (De Becker & Raucq 2013). The relativistic particles (mostly electrons) may inverse comptonize the photospheric stellar light to X-rays or even soft γ-rays. This opens up the possibility that some non-thermal X-ray emission might be measured in CWBs above 10 keV. However, no significant X-ray emission above 10 keV was observed from WR 25 by *NuSTAR* which provides evidence that no inverse compton scattering emission is produced by WR 25 above the background level. The upper limit, that we derived on the putative non-thermal X-ray luminosity, is of the order of $10^{32}\,\mathrm{erg\,s^{-1}}$. We argue that a sensitivity improvement of at least one order of magnitude is needed to access more constraining limits on the putative IC emission, or even have a chance to detect it for massive star systems with the most powerful winds.

References

Arora, B., Pandey, J. C., & De Becker, M. 2019, MNRAS, 487, 2624
Crowther, P. A. & Walborn, N. R. 2011, MNRAS, 416, 1311
De Becker, M. & Raucq, F. 2013, A&A, 558, A28
Gamen, R., Gosset, E., Morrell, N., et al. 2006, A&A, 460, 777
Pandey, J. C., Pandey, S. B., & Karmakar, S. 2014, ApJ, 788, 84

Winds of Stars and Exoplanets
Proceedings IAU Symposium No. 370, 2023
A. A. Vidotto, L. Fossati & J. S. Vink, eds.
doi:10.1017/S1743921323000340

Double tail structure in escaping atmospheres of magnetised close-in planets

A. A. Vidotto[1], S. Carolan[2], G. Hazra[1,3], C. Villarreal D'Angelo[4] and D. Kubyshkina[5]

[1]Leiden Observatory, Leiden University, PO Box 9513, 2300 RA Leiden, The Netherlands

[2]School of Physics, Trinity College Dublin, College Green, D02 PN40 Dublin 2, Ireland

[3]Dept. of Astrophysics, University of Vienna, Türkenschanzstrasse 17, A-1180 Vienna, Austria

[4]Inst. Astronomía Téorica y Exp. (CONICET-UNC), Laprida 854, Córdoba, Argentina

[5]Space Research Inst., Austrian Academy of Sciences, Schmiedlstrasse 6, A-8042 Graz, Austria

Abstract. High-energy stellar irradiation can photoevaporate planetary atmospheres, which can be observed in spectroscopic transits of hydrogen lines. Here, we investigate the effect of planetary magnetic fields on the observational signatures of atmospheric escape in hot Jupiters.

Keywords. MHD, planets and satellites: atmospheres, magnetic fields, planet-star interactions

---

In this work, we use our newly developed 3D self-consistent radiative magnetohydrodynamic (MHD) simulations (Carolan et al. 2021; Hazra et al. 2022) to study the effects of planetary magnetic fields on the dynamics of escaping atmosphere. To investigate the resulting observational signature, we couple the results of our 3D models to Lyman-α transit calculations (Vidotto et al. 2018; Allan & Vidotto 2019). In our models, we account for high energy stellar photons ionising the atmospheric neutral hydrogen atoms, which affects both the heating deposition in the atmosphere of the planet and its ionisation state. In addition to photoionisation, we also include collisional ionisation and recombination and track the proton and neutral components of the flow. Additionally, we also consider Lyman-α cooling and collisional cooling. The reader will be able to find further details of the model and this work in Carolan et al. (2021).

In our model, the stellar wind is injected in the numerical grid through an external boundary (see Figure 1). Using the same stellar wind property, we vary the planet's dipole field strength from 0 to 10 G. In Figure 1, we show a typical structure of the escaping atmosphere in magnetised planets (>3 G). Escape occurs through polar outflows, as opposed to the predominantly comet-like tail from non-magnetised models (see comparison in Figure 2). We find a small increase in evaporation rate with planetary field, though this should not affect the timescale of atmospheric loss (Figure 3a).

The double-tail structure has some key effects in Lyman-α transit signatures, as summarised in Figure 3b. When considering magnetic fields, we see an increase in line centre absorption due to an increase of the size of dead-zones. Additionally, we also see an increase in redshifted absorption with increase in planetary field. Most of the red shifted material exists around the night-side orbital plane, as some material falls from the comet-like tails back towards the planet. Finally, we also see that with increased magnetic field, there is an initial decrease in blueshifted absorption, as planetary material begins to be

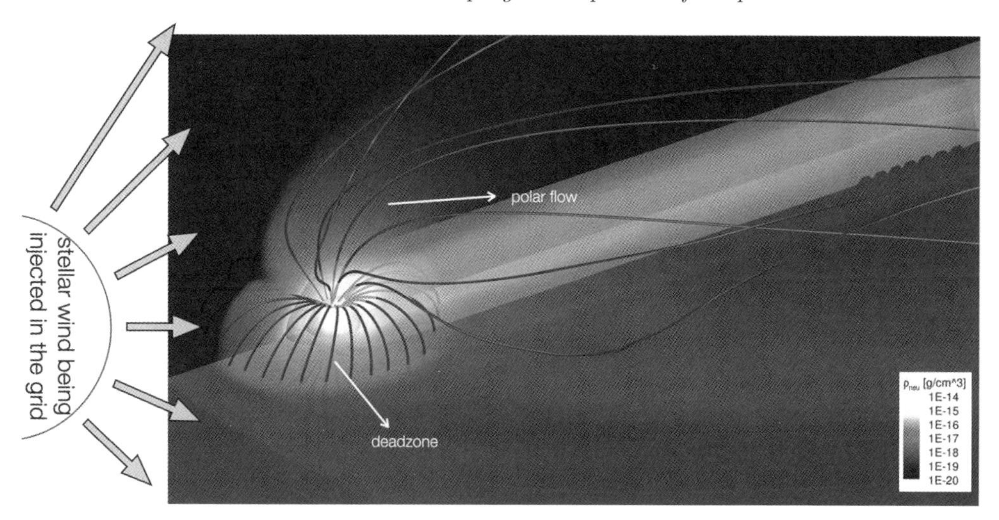

Figure 1. The polar outflows seen in magnetised planets lead to the formation of a double tail structure, above and below the orbital plane, and a dead-zone around the equator.

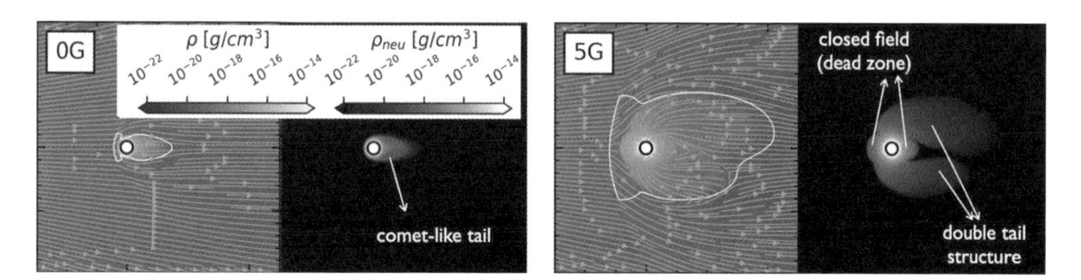

Figure 2. Side view of the planet orbit showing the total (ρ) and neutral Hydrogen ($\rho_{\rm neu}$) density for unmagnetised (left panels) and magnetised (right panels) Hot Jupiters.

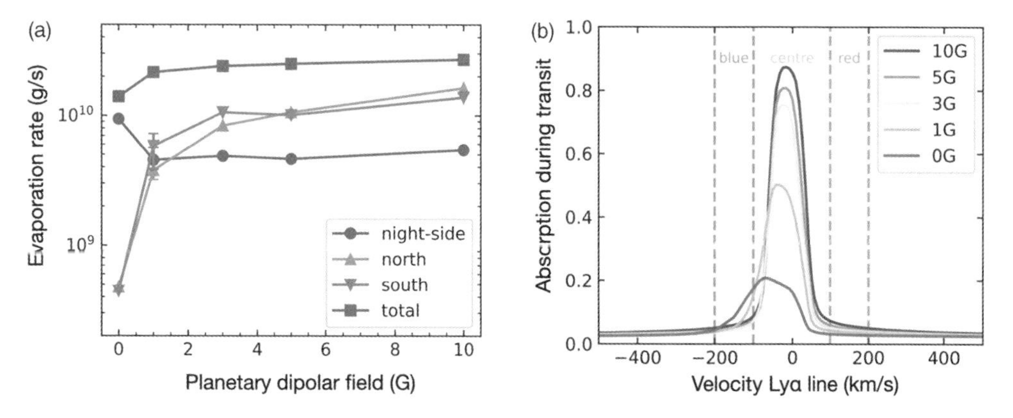

Figure 3. In spite of the small increase in escape rate with planetary field (red line, panel a), we find that Ly-α transit observations are strongly affected by planetary magnetism (panel b).

launched above and below the orbital plane, instead of being fully funnelled on to the orbital plane by the stellar wind, as seen in the 0-G model.

For further details of this work, please see Carolan et al. (2021).

References

Allan, A. & Vidotto, A. A. 2019, MNRAS, 490, 3760
Carolan, S., Vidotto, A. A., Hazra, G., Villarreal D'Angelo, C., & Kubyshkina, D. 2021, MNRAS, 508, 6001
Hazra, G., Vidotto, A. A., Carolan, S., Villarreal D'Angelo, C., & Manchester, W. 2022, MNRAS, 509, 5858
Vidotto, A. A., Lichtenegger, H., Fossati, L., et al. 2018, MNRAS, 481, 5296

Winds of Stars and Exoplanets
Proceedings IAU Symposium No. 370, 2023
A. A. Vidotto, L. Fossati & J. S. Vink, eds.
doi:10.1017/S1743921322004471

Shock breakout in winds of red supergiants: Type IIP supernovae

Alak Ray[1], **Harita Palani Balaji**[2], **Adarsh Raghu**[3] **and Gururaj Wagle**[4]

[1]Tata Institute of Fundamental Research, Mumbai 400005, India

[2]IISER, Pune 411008, India

[3]IISER, Kolkata 741246, India

[4]Louisiana State University, Baton Rouge, LA 70803, USA

Abstract. When a supernova shockwave launched deep inside the star exits the surface, it probes the circumstellar medium established by prior mass loss from the pre supernova star. The bright electromagnetic display accompanying the shock breakout is influenced by the properties of the star and scripts the history of the stellar mass loss. We investigate with MESA and STELLA codes the radiative display resulting from a set of progenitors that we evolved to core collapse. We simulate with different internal convective overshoot and compositional mixing and two sets of mass loss schema, one the standard "Dutch" scheme and another, an enhanced, episodic mass loss at a late stage. Shock breakout from the star shows double peaked bolometric light curves for the Dutch wind, as well as high velocity ejecta accelerated during shock breakout. We contrast the breakout flash from an optically thick CSM with that of the rarified medium.

Keywords. supernovae: general, shock waves, stars: mass loss, convection

---

Type IIP supernovae constitute nearly 48% of all core-collapse supernovae in volume limited samples in the local universe (Smith et al. 2011). Their progenitors, red supergiants, lose a lot of mass in slow moving winds that form the circumstellar medium (CSM) around the star. This CSM is probed rapidly by a fast moving hydrodynamic shock as it first sweeps across the star and then the CSM. A short span of the supernova's display thus reveals a long term history of mass loss from the star.

We investigate the radiative display of Type IIP SN resulting from a set of progenitors evolved, in different possible ways, from the central hydrogen burning stages all the way to the end of most advanced nuclear fuel burning and subsequent core collapse. We utilise in these works our suite of progenitor models that were evolved employing a variety of input physics and computational control that regulate the evolutionary sequence of these stars (see Das & Ray (2017); Wagle et al. (2019); Wagle & Ray (2020); Wagle et al. (2020)). This was undertaken with the Modules for Experiments in Stellar Astrophysics (MESA code r-10398) Paxton et al. (2018). While we have calculated stars of ZAMS 12 to 26 MSun with Z= 0.006 (very similar to that of LMC) with MESA, in this work we use models mainly in the range 12-14 Msun.

During the breakout of the star, the shock dominated by radiation pressure, makes the stellar envelope expand outward. With time, the optical depth of the material ahead of the shock decreases due to this expansion. Radiation begins to leak out of the shocked zone once its overlying optical depth falls below $c/v(r)$ (Ohyama 1963). Here c and $v(r)$ are the velocities of light and the fluid respectively. However, in many cases, a significant

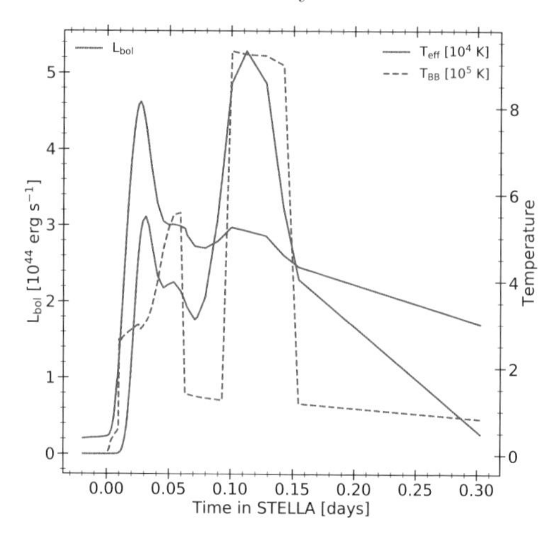

Figure 1. L, T vs time in 12 $M_{\odot}$, f = 0.025, EDEP = 1.0 FOE, Dutch mass loss.

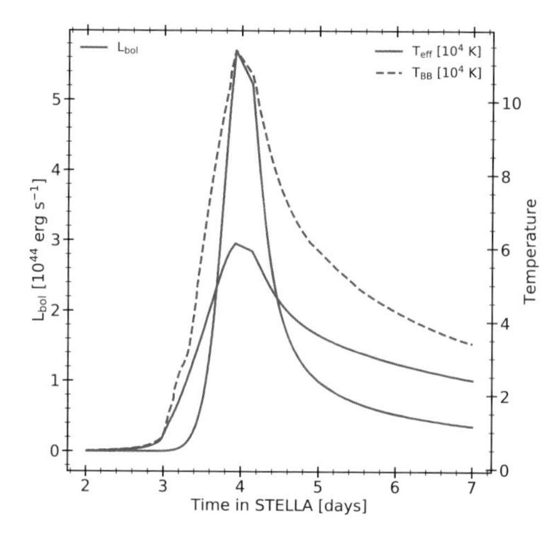

Figure 2. Case of 12 $M_{\odot}$, f = 0.020, EDEP = 1.0 FOE, Enhanced mass loss.

circumstellar medium (CSM) forms due to the ejection of mass from the progenitor star prior to its explosion either in a steady wind or in episodic, massive outflows. For explosions inside dense CSMs the shock breakout may take place at a larger distance and at later times (Chevalier & Irwin 2011, 2021; Waxman & Katz 2017). This early radiation carry signatures of the progenitor star (e.g. its radius) and its mass loss history a short time before its explosion. Study of the characteristics of this early radiative emission of type IIP SNe for a given set of progenitors and their mass loss history therefore sheds light on the connections between different types of SNe and their much fainter progenitors.

In Figs. 1 and 2, we plot L_{bol}, T_{eff}, T_{BB} as a function of STELLA time for our prototypical Dutch mass loss, and an episodic, late time enhanced mass loss model. We find multiple peaks in L_{bol} for all of the Dutch mass loss models. Note that the first and the last of these peaks in Fig. 1 are well separated compared to the light travel time ("retardation" time) which is typically 0.016 d. We explain these as a combination of SBO flash and the subseuent interaction of the shock wave with shells of gas in the CSM ejected earlier in pre-SN stage.

High-velocity, very low-mass shocked ejecta is present at early times after SBO in Dutch cases. The ejecta was moving slowly in the pre-SN phase but was accelerated during the

SBO flash. The ejecta velocity structure is qualitatively different from that of the bare star SBO. Photon spectra of the breakout flash from a dense CSM differs substantially from that of the optically thin CSM case fed by pre-SN Dutch wind. Initially the photon spectrum for the latter CSM has an approximate thermal shape but soon a large, disjoint high-energy bump develops. In contrast the SBO flash spectrum in a dense CSM lacks the disjoint peak. Details of this work is found in Palani Balaji et al. (2022).

Alak Ray acknowledges a DAE Raja Ramanna Fellowship during this work.

References

Chevalier, R. A. & Irwin, C. M. 2021, ApJ Letters, 914, L25. doi:10.3847/2041-8213/ac0884

Chevalier, R. A. & Irwin, C. M. 2011, ApJ Letters, 729, L6. doi:10.1088/2041-8205/729/1/L6

Das, S. & Ray, A. 2017, ApJ, 851, 138. doi:10.3847/1538-4357/aa97e1

Ohyama, N. 1963, Progress of Theoretical Physics, 30, 170. doi:10.1143/PTP.30.170

Palani Balaji, H., Ray, A., Wagle, G. A., et al. 2022, ApJ, 933, 194. doi:10.3847/1538-4357/ac7528

Paxton, B., Schwab, J., Bauer, E. B., et al. 2018, ApJS, 234, 34. doi:10.3847/1538-4365/aaa5a8

Smith, N., Li, W., Filippenko, A. V., et al. 2011, MNRAS, 412, 1522. doi:10.1111/j.1365-2966.2011.17229.x

Wagle, G. A., Ray, A., & Raghu, A. 2020, ApJ, 894, 118. doi:10.3847/1538-4357/ab8bd5

Wagle, G. A. & Ray, A. 2020, ApJ, 889, 86. doi:10.3847/1538-4357/ab5d2c

Wagle, G. A., Ray, A., Dev, A., et al. 2019, ApJ, 886, 27. doi:10.3847/1538-4357/ab4a19

Waxman, E. & Katz, B. 2017, Handbook of Supernovae, 967. doi:10.1007/978-3-319-21846-5_33

Winds of Stars and Exoplanets
Proceedings IAU Symposium No. 370, 2023
A. A. Vidotto, L. Fossati & J. S. Vink, eds.
doi:10.1017/S1743921323000856

On the making of a PN: the interaction of a multiple stellar wind with the ISM

Arturo Manchado[1,2,3], **Eva Villaver**[4], **G. García-Segura**[5] and **Luciana Bianchi**[6]

[1]Instituto de Astrofísica de Canarias, Vía Láctea S/N, E-38200 La Laguna, Tenerife, Spain

[2]Departmento de Astrofísica, Universidad de La Laguna (ULL), E-38206 La Laguna, Tenerife, Spain

[3]Consejo Superior de Investigaciones Científicas, Spain

[4]Centro de Astrobiología (CAB, CSIC-INTA), ESAC Campus Camino Bajo del Castillo, s/n, Villanueva de la Cañada, 28692, Madrid, Spain

[5]Instituto de Astronomía-UNAM, Apartado postal 877, Ensenada, 22800 Baja California, México

[6]Department of Physics and Astronomy, Johns Hopkins University, Baltimore, MD 21218, USA

Abstract. NGC 7293, the Helix nebula, represents one of the rare instances in which theoretical predictions of stellar evolution can be accurately tested against observations since the precise parallax distance and the velocity and proper motion of the star are well known. We present numerical simulations of the formation of the Helix PN that are fully constrained by the progenitor stellar mass, stellar evolution history, and star-interstellar medium (ISM) interaction. In the simulations, multiple bow-shock structures are formed by fragmentation of the shock front where the direct interaction of the stellar wind with the ISM takes place.

Keywords. hydrodynamics–ISM: planetary nebula: general; planetary nebula: individual (NGC 7293) – stars: AGB and post-AGB.

---

1. Introduction

NGC 7293 (a.k.a The Helix, PN G036.1 − 57.1) is one of the planetary nebula (PN) with a very good determination of its distance; 202 pc from Gaia DR3 (González-Santamaría et al. 2021). The central star has a mass of 0.60±0.02 $M_\odot$ so (Benedict et al. 2009) and a high temperature T_{eff} = 104 000 K (Guerrero & De Marco 2013) implying that the star+nebula system is in a rather evolved stage. The *Galaxy Evolution Explorer* (*GALEX*; Martin et al. 2005) wide field-of-view (26 arcminutes) revealed for the first time striking morphological features in the form of extended bow-shocks beyond the nebular halo (Bianchi et al. 2012).

2. Numerical simulations

The numerical simulations have been performed with the fluid solver ZEUS-3D (Stone, Mihalas, & Norman 1992), developed by M. L. Norman and the Laboratory for Computational Astrophysics. The computations have been carried out on a 2D spherical polar grid with the angular coordinate ranging from 0° to 180° and a physical radial extension of 3 pc. They have been run at a resolution of 1600 × 1440 zones in the radial and angular coordinates of the grid respectively (or equivalently ∼ 388 au × 0.125°).

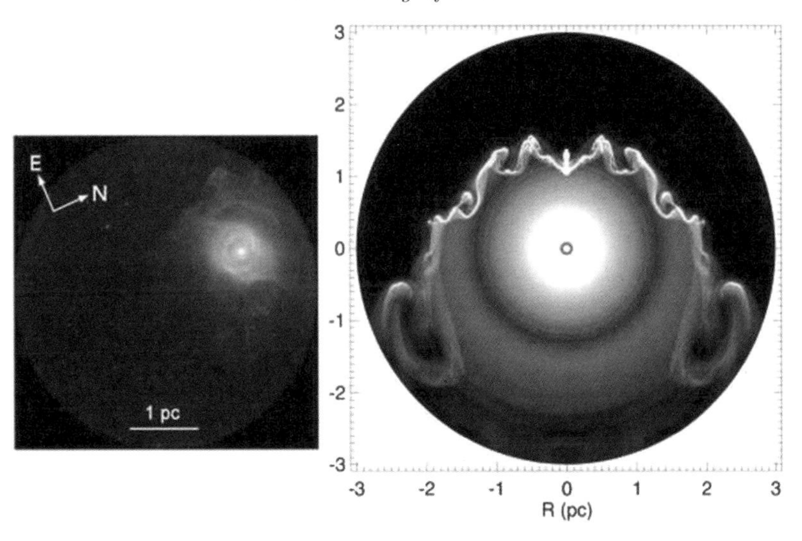

Figure 1. Left, *GALEX* FUV image of NGC 7293. At a distance of 202 pc, the FOV is 4.23 pc. Right, density map from our simulations, the snapshot corresponds to a ~1 000 yr old PN. The morphology of the UV features is reproduced by the simulations; more important, the physical scale of the predicted structures matches the observations.

Our boundary conditions are the AGB stellar wind of a 1.5 $M_{\odot}$ star and a ISM density of $n_{\rm ISM}$= 0.06 cm^3, and a relative velocity respect to its ambient medium of 40 km s^{-1}. For the post-AGB stage and PN evolution we follow the stellar wind according to the prescription given in Villaver et al. (2002a) by using the post-AGB evolutionary sequence given by Vassiliadis & Wood (1994) for a hydrogen burner with solar metallicity for the assumed stellar mass.

3. RESULTS

In Fig.1, (right panel) we show the result of the numerical simulation after 819000 yr in the AGB, and 1000 yr, after the onset of the photoionization. In the left panel we show the *GALEX* FUV filter image, which at a distance of 202 pc has a FOV of 4.23 pc. It is remarkable that both figures show bow shock structures in the direction of the movement, and that the actual size in parsec is quite similar. We conclude that the morphology of the Helix can be explained by the evolution of a 1.5 $M_{\odot}$ star interacting with an ISM with a relative velocity of 40 km s^{-1}. apparent multiple bow-shocks.

Acknowledgments

A. M. acknowledge support from the ACIISI, Gobierno de Canarias and the European Regional Development Fund (ERDF) under grant with reference PROID2020010051 as well as from the State Research Agency (AEI) of the Spanish Ministry of Science and Innovation (MICINN) under grant PID2020-115758GB-I00.

References

Benedict, G. F., McArthur, B. E., Napiwotzki, R., et al. 2009, AJ, 138, 1969
Bianchi, L., Manchado, A., Forster, K. 2012, IAU Symp. 283, p. 308
González-Santamaría, I., Manteiga, M., Manchado, A., et al. 2021, A&A, 656, A51. doi:10.1051/0004-6361/202141916
Guerrero, M. A., & De Marco, O. 2013, A&A, 553, A126
Martin, D. C., Fanson, J., Schiminovich, D., et al. 2005, ApJ Letters, 619, L1

Martin, D. C., et al. 2007, Nature, 448, 780
Stone, J. M., Mihalas, D., & Norman, M. L. 1992, ApJS, 80, 819
Villaver, E., García-Segura, & Manchado, A. 2002a, ApJ, 571, 880
Villaver, E., García-Segura, G., & Manchado, A. 2003, ApJ Letters, 585, L49
Vassiliadis, E., & Wood, P. R. 1994, ApJS, 92, 125

Part 5:
Relevance of winds on stellar/planetary evolution

Winds of Stars and Exoplanets
Proceedings IAU Symposium No. 370, 2023
A. A. Vidotto, L. Fossati & J. S. Vink, eds.
doi:10.1017/S1743921322004446

Role of planetary winds in planet evolution and population

D. Modirrousta-Galian

Yale University Department of Earth and Planetary Sciences, 210 Whitney Ave., New Haven, CT 06511, USA

Abstract. The role of atmospheric evaporation in shaping exoplanet populations remains a major unsolved problem in the literature. Observational evidence, like the bimodal distribution of exoplanet radii, suggests a catastrophic past in which exoplanets with masses of approximately $1-10M_{\oplus}$ often lose their primordial envelopes and experience a drastic reduction in their radii. Our knowledge of the mechanisms behind atmospheric evaporation remains nebulous, with new models regularly introduced in the literature. Understanding the principles behind these models and knowing when to apply them is essential for constraining how planets evolve. This communication reviews the mechanisms behind atmospheric evaporation by exploring observations and theory, as well as introducing some of the principles in the forthcoming paper Modirrousta-Galian & Korenaga (in press).

Keywords. Mini Neptunes(1063) — Super Earths(1655) — Star-planet interactions(2177) — Exoplanet evolution(491) — Exoplanet atmospheres(487) — Planetary interior(1248)

1. Introduction

The first exoplanet was discovered in 1995 (Mayor & Queloz 1995) and, since then, the total number of known exoplanets has surpassed five thousand as of writing. The flourishing of the observational exoplanet sector has led to numerous surprising discoveries. One such finding is that atmospheric evaporation shapes the histories and evolutions of exoplanets in their totality. This discovery led to a surge in the popularity of atmospheric evaporation research and the generation of various mass loss models. In this communication, I focus on the two main sources of atmospheric evaporation: core powered mass loss (e.g., Ginzburg et al. 2018) and X-ray and ultraviolet irradiation (e.g., Micela et al. 2022). Other mass loss mechanisms, such as mechanical impact erosion (e.g., Cameron 1983; Ahrens 1993; Genda & Abe 2005; Schlichting et al. 2015) and coronal mass ejections (e.g., Cohen et al. 2011; Hazra et al. 2022), will not be discussed. This communication provides a short review of the data and theory of atmospheric evaporation for a nonspecialist audience. I will first discuss the available data and theory before the Second Light (K2) mission in 2014. I will then summarize the current state of research in the atmospheric evaporation literature. In the last section, I will explore possible future directions and conclude. Throughout this writing, the structure of planets will be categorized in two parts: (1) The central condensed section of the planet will be called the embryo (i.e., the silicate mantle with the central metallic core) and (2) the primordial atmosphere. A full theoretical framework is provided in the forthcoming paper Modirrousta-Galian & Korenaga (in press).

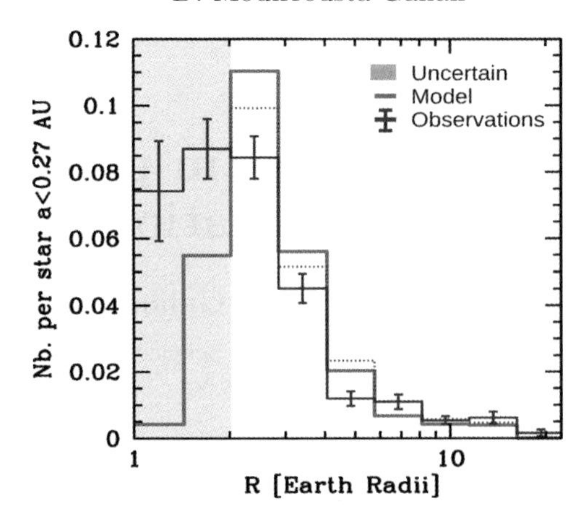

Figure 1. The radius distribution of exoplanets with periods less than 50 days (a<0.27 AU for a $1M_*$ star). The line without error bars shows their synthetic population, the line with error bars marks the data from Howard et al. (2010), Borucki et al. (2011), and Batalha et al. (2011), and the shaded region is uncertain. The age of the synthetic population is 5 Gyr. Figure adapted from Mordasini et al. (2012).

1.1. *The past*

Data on exoplanet population trends was limited before the launch of the Kepler mission in 2009 because most known exoplanets were hot Jupiters, so our understanding was based mostly on theoretical principles with origins in the Earth and planetary science field (e.g., Armitage 2014). Formation simulations indicate that the occurrence rate of planetary embryos inversely scales with their masses, that is, smaller mass embryos are more abundant than heavier ones (e.g., Ida & Lin 2004, 2005; Baraffe et al. 2006). In addition, gas accretion simulations suggest that exoplanets usually acquire atmospheres that are ~1% of their total masses (e.g., Ikoma & Hori 2012; Lee et al. 2014), though there is a large variance (e.g., Ikoma et al. 2000). By combining the predicted embryo mass distribution with their expected atmospheric masses, it was thought that the radius distribution of exoplanet radii would follow a skewed Gaussian (e.g., Mordasini et al. 2012). Figure 1 shows the predicted radius distribution according to the simulations of Mordasini et al. (2012) plotted with the available data at the time (Howard et al. 2010; Borucki et al. 2011; Batalha et al. 2011).

The number of known exoplanets grew significantly after the activation of the Second Light (K2) mission in 2014 (e.g., Burke et al. 2014). It was found that, rather than being Gaussian, the radius distribution was bimodal with one peak at $1.3R_{\oplus}$, the other peak at $2.4R_{\oplus}$, and the minimum point at $1.75R_{\oplus}$ (Fulton et al. 2017). The statistical robustness of the distribution suggested that the peaks were the manifestations of two distinct exoplanet populations: one without primordial atmospheres called super-Earths and the other with primordial atmospheres called sub-Neptunes (see Figure 2). This was further evidenced by the paucity of planets with large radii and orbital periods less than three days (i.e., the sub-Jovian desert; Fulton et al. 2017), indicating a hidden mechanism that causes planets to experience a significant reduction in their radii through their lifetimes (e.g., Owen & Lai 2018).

Three mechanisms have been proposed to explain the bimodal distribution and sub-Jovian desert. The first is core powered mass loss, which suggests that a planet acquires a large amount of energy after its giant impact phase of formation that triggers extreme mass outflow (e.g., Ginzburg et al. 2018; Biersteker & Schlichting 2019). The second argument is that X-ray and ultraviolet (XUV) irradiation from stars ionizes and heats

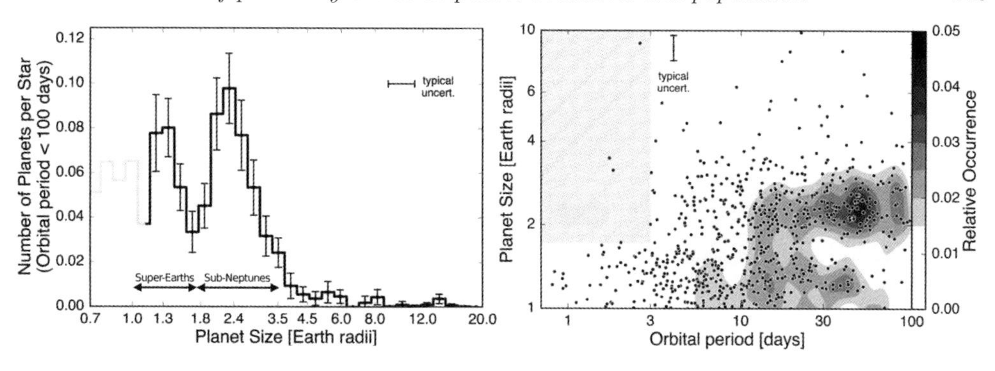

Figure 2. *Left:* The bimodal distribution of exoplanet radii. The data has been corrected for completeness. The lighter grey region for $R<1.14R_{\oplus}$ suffers from low completeness. The $1-1.75R_{\oplus}$ and $1.75-3.5R_{\oplus}$ regions are super-Earths and sub-Neptunes, respectively. *Right:* The sub-Jovian desert (the dark shaded area for periods less than three days and radii larger than $1.75R_{\oplus}$). Both figures have been adapted from Fulton et al. (2017).

the thermospheres of planetary atmospheres, causing their loss (e.g., Owen & Wu 2017; Modirrousta-Galian et al. 2020). The final argument proposes that the bimodal distribution is a natural outcome of planet formation in which the two peaks are comprised of rocky and ice-rich planets, respectively (e.g., Zeng et al. 2019; Venturini et al. 2020). The following section explores the core powered and XUV evaporation models. The formational argument for the bimodal distribution and sub-Jovian desert will not be further discussed because this communication focuses on the role of planetary winds and not cosmochemistry or planetary formation.

1.2. *The present*

There is an ongoing discussion on whether core powered mass loss or X-ray and ultraviolet irradiation are responsible for the bimodal distribution and sub-Jovian desert. Both models predict greater mass loss rates in the early stages of a planet's life because stars and planets are more energetic immediately after formation. The core powered and XUV-induced mass loss models will be qualitatively discussed below.

1.2.1. Core powered mass loss

The core powered mass loss model builds on concepts that originated in the Earth and planetary science field (e.g., Armitage 2014), where it has long been suggested that planets the size of Earth and larger experience a giant impact phase in which planetesimals collide and merge, leading to extreme surface temperatures. Simulations suggest that Earth could have had temperatures above 10,000 K (e.g., Karato 2014; Nakajima & Stevenson 2015; Lock et al. 2018) and, because super-Earths and sub-Neptunes have greater masses than Earth, they probably experienced more giant impacts and had higher surface temperatures. There is no simple analytic formula for estimating the mass loss rate of a planet experiencing core powered mass loss because the atmospheric dynamics will change depending on the internal state of the embryo. Only by creating a fully self-consistent planetary model can core powered mass loss be estimated.

Ginzburg et al. (2018) was the first to propose that a super-Earth or sub-Neptune could experience an extreme loss of primordial gases after a giant impact (see Figure 3). They suggest that the core powered mass loss mechanism can remove the light envelopes of small planets but not the heavy envelopes of larger ones, leading to the bimodal distribution of exoplanet radii. Their numerical framework assumes that the magma

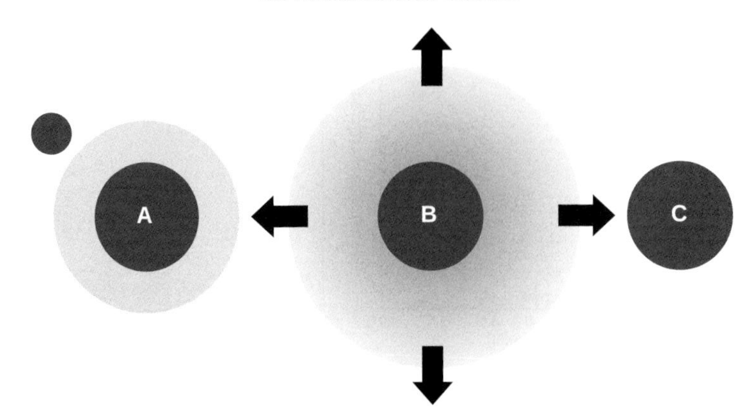

Figure 3. Schematic diagram showing the core powered mass loss mechanism. A shows a protoplanet with a primordial atmosphere immediately before a giant impact. B shows how the atmosphere expands because of the large amount of internal energy deposited from the giant impact. C shows a possible final outcome in which the entire atmosphere is lost.

ocean is isothermal, which is adequate as a first-order approximation, but a more realistic treatment would be needed if a detailed analysis is desired (e.g., Miyazaki & Korenaga 2019). Magma is composed of a mixture of minerals, so rather than crystallizing at a single temperature like water, it has a range of temperatures over which crystallization occurs (e.g., Solomatov 2015). Three points in this temperature range are of notable importance: (1) The liquidus is defined as the lowest temperature at a given pressure where magma is fully molten, (2) The solidus is the highest temperature at a given pressure where magma is fully crystallized, and (3) the rheological transition is the temperature at a given pressure where magma transitions from behaving rheologically like a liquid to a solid. A magma ocean can cool efficiently only when its surface temperature exceeds the rheological transition; convection will be sluggish and cooling inefficient at temperatures below it. By not including the effects of rheology in their magma ocean model, the internal energy available to drive atmospheric evaporation was overestimated.

Regarding observational data, the core powered mass loss model cannot explain the sub-Jovian desert because it defines mass loss only by the internal conditions of the embryo. In other words, this model is independent of the orbital period of the planet or the type of star it orbits. The inability of the core powered mass loss model to explain the sub-Jovian desert has been used as a justification for ignoring the internal heat flux as a source of mass loss. In the next section, the XUV-induced photoevaporation mechanism is examined in the absence of interior energy.

1.2.2. X-ray and ultraviolet mass loss

X-ray and ultraviolet induced photoevaporation (see Figure 4) has long been discussed in the Earth and planetary science literature (e.g., Watson et al. 1981; Hunten et al. 1987). Its application to exoplanets was popularized by Owen & Wu (2013, 2017), who suggested that the bimodal distribution can be explained by a population of exoplanets with masses of approximately $3M_{\oplus}$, where half have primordial atmospheres and the other half do not. The sub-Jovian desert is straightforward to explain with XUV-induced photoevaporation because closely orbiting planets are more highly irradiated than further orbiting ones, so they are more prone to total atmospheric loss. Owen & Wu (2013, 2017) used the energy limited approximation (Watson et al. 1981; Erkaev et al. 2007) for estimating mass loss, which does not incorporate thermal heating from the star's bolometric flux because it assumes that only XUV irradiation drives outflow. Indeed, fluid dynamical simulations

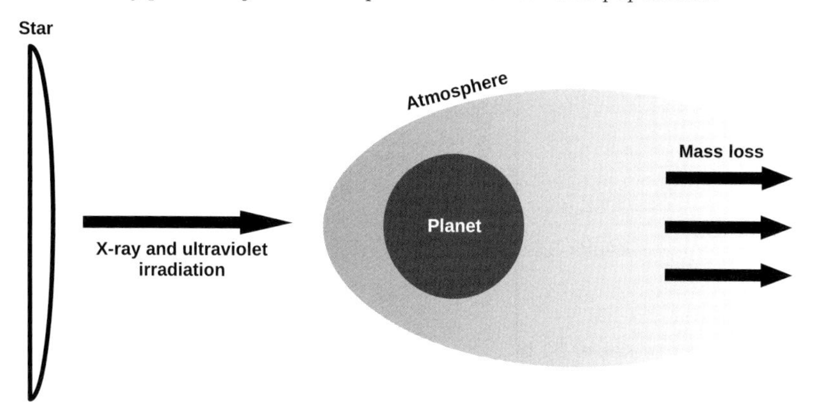

Figure 4. Schematic diagram showing the configuration of a planet experiencing atmospheric evaporation because of X-ray and ultraviolet irradiation.

show that the energy limited approximation underestimates mass loss rates by several orders of magnitude (e.g., García Muñoz 2007; Tian 2015; Caldiroli et al. 2021) and should not be used for highly irradiated super-Earths and sub-Neptunes.

Because the mass loss rates predicted by hydrodynamic models are substantially different from those of the energy limited approximation, it was not obvious if the bimodal distribution and sub-Jovian desert were still reproducible through XUV-induced photoevaporation alone. Modirrousta-Galian et al. (2020) tackled this problem by running a population simulation using the models of Kubyshkina et al. (2018a,b). Their findings suggested that the masses of exoplanets had to be larger than those predicted by Owen & Wu (2017) for the bimodal distribution to form. Their best fit mass distribution had few planets with masses less than three Earth masses, a uniform distribution between three and eight Earth masses, and a decrease in the occurrence rate of exoplanets with masses above eight Earth masses. The simulations of Modirrousta-Galian et al. (2020) implied that XUV irradiation alone could explain observations without requiring other mass loss mechanisms, such as the core powered mass loss model. However, the discovery of super-Earths and sub-Neptunes with densities lower than cold hydrogen gas (called super-puffs) provided renewed support for the core powered mass loss mechanism because such enlarged radii can be attained only with high internal luminosities. Figure 5 shows the mass and radius plots of DS Tuc A b and the planets in the Kepler-51 system (Libby-Roberts et al. 2020; Benatti et al. 2019, 2021). In other words, a combination of photoevaporation and core powered mass loss is required to explain the bimodal distribution, sub-Jovian desert, and super-puff presence.

1.3. *The future*

Core powered mass loss and XUV-induced photoevaporation are endmember cases for atmospheric evaporation. The first models mass loss arising from the internal luminosity of the planet, and the other models the mass loss from incoming stellar irradiation. Fluids behave very differently depending on whether they are heated from above or below, so the incoming and outgoing energy fluxes need to be considered concurrently when modeling mass loss. For example, fluids are prone to Rayleigh-Bénard convection when heated from below, whereas heating from above causes them to become stably stratified. The balance between the outgoing and incoming energy fluxes will influence the thermal structure of the entire planet. By considering the internal and incoming luminosity, it can be shown that mass loss is neither purely XUV driven nor purely core powered but a mixture of both.

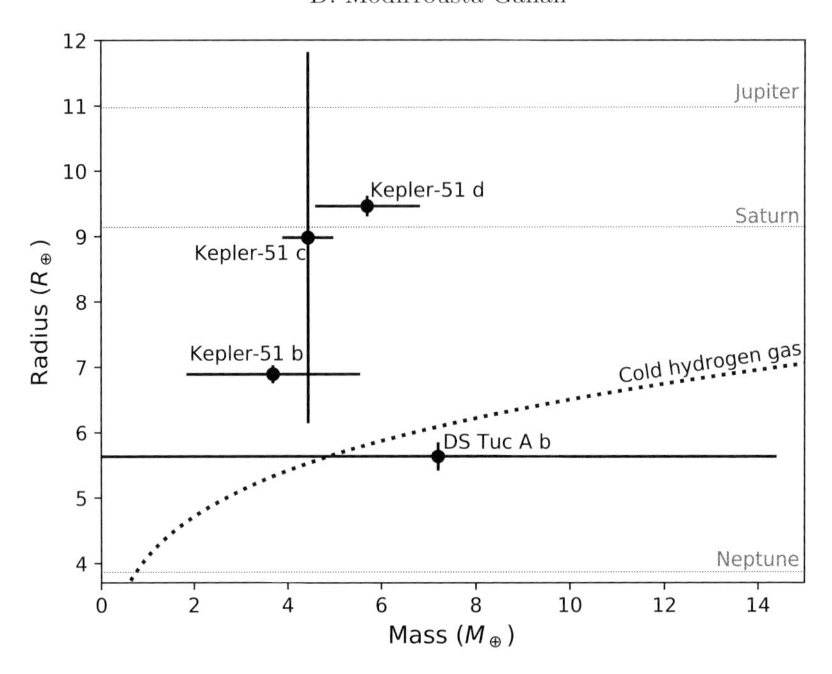

Figure 5. Mass and radius of Kepler-51 b, c, d (Libby-Roberts et al. 2020), and DS Tuc A b (Benatti et al. 2019, 2021). The cold hydrogen gas curve is from Zeng et al. (2019) who used the equation of state data of Becker et al. (2014).

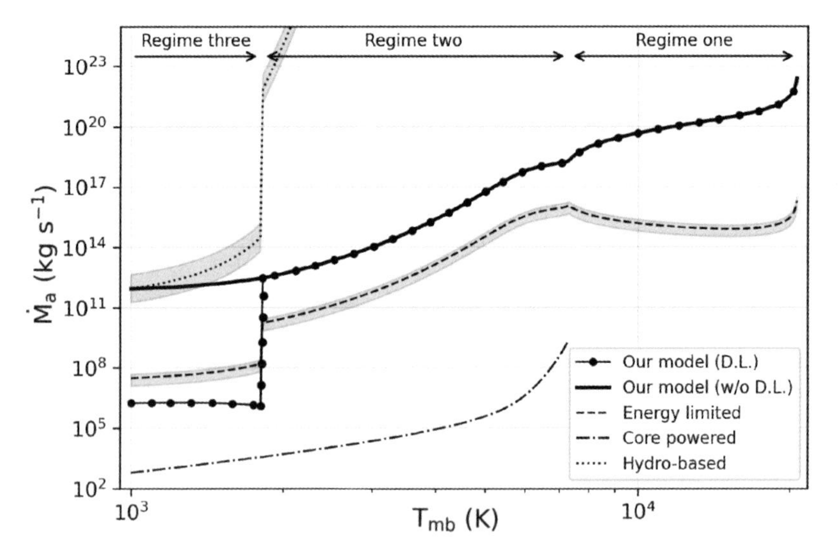

Figure 6. The predicted atmospheric evaporation rates as a function of the temperature of the boundary layer of the magma ocean ($T_{\rm mb}$) for an exoplanet that is three times the mass of Earth with a core-mass fraction of 26%, a primordial (hydrogen-rich) atmosphere that is 1% of the total planetary mass, an equilibrium temperature of 500 K, and experiencing an XUV radiant flux of 0.1 W m^{-2}. The black solid with circles, thick black solid, dashed, dash-dotted, and dotted lines are for our model with and without the diffusion limited mass loss included, the energy limited model (Watson et al. 1981), the core powered model (Biersteker & Schlichting 2019, 2021), and the hydro-based model (Kubyshkina et al. 2018a,b). The gray regions mark the typical uncertainty of the hydro-based and energy limited models. Figure from Modirrousta-Galian & Korenaga (in press).

In a forthcoming paper, Modirrousta-Galian & Korenaga (in press) self-consistently incorporate stellar irradiation with atmospheric and geodynamical principles for estimating the mass loss rate of super-Earth and sub-Neptune exoplanets. This combination gives rise to unforeseen emergent properties, such as the three regimes of atmospheric evaporation (see Figure 6). In the first regime, a planet has very high internal temperatures from its high-energy formation processes. These high temperatures give rise to a fully convecting atmosphere that efficiently loses mass without much internal cooling. The second regime applies to planets with lower internal temperatures, so a radiative region forms, but the photosphere still remains outside the Bondi radius. Hence, mass loss continues to depend only on internal temperatures. Planets with the lowest internal temperatures are in the third regime when the photosphere forms below the Bondi radius and mass is lost primarily because of X-ray and ultraviolet irradiation. Modirrousta-Galian & Korenaga (in press) provide the first unifying framework for modeling mass loss through the lifespan of super-Earth and sub-Neptune exoplanets that describes the circumstances at which photoevaporation and core powered mass loss occur. Indeed, the framing of the question should not be whether one mechanism is responsible or the other, but rather when one mechanism is active versus the other.

References

Ahrens, T. J. 1993, Annual Review of Earth and Planetary Sciences, 21, 525

Armitage, P. J. 2014, Lecture Notes on the Formation And Early Evolution of Planetary Systems (CreateSpace Independent Publishing Platform)

Baraffe, I., Alibert, Y., Chabrier, G., & Benz, W. 2006, A&A, 450, 1221

Batalha, N. M., Borucki, W. J., Bryson, S. T., et al. 2011, ApJ, 729, 27

Becker, A., Lorenzen, W., Fortney, J. J., et al. 2014, ApJS, 215, 21

Benatti, S., Damasso, M., Borsa, F., et al. 2021, A&A, 650, A66

Benatti, S., Nardiello, D., Malavolta, L., et al. 2019, A&A, 630, A81

Biersteker, J. B. & Schlichting, H. E. 2019, MNRAS, 485, 4454

Biersteker, J. B. & Schlichting, H. E. 2021, MNRAS, 501, 587

Borucki, W. J., Koch, D. G., Basri, G., et al. 2011, ApJ, 736, 19

Burke, C. J., Bryson, S. T., Mullally, F., et al. 2014, ApJS, 210, 19

Caldiroli, A., Haardt, F., Gallo, E., et al. 2021, A&A, 655, A30

Cameron, A. G. W. 1983, Icarus, 56, 195

Cohen, O., Kashyap, V. L., Drake, J. J., Sokolov, I. V., & Gombosi, T. I. 2011, ApJ, 738, 166

Erkaev, N. V., Kulikov, Y. N., Lammer, H., et al. 2007, A&A, 472, 329

Fulton, B. J., Petigura, E. A., Howard, A. W., et al. 2017, AJ, 154, 109

García Muñoz, A. 2007, Planetary and Space Science, 55, 1426

Genda, H. & Abe, Y. 2005, Nature, 433, 842

Ginzburg, S., Schlichting, H. E., & Sari, R. 2018, MNRAS, 476, 759

Hazra, G., Vidotto, A. A., Carolan, S., Villarreal D'Angelo, C., & Manchester, W. 2022, MNRAS, 509, 5858

Howard, A. W., Marcy, G. W., Johnson, J. A., et al. 2010, Science, 330, 653

Hunten, D. M., Pepin, R. O., & Walker, J. C. G. 1987, Icarus, 69, 532

Ida, S. & Lin, D. N. C. 2004, ApJ, 604, 388

Ida, S. & Lin, D. N. C. 2005, ApJ, 626, 1045

Ikoma, M. & Hori, Y. 2012, ApJ, 753, 66

Ikoma, M., Nakazawa, K., & Emori, H. 2000, ApJ, 537, 1013

Karato, S.-I. 2014, Proceedings of the Japan Academy, Series B, 90, 97

Kubyshkina, D., Fossati, L., Erkaev, N. V., et al. 2018a, ApJ Letters, 866, L18

Kubyshkina, D., Fossati, L., Erkaev, N. V., et al. 2018b, A&A, 619, A151

Lee, E. J., Chiang, E., & Ormel, C. W. 2014, ApJ, 797, 95

Libby-Roberts, J. E., Berta-Thompson, Z. K., Désert, J.-M., et al. 2020, AJ, 159, 57

Lock, S. J., Stewart, S. T., Petaev, M. I., et al. 2018, Journal of Geophysical Research (Planets), 123, 910
Mayor, M. & Queloz, D. 1995, Nature, 378, 355
Micela, G., Cecchi-Pestellini, C., Colombo, S., Locci, D., & Petralia, A. 2022, Astronomische Nachrichten, 343, e10097
Miyazaki, Y. & Korenaga, J. 2019, Journal of Geophysical Research (Solid Earth), 124, 3399
Modirrousta-Galian, D., Locci, D., & Micela, G. 2020, ApJ, 891, 158
Mordasini, C., Alibert, Y., Klahr, H., & Henning, T. 2012, A&A, 547, A111
Nakajima, M. & Stevenson, D. J. 2015, Earth and Planetary Science Letters, 427, 286
Owen, J. E. & Lai, D. 2018, MNRAS, 479, 5012
Owen, J. E. & Wu, Y. 2013, ApJ, 775, 105
Owen, J. E. & Wu, Y. 2017, ApJ, 847, 29
Schlichting, H. E., Sari, R., & Yalinewich, A. 2015, Icarus, 247, 81
Solomatov, V. 2015, in Treatise on Geophysics (Second Edition), second edition edn., ed. G. Schubert (Oxford: Elsevier), 81–104
Tian, F. 2015, Annual Review of Earth and Planetary Sciences, 43, 459
Venturini, J., Guilera, O. M., Haldemann, J., Ronco, M. P., & Mordasini, C. 2020, A&A, 643, L1
Watson, A. J., Donahue, T. M., & Walker, J. C. G. 1981, Icarus, 48, 150
Zeng, L., Jacobsen, S. B., Sasselov, D. D., et al. 2019, PNAS, 116, 9723

Winds of Stars and Exoplanets
Proceedings IAU Symposium No. 370, 2023
A. A. Vidotto, L. Fossati & J. S. Vink, eds.
doi:10.1017/S1743921322003970

Size Evolution and Orbital Architecture of KEPLER Small Planets through Giant Impacts and Photoevaporation

Gu Pin-Gao[1], Matsumoto Yuji[2], Kokubo Eiichiro[2] and Kurosaki Kenji[3]

[1]Institute of Astronomy and Astrophysics, Academia Sinica, Taipei 10617, Taiwan

[2]National Astronomical Observatory of Japan, 2-21-1, Osawa, Mitaka, 181-8588 Tokyo, Japan

[3]Department of Planetology, Kobe University, Nada-ku, Kobe, Hyogo, 657-8501, Japan

Abstract. The KEPLER transit survey with follow-up spectroscopic observations has discovered numerous small planets (super-Earths/sub-Neptunes) and revealed intriguing features of their sizes, orbital periods, and their relations between adjacent planets. The planet size distribution exhibits a bimodal distribution separated by a radius gap at around 1.8 Earth radii. Besides, these small planets within multiple planetary systems show that adjacent planets are similar in size and their period ratios of adjacent planet pairs are similar as well, a phenomenon often dubbed as peas-in-a-pod in the exoplanet community. While the radius gap has been predicted and theorized for years, whether it can be relevant to the orbital architecture peas-in-a-pod is physically unknown. For the first time, we attempted to model both features together through planet formation and evolution processes involving giant impacts and photoevaporation. We showed that our model is generally consistent with the KEPLER results but with a smaller radius gap. The impact of Kubyshikina's model for photoevaporation on our model is discussed.

Keywords. Exoplanet formation, Exoplanet atmospheres

---

1. Introduction

This study was motivated by two key results from the KEPLER transit mission and the follow-up California-Kepler Survey. First of all, the radius gap separates two populations of small planets, sub-Neptunes (with size $\gtrsim$ 1.8 earth radii) and suer-Earths (with size $\lesssim$ 1.8 earth radii), for the orbital period $<$ 100 days (Fulton et al. 2017; Fulton & Petigura 2018). The presence of the bimodal distribution of close-in small planets separated by the radius gap was predicted a long time ago as a result of XUV photoevaporation (Owen & Wu 2013), and more other models have been proposed ever since (e.g., Ginzburg et al. 2018). In some of these models, such as the XUV photoevaporation, the radius gap can provide a hint about the interior structure of the two populations. The sub-Neptunes consist of a rocky core surrounded by a Hydrogen-rich gas envelope. The envelope mass is a few precedents of the core mass of the planet. In contrast, the super-Earths are just a bare rocky core. Their gas envelopes probably have been stripped away by a few mass loss mechanisms.

The second piece of observational evidence is the size and spacing distributions in multiple-planet systems. This has been referred to as "peas-in-a-pod", meaning that they are similar in size and are regularly spaced for adjacent pairs of small planets (Weiss et al. 2018; Weiss & Petigura 2020). The distribution of the radius ratio of the adjacent planet pairs in a multiple planet system tends to peak at 1, with the standard deviation smaller than those by drawing planet radii at random from the distribution of observed planet radii. A similar distribution is present for the spacing, which is represented by the period ratio of three adjacent planets.†

As a result, an outstanding question can be raised – is the radius gap physically relevant to peas-in-a-pod?

2. Model

We studied this issue from the viewpoint of planet formation and evolution. We summarize our model below and refer the readers to Matsumoto et al. (2021) for details. In proto-planetary disks, planetary embryos can form with gas accretion onto a core based on the core-accretion model. After the disk dissipates, the protoplanets, i.e., planetary embryos, go through giant impacts to assemble the final planets. The gas envelope can be lost via giant impact shock, XUV photoevaporation, or core-powered mass loss, each of which occurs on different timescales. The giant impact proceeds on the timescale of $t \sim 10^7$ years (Genda & Abe 2003; Keszthelyi et al. 2020), the XUV photoevaporation occurs on $t \sim 10^9$ years (Owen & Wu 2017; Rogers et al. 2021), and the core-powered mass loss may take $t \sim 10^9$ years (Gupta & Schlichting 2019; Rogers et al. 2021). In this study, We considered the planetary winds driven only by the giant impacts and XUV photoevaporation. The two different timescales of the mass-loss processes enable us to perform two-stage mass loss – impact erosion followed by the XUV photoevaporation.

We considered the standard model for core formation through planetesimal accretion, with the initial condition given by the minimum-mass extrasolar nebula disk profile following a power law (Dai et al. 2020). We started with the protoplanets deployed from 0.05 to 1 au in a gas-free disk. The core mass of each protoplanet is given by the isolation mass with the mutual separation of 10 Hill radii, which are standard outcomes from N-body simulations for the planetesimal accretion model (Kokubo & Ida 2002). The initial envelope mass fraction X_{init} and the power index of the density power law p_{init} are currently unknown. Therefore, we parameterized them with various values.

We adopted the two-stage mass loss: impact erosion followed by EUV photoevaporation. The first stage of the evolution was performed by the N-body simulation for giant impact growth. In the N-body simulation, the impact energy can be computed. The mass loss of the envelope driven by an impact event was subsequently calculated based on the SPH simulation in Keszthelyi et al. (2020), in which the impact mass loss is a power law of the impact energy. After we updated the envelope mass and core mass due to a merger, we computed the planetary size based on the adiabatic profile of the convective atmosphere (Owen & Wu 2017), and continued the N-body part of the simulation. After the orbits of final assembled planets settle at $t = 10^7$ years, N-body simulations for giant impacts were terminated. The EUV photoevaporation takes over the impact erosion for mass loss. We adopted the analytical model for energy-limited photoevaporation (Erkaev et al. 2007), with the XUV flux decreasing with time after $t = 10^8$ years (Owen & Wu 2017). Meanwhile, the convective atmosphere cools and contracts. We evolved the atmospheres and hence the planet sizes up to 10^9 years.

3. Results

Fig. 1 shows the mass and size evolution from $t = 10^7$ years (right after the giant impact stage) to $t = 10^9$ years in the case of $X_{init} = 0.1$ and $p_{init} = 3/2$. The dashed lines indicate the initial mass and size distribution of the protoplanets as a function

† The "peas-in-a-pod" orbital architecture remains controversial due to the KEPLER observational bias (Zhu 2020).

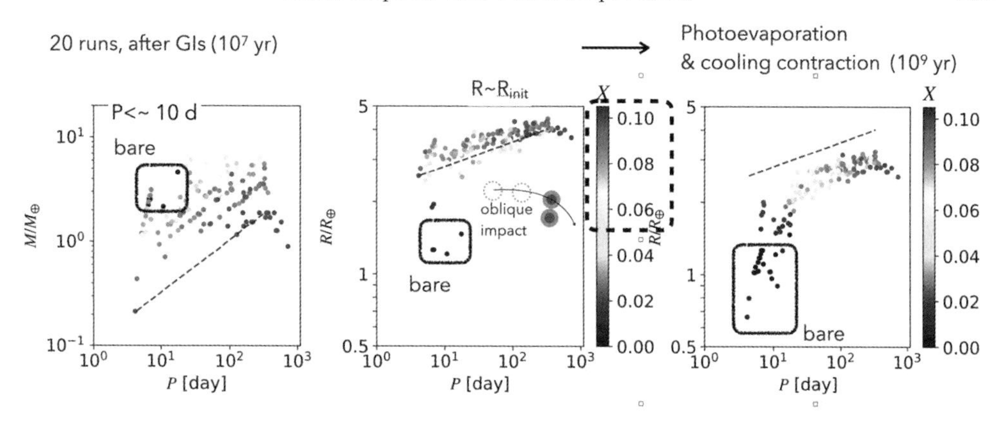

Figure 1. Size and mass evolutions from $t = 10^7$ years (after the giant-impact stage) to $t = 10^9$ years (after the photoevaporation stage) in the case of $X_{init} = 0.1$ and $p_{init} = 3/2$. The dashed lines indicate the initial spatial distribution of mass and size. X, presented by colors, is the mass fraction of gas envelope. "bare" means bare cores. An inset sketches an oblique impact between two protoplanets, which has a rocky core (in brown) surrounded by a gas envelope (in blue).

of the orbital period. The color denotes the mass fraction of the gas envelope X of each planet (i.e., each dot), which is X_{init} initially. The left panel of Fig. 1 shows the planet masses after the giant impacts ($t = 10^7$ years), which are larger than the initial mass of the protoplanets due to impact assembly. In contrast, the final size distribution of assembled planets does not significantly deviate from the initial size distribution, as illustrated in the middle panel of Fig. 1. It arises because most of the planets almost inherit the gas envelope from the protoplanets from which they are assembled, except for the extremely close-in planets that become bare cores. Most of the collisions occur as oblique impacts, hence driving little mass loss as illustrated by the inset in the middle panel of Fig. 1. After the giant impact phase ($t > 10^7$), the photoevaporation is turned on to further evolve the planetary atmospheres, which also cool and contract. At the end of the evolution ($t = 10^9$) years, the planetary sizes become smaller than the initial sizes of the protoplanets as shown in the right panel of Fig. 1. Moreover, a distinct population of the bare cores appears at smaller orbital periods as a result of stronger stellar EUV irradiation. A bimodal size distribution has emerged at the end of the evolution.

Fig. 2 shows the fraction distribution of the size ratio between adjacent planet pairs right after the giant impact phase ($t = 10^7$ years) and at the end of the evolution ($t = 10^9$ years). The averaged size ratios right after the giant impact phase are closer to unity than those after the photoevaporation stage. This is because the assembled planets do not further evolve their orbits and the inner planets lose more gas envelopes through photoevaporation. Nonetheless, the distribution of the size ratio does not evolve significantly after 10^7 years. Therefore, our model indicates that the size and spacing of adjacent planets are correlated, consistent with the peas-in-a-pod pattern, which is primarily caused by giant impacts. Note that the standard deviations of the simulated planet pairs are smaller than $\langle R_{i+1}/R_i \rangle = 1.29 \pm 0.63$ from the KEPLER results (Weiss et al. 2018). The giant impacts yield a stronger correlation between adjacent planets than those in the KEPLER multiple-planet systems.

On the other hand, the bimodal size distribution does not appear until 10^9 years. Fig. 3 shows that there are no clear radius gaps right after giant impacts. This is because less massive planets are not bare after giant impacts. Similar to previous models, the envelope stripping by the XUV photoevaporation transforms some of the sub-Neptunes into super-Earths, hence producing the bimodal size distributions and radius gaps. However, the

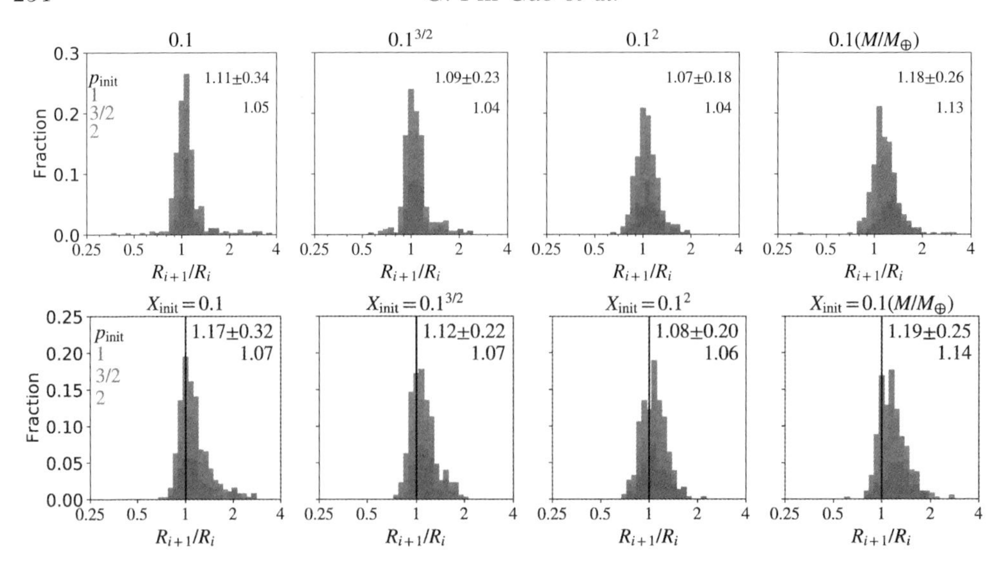

Figure 2. The fraction distribution of the size ratio between adjacent planet pairs with period < 100 days at $t \sim 10^7$ (top) and 10^9 (bottom panel) years. The various initial mass distributions of the protoplanets (described by p_{init} and presented as colored histograms) and initial mass fraction of the gas envelope of protoplanets (denoted by X_{init}) are displayed.

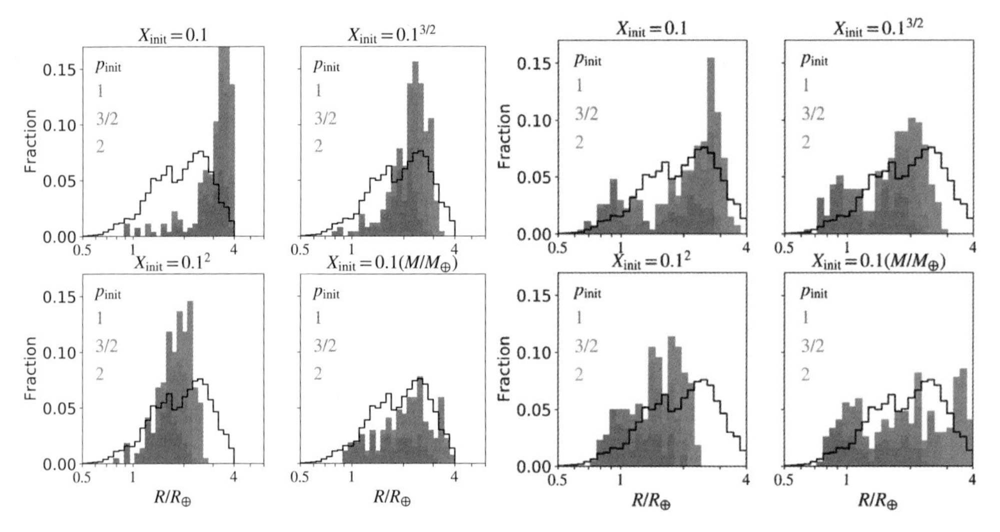

Figure 3. Fraction distribution of planet size in our model (colored histogram) compared with that for the KEPLER planets (black histogram) at $t = 10^7$ (left four panels) and 10^9 (right four panels) years. The cases of various X_{init} and p_{init} are presented.

simulated radius gap is smaller than the observed radius gap, i.e., our model produces overabundant sub-Neptunes and smaller super-Earths. It may imply that more envelope-loss mechanisms, such as the core-powered mass loss, are required to evolve some of the simulated sub-Neptunes to super-Earths, possibly helping to increase the simulated radius gap toward the observed one in our framework. Our result could be also remedied by considering a mass distribution with a continuous profile of p_{init} with $X_{init} = 0.1$. We refer the readers to Matsumoto et al. (2021) for more details.

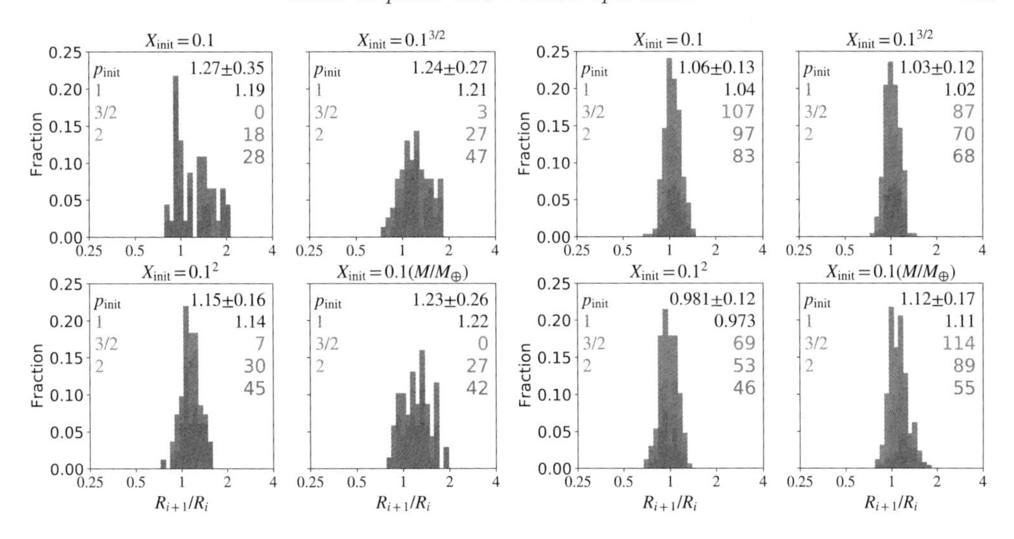

Figure 4. Fraction distribution of planet size for SE-SE pairs (left four panels) and for SN-SN pairs (right four panels) at the end of evolution ($t = 10^9$ years) in our model. The cases of various X_{init} and p_{init} are presented.

4. Discussions

Millholland & Winn (2021) divided the KEPLER small planet pairs into the superEarth-superEarth (SE-SE) pairs and the subNeptune-subNeptune (SN-SN) pairs. They found that the tendency for intra-system uniformity is twice as strong when considering planets within the same size category than it is when combining all planets together. Motivated by this recent finding, we show the separate fraction distributions for SE-SE pairs and SN-SN pairs from our model in Fig. 4. The colored numbers on the right-hand sides of each panel indicate the number of the plotted pairs for each $p_{\rm init}$. For SN-SN pairs, the peak values become large, as suggested by Millholland & Winn (2021). In contrast, the peaks in the SE-SE pairs are not always larger. This would be attributed to the initial mass distribution of the protoplanets as a function of the orbital distance in our model.

Several speakers including Daria Kubyshikina at this conference highlighted the revised mass loss rate driven by XUV photoevaporation in a hydrogen-only atmosphere. Kubyshikina et al. (2018a,b) derived an analytic fit based on a grid of hydrodynamic models. They demonstrated that the mass loss rate based on the energy-limit model is underestimated by several orders of magnitude in the regime of the low Jeans escape parameter Λ, which is a ratio of the gravitation energy to the thermal energy for a hydrogen atom in an atmosphere. The bimodal distribution of small planets can be reproduced using the revised version of photoevaporation (Modirrousta-Galian et al. 2020). The enhanced mass loss rate may improve our model, which needs more mass loss to be more consistent with the bimodal size distribution. It may also strengthen the correlation of SE-SE pairs in our model. However, the incredibly high mass loss in the low gravity regime implies that the close-in protoplanets could completely lose their hydrogen-rich envelope during the giant impact phase. Nevertheless, the cooling due to additional emissions in the presence of other molecules in the atmosphere can reduce the mass loss rate (Yoshida et al. 2022). Additionally, dust generated through impact events in a natal debris disk (Genda et al. 2015; Kenyon & Bromley 2016) may shield XUV and thus weaken the mass loss of the gas envelope of close-in protoplanets during the giant-impact phase.

5. Summary

Motivated by the KEPLER results, we attempted to simultaneously model the size distribution and orbital architecture for close-in small planets. We employed N-body simulations for giant impacts including an analytical structure model for a planetary gas envelope. A two-stage mass loss was considered – giant-impact shock followed by XUV photoevaporation. Our results are summarized below:

• the observed size-ratio and period-ratio distributions (i.e., peas in a pod) are generally reproduced. They are primarily caused by giant impacts of protoplanets, although the adjacent planet pairs are more correlated in our model.

• the simulated radius gap is smaller – too many sub-Neptunes and too small super-earths. The issue may be mitigated by invoking more mass loss mechanisms or considering a wide range of initial mass distributions of protoplanets. Kubyshikina's analytic model for XUV photoevaporation could improve our model.

• is the bimodal size distribution relevant to the orbital architecture? Our model is not completely consistent with the observations, and a few concerns related to new developments in the field are raised in the discussion section, so our work has not yet answered the outstanding question. Nevertheless, our study provides hints and suggests hopes for the challenges that the community faces.

We would like to thank the organizers for giving us an opportunity to present the work online. P.G. Gu would like to thank Aline Vidotto for the helpful discussions and acknowledge the support from MOST through the grant number 111-2112-M-001-037.

References

Dai, F., Winn, J. N., Schlaufman, K., et al. 2020, AJ, 159, 247
Erkaev, N. V., Kulikov, Y. N., Lammer, H., et al. 2007, A&A, 472, 329
Fulton, B. J., Petigura, E. A. 2018, AJ, 156, 264
Fulton, B. J., Petigura, E. A., Howard, A. W., et al. 2017, AJ, 154, 109
Genda, H., Kobayashi, H., Kokubo, E. 2015, ApJ, 810, 136
Genda, H., Abe, Y. 2003, Icar, 164, 149
Ginzburg, S., Schlichting, H. E., Sari, R. 2018, MNRAS, 476, 759
Gupta, A., Schlichting, H. E. 2019, MNRAS, 487, 24
Kegerreis, J. A., Eke, V. R., Massey, R. J., Teodoro, L. F. A. 2020, ApJ, 897, 161
Kenyon, S. J., Bromley, B. C. 2016, ApJ, 817, 51
Kokubo, E., Ida, S. 2002, ApJ, 581, 666
Kubyshikina, D., Fossati, L., Erkaev, N. V., et al. 2018a, ApJ, 619, 151
Kubyshikina, D., Fossati, L., Erkaev, N. V., et al. 2018b, ApJ Letters, 886, 18
Matsumoto, Y., Kokubo, E., Gu, P.-G., Kurosaki, K. 2021, ApJ, 923, 81
Millholland, S. C., Winn, J. N. 2021, ApJ Letters, 920, 34
Mordirrousta-Galian, D., Locci, D., Micela, G. 2021, ApJ, 891, 158
Owen, J. E., Wu, Y. 2013, ApJ, 775, 105
Owen, J. E., Wu, Y. 2017, ApJ, 847, 29
Rogers, J. G., Gupta, A., Owen, J. E., Schlichting, H. E. 2021, MNRAS, 508, 5886
Weiss, L. M., Marcy, G. W., Petigura, E. A., et al. 2018, AJ, 155, 48
Weiss, L. M., Petigura, E. A. 2020, ApJ Letters, 893, L1
Yoshida, Y., Terada, N., Ikoma, M., Kuramoto, K. 2022, ApJ, 934, 137
Zhu, W. 2020, AJ, 159, 188

Winds of Stars and Exoplanets
Proceedings IAU Symposium No. 370, 2023
A. A. Vidotto, L. Fossati & J. S. Vink, eds.
doi:10.1017/S1743921322004227

Spin-down and reduced mass loss in early-type stars with large-scale magnetic fields

Z. Keszthelyi[1,2], A. de Koter[1,3], Y. Götberg[4], G. Meynet[5], S.A. Brands[1], V. Petit[6], M. Carrington[7], A. David-Uraz[8,9], S.T. Geen[1], C. Georgy[5], R. Hirschi[10,11], J. Puls[12], K.J. Ramalatswa[13,14], M.E. Shultz[6] and A. ud-Doula[15]

[1]Anton Pannekoek Institute for Astronomy, University of Amsterdam, Science Park 904, 1098 XH, Amsterdam, The Netherlands

[2]Center for Computational Astrophysics, Division of Science, National Astronomical Observatory of Japan, 2-21-1, Osawa, Mitaka, Tokyo 181-8588, Japan

[3]Institute of Astronomy, KU Leuven, Celestijnenlaan 200D, 3001 Leuven, Belgium

[4]The observatories of the Carnegie institution for science, 813 Santa Barbara Street, Pasadena, CA 91101, USA

[5]Geneva Observatory, University of Geneva, Maillettes 51, 1290 Sauverny, Switzerland

[6]Dept. of Physics and Astronomy, Bartol Research Institute, University of Delaware, 217 Sharp Lab, Newark, DE 19716, USA

[7]Dept. of Physics and Space Science, Royal Military College of Canada, PO Box 1700, Station Forces, Kingston, ON K7K 0C6, Canada

[8]Dept. of Physics and Astronomy, Howard University, Washington, DC 20059, USA

[9]Center for Research and Exploration in Space Science and Technology, and X-ray Astrophysics Laboratory, NASA/GSFC, Greenbelt, MD 20771, USA

[10]Astrophysics Group, Keele University, Keele, Staffordshire ST5 5BG, UK

[11]Institute for Physics and Mathematics of the Universe (WPI), University of Tokyo, 5-1-5 Kashiwanoha, Kashiwa 277-8583, Japan

[12]LMU München, Universitätssternwarte, Scheinerstr. 1, 81679 München, Germany

[13]Dept. of Astronomy, University of Cape Town, Private Bag X3, Rondebosch 7701, South Africa

[14]South African Astronomical Observatory, PO Box 9, Observatory, 7935, South Africa

[15]Dept. of Physics, Penn State Scranton, 120 Ridge View Drive, Dunmore, PA 18512, USA

Abstract. Magnetism can greatly impact the evolution of stars. In some stars with OBA spectral types there is direct evidence via the Zeeman effect for stable, large-scale magnetospheres, which lead to the spin-down of the stellar surface and reduced mass loss. So far, a comprehensive grid of stellar structure and evolution models accounting for these effects was lacking. For this reason, we computed and studied models with two magnetic braking and two chemical mixing schemes in three metallicity environments with the MESA software instrument. We find notable differences between the subgrids, which affects the model predictions and thus the detailed characterisation of stars. We are able to quantify the impact of magnetic fields in terms of preventing quasi-chemically homogeneous evolution and producing slowly-rotating, nitrogen-enriched ("Group 2") stars. Our model grid is fully open access and open source.

Keywords. stars: evolution — stars: massive — stars: magnetic field — stars: rotation — stars: abundances

---

1. Introduction

Magnetism is a key component in several astrophysical phenomena. For example, magnetic fields play a crucial role in regulating star formation and controlling the formation of neutron stars (e.g., Commerçon et al. 2011; Takiwaki & Kotake 2011). A fraction of massive stars (initially $M_\star > 8$ M$_\odot$) and intermediate-mass stars (initially 3 M$_\odot$ $< M_\star < 8$ M$_\odot$) show evidence of stable, globally organised, large-scale magnetic fields that are understood to be of fossil origin. The exact origin of such fossil fields is unclear; however, they may result from the pre-main sequence evolution of the star or from stellar merger events (Schneider et al. 2019). Fossil fields form a magnetosphere around the star, affecting the wind and stellar rotation. For their stability, they must be anchored in deep stellar layers. While it is known that magnetic fields can significantly impact the physics and evolution of stars, there has been no comprehensive grid of stellar evolution models of massive stars that would take into account the effects of surface fossil magnetic fields. We use the MESA software instrument (Paxton et al. 2019) to compute models and map out a large parameter space. The main new additions in our calculations are magnetic mass-loss quenching, magnetic braking, and efficient angular momentum transport (see next section).

2. Background

2.1. *Alfvén radius*

The Alfvén radius characterises a critical distance at which the magnetic energy density and the gas kinetic energy density are equal. ud-Doula et al. (2009) use a numerical fitting to quantify the Alfvén radius as:

$$\frac{R_{\rm A}}{R_\star} \approx 1 + (\eta_\star + 0.25)^x - (0.25)^x \,, \tag{2.1}$$

with $x = 1/4$ and $1/6$ for dipolar and quadrupolar field geometries, which we assume in our INT and SURF models (see next section), respectively. $R_\star$ is the stellar radius. The equatorial magnetic confinement parameter $\eta_\star$ is defined as:

$$\eta_\star = \frac{B_{\rm eq}^2 R_\star^2}{\dot{M}_{B=0} \cdot v_\infty} \,, \tag{2.2}$$

where $B_{\rm eq}$ is the equatorial magnetic field strength, $\dot{M}_{B=0}$ is the mass-loss rate in absence of a magnetic field, and v_∞ is the terminal velocity (ud-Doula et al. 2009).

2.2. *Mass-loss quenching*

Large-scale magnetic fields lead to channelling and trapping the wind plasma within the magnetosphere. To account for the global, time-averaged effect of this process, the adopted mass-loss rates in our models are reduced by a parameter $f_{\rm B}$, which is defined following the works of ud-Doula et al. (2008, 2009):

$$f_{\rm B} = \frac{\dot{M}}{\dot{M}_{B=0}} = 1 - \sqrt{1 - \frac{1}{R_{\rm c}}} \quad \text{if} \quad R_{\rm A} < R_{\rm K} \tag{2.3}$$

and

$$f_{\rm B} = \frac{\dot{M}}{\dot{M}_{B=0}} = 2 - \sqrt{1 - \frac{1}{R_{\rm c}}} - \sqrt{1 - \frac{0.5}{R_{\rm K}}} \quad \text{if} \quad R_{\rm K} < R_{\rm A} \tag{2.4}$$

where $R_{\rm A}$, $R_{\rm K}$, and $R_{\rm c}$ are the Alfvén radius, the Kepler co-rotation radius, and the closure radius in units of the stellar radius, respectively (see Keszthelyi et al. 2019, 2020, 2021, and references therein).

2.3. *Magnetic braking*

Stellar rotation exerts a force on magnetic field lines and causes the field to bend in the azimuthal direction. The associated Maxwell stresses are very efficient at transferring angular momentum to the surrounding wind plasma, which results in slowing the spin of the star. We quantify this process following the work of ud-Doula et al. (2009), where the total wind and magnetic field-induced loss of angular momentum can be expressed via a Weber-Davis (Weber & Davies 1967) scaling relation:

$$\frac{{\rm d}J_{\rm B}}{{\rm d}t} = \frac{2}{3}\dot{M}_{B=0}\,\Omega_\star R_A^2\,, \tag{2.5}$$

with ${\rm d}J_{\rm B}/{\rm d}t$ the rate of angular momentum loss from the system, $\Omega_\star$ the surface angular velocity, and R_A the Alfvén radius (defined in Equation 2.1).

Based on how magnetic braking is applied, we split the model calculations into two branches. In one case, we assume internal magnetic braking (INT models). These models are solid-body rotating and specific angular momentum is extracted from each layer of the stellar model. In the other case, we only allow the model to directly remove specific angular momentum from the upper envelope of the star (SURF models). Thus, radial differential rotation can develop in deeper layers in the SURF models.

2.4. *Chemical mixing*

Due to stellar rotation, chemical elements are also mixed in radiative regions of stars. Consequently, main sequence massive stars can replenish their core with more hydrogen, and also enrich their surface with core-produced materials (most importantly, nitrogen). We adopt a diffusive scheme, following the work of Pinsonneault et al. (1989), to account for these effects.

$$\frac{\partial X_i}{\partial t} = \frac{\partial}{\partial m}\left[(4\pi r^2 \rho)^2\, D_{\rm chem}\,\frac{\partial X}{\partial m}\right] + \left(\frac{{\rm d}X_i}{{\rm d}t}\right)_{\rm nuc}, \tag{2.6}$$

where X_i is the mass fraction of a given element i, t is the time, m and ρ are the mass coordinate and mean density at a given radius r, $D_{\rm chem}$ is the sum of individual diffusion coefficients contributing to chemical mixing, and the final term accounts for nuclear burning.

The detailed mixing processes remain highly uncertain in massive star evolutionary models. For this reason, we also split the model calculations to construct $D_{\rm chem}$ in two different ways. In one case (Mix1), we assume chemical mixing equations that are commonly adopted in MESA. We thus adopt $D_{\rm chem}$ as a sum of individual diffusion coefficients describing meridional circulation, shear, and GSF instabilities. We then scale $D_{\rm chem}$ with commonly used factors in this approach. In the other case (Mix2), we implement the mixing equations of Zahn (1992) into the MESA code. Here we use the vertical shear mixing and meridional circulation (both defined differently than in the above approach) to construct $D_{\rm chem}$. In this case, scaling factors are not used. We do not change the equations for angular momentum transport for the sake of a consistent model-to-model comparison.

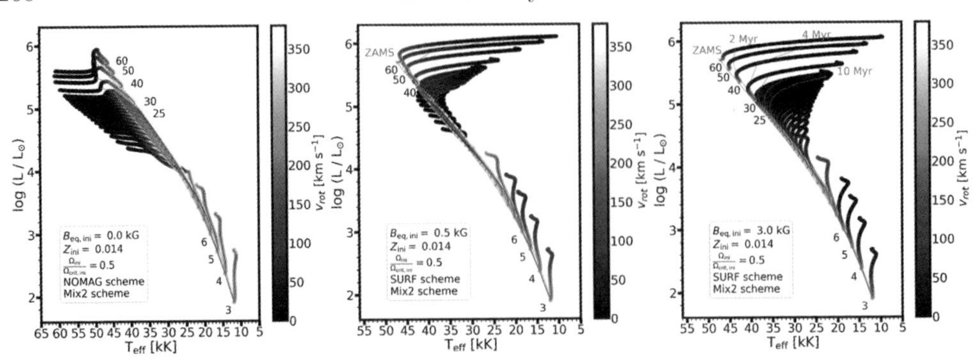

Figure 1. HRDs of the computed models at solar metallicity with an initial ratio of 0.5 critical angular velocity within the efficient Mix2 chemical mixing scheme. Panels from left to right show models with initial equatorial magnetic field strengths of 0, 0.5, 3 kG, respectively. The colour-coding corresponds to the surface rotational velocity. The initial masses in solar units are indicated next to the ZAMS. Between 6 and 25 $M_\odot$, the increment is 1 $M_\odot$. From Keszthelyi et al. (2022, accepted in MNRAS).

3. Summary of computed models

In total, we computed 8,748 main sequence stellar evolution models, with 3 stellar structure models for each of these corresponding to ZAMS, mid-MS, and TAMS evolutionary stages. We also generated isochrones. All the computed models are open access and open source and available on Zenodo at `https://doi.org/10.5281/zenodo.7069766`. The initial mass and equatorial magnetic field strength spans from 3 to 60 $M_\odot$ and from 0 to 50 kG in our grid of models. Four sub-grids are available, accounting for two magnetic braking and two chemical mixing schemes, introduced above. Hot star mass-loss rates in our models are adopted from Vink et al. (2001) and are reduced by a factor of 2 to account for the general trend evidenced from recent clumping-corrected mass-loss rate determinations. Mass-loss rates lower than the predictions of Vink et al. (2001) are also expected from new numerical simulations (Björklund et al. 2020; Krtička et al. 2021). The metallicity in our model grid corresponds to Solar, LMC, and SMC abundances, respectively.

4. Results & Discussion

4.1. *Quasi-chemically homogeneous evolution*

When internal chemical mixing is efficient, our models at solar metallicity with an initial rotation rate of 0.5 critical angular velocity show a blueward evolution on the HRD (left panel of Figure 1). This is because all stellar layers have a nearly homogeneous chemical composition. Given the blueward evolution, the increase in effective temperature will lead to Wolf-Rayet type mass-loss rates in our models, which can help i) spin down the stellar models and ii) decrease the surface hydrogen abundance. With an initial equatorial magnetic field strength of 0.5 kG (middle panel), the models experience a short blueward evolution. However, within a spin-down timescale, the decrease of rotational velocity results in less efficient chemical mixing. Therefore, the models turn to a redward evolution on the main sequence. With $B_{\rm eq,ini} = 3$ kG, the initial blueward evolution is shorter, and most of the main sequence evolution follows a classical path. We can thus conclude that surface fossil magnetic fields may play an important role in preventing quasi-chemically homogeneous main sequence evolution, which may be an important channel for several astrophysical phenomena (e.g., Yoon et al. 2006).

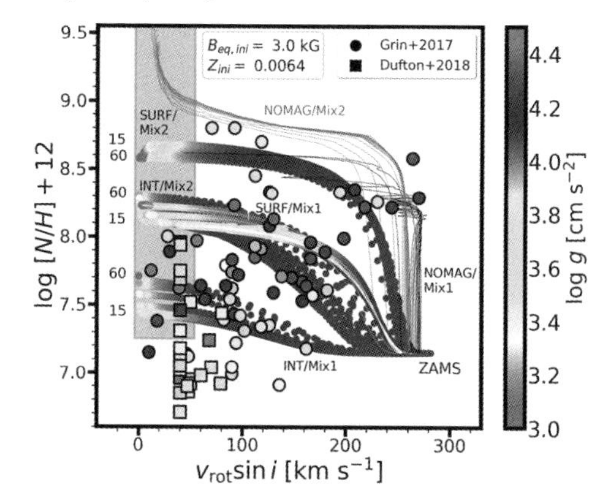

Figure 2. Hunter diagram of magnetic single-star evolutionary models with $B_{\mathrm{eq,ini}} = 3$ kG, $\Omega_{\mathrm{ini}}/\Omega_{\mathrm{ini,crit}} = 0.5$ at LMC ($Z_{\mathrm{ini}} = 0.0064$) metallicity within two magnetic braking and two chemical mixing schemes. Models within the SURF/Mix1 scheme are also shown with white lines since they overlap with the INT/Mix2 models. Additionally, two sets of non-magnetic (NOMAG) models are shown within the Mix1 (grey) and Mix2 (black) schemes. Models with initial masses from 15 to 60 $M_{\odot}$ are shown. The coloured area corresponds to our definition of Group 2 stars ($v \sin i < 50$ km s^{-1} and at least 0.1 dex nitrogen enrichment in spectroscopic units). The colour-coding of the models shows the logarithmic surface gravity. Observations are shown with circles and squares. A typical reported uncertainty in the observed nitrogen abundances is about 0.1 dex. From Keszthelyi et al. (2022, accepted in MNRAS).

4.2. *Magnetic stellar evolution models on the Hunter diagram*

The Hunter diagram (Hunter et al. 2008), showing surface nitrogen abundance as a function of projected rotational velocity, is an important diagnostic tool to gain insights into the chemical and rotational evolution of stars. The efficiency of chemical mixing and magnetic braking play an important role in shaping the quantitative evolution of the models on this diagram. For LMC metallicity, we assumed a baseline abundance of $\log[N/H] = 7.15$. The uncertainties in the magnetic models may lead to an order of magnitude difference in the predicted nitrogen abundances for a typical 3 kG initial equatorial magnetic field strength (Figure 2). Nonetheless, we find that these magnetic models, regardless of the uncertainties related to mixing and braking schemes, produce slowly-rotating, nitrogen-enriched stars. These so-called “Group 2” stars could not be explained with typical single star evolution thus far, though they are commonly found in spectroscopic samples. We note however that spectropolarimetric observations and comprehensive magnetic characterisation of Group 2 stars are largely lacking in the Galaxy (see however, e.g., Aerts et al. 2014; Martins et al. 2012, 2015) and are practically unavailable in the Magellanic Clouds. Therefore Group 2 stars might be prime targets to detect magnetic fields.

5. Conclusions

We computed and studied an extensive grid of stellar evolution and structure models, incorporating the effects of surface fossil magnetic fields. We can quantitatively demonstrate that i) quasi-chemically homogeneous evolution could be mitigated and prevented for increasing magnetic field strength, and ii) slowly-spinning, nitrogen-enriched Group 2 stars on the Hunter diagram could be produced by magnetic massive stars. The library of stellar models is available to the community via Zenodo at `https://doi.org/10.5281/zenodo.7069766`.

References

Aerts, C., Molenberghs, G., Kenward, M. G., Neiner, C., 2014, *ApJ* 781, 88
Björklund, R., Sundqvist, J. O., Puls, J., Najarro, F., 2020, *A&A* 648, A36
Commerçon, B., Hennebelle, P., Henning, T., 2011, *ApJL* 742, L9
Hunter, I., Brott, I., Lennon, D. J., Langer, N., Dufton, P. L., et al., 2008, *ApJL* 676, L29
Keszthelyi, Z., Meynet, G., Georgy, C., Wade, G. A., Petit, V., David-Uraz, A., 2019, *MNRAS* 485, 5843
Keszthelyi, Z., Meynet, G., Shultz, M. E., David-Uraz, A., ud-Doula, A., et al., 2020, *MNRAS* 493, 518
Keszthelyi, Z., Meynet, G., Martins, F., de Koter, A., David-Uraz, A., 2021, *MNRAS* 504, 2474
Keszthelyi et al. 2022, *MNRAS*, 517, 2028
Krtička, J., Kubát, J., Krtičková, I., 2021, *A&A* 647, A28
Martins, F., Hervé, A., Bouret, J.-C., Marcolino, W., Wade, G. A., et al., 2015, *A&A* 575, A34
Martins, F., Escolano, C., Wade, G. A., Donati, J. F., Bouret, 2012, *A&A* 538, A29
Paxton, B., Smolec, R., Schwab, J., Gautschy, A., Bildsten, L., et al., 2019, *ApJS* 243, 10
Pinsonneault, M. H., Kawaler, S. D., Sofia, S., Demarque, P., 1989, *ApJ* 338, 424
Schneider, F. R. N., Ohlmann, S. T., Podsiadlowski, P., 2019, *Nat* 574, 211
ud-Doula, A., Owocki, S. P., Townsend, R. H. D., 2008, *MNRAS* 385, 97
ud-Doula, A., Owocki, S. P., Townsend, R. H. D., 2009, *MNRAS* 392, 1022
Takiwaki, T., Kotake, K, 2011, *ApJ* 743, 30
Vink, J. S., de Koter, A., Lamers, H. J. G. L. M., 2001, *A&A* 369, 574
Weber, E. J., Davis, Jr., L., 1967, *ApJ* 148, 217
Yoon, S. -C., Langer, N., Norman, C., 2006, *ApJ* 460, 199
Zahn, J.-P., 1992, *A&A* 265, 115

Winds of Stars and Exoplanets
Proceedings IAU Symposium No. 370, 2023
A. A. Vidotto, L. Fossati & J. S. Vink, eds.
doi:10.1017/S1743921322003623

Mass loss implementation and temperature evolution of very massive stars

Gautham N. Sabhahit[1], Jorick S. Vink[1], Erin R. Higgins[1] and Andreas A.C. Sander[1,2]

[1]Armagh Observatory and Planetarium, College Hill, Armagh BT61 9DG, N. Ireland
E-mail: gautham.sabhahit@armagh.ac.uk

[2]Zentrum für Astronomie der Universität Heidelberg, Astronomisches Rechen-Institut, Mönchhofstr. 12-14, 69120 Heidelberg, Germany

Abstract. Very massive stars (VMS) dominate the physics of young clusters due to their extreme stellar winds. The mass lost by these stars in their winds determine their evolution, chemical yields and their end fates. In this contribution we study the main-sequence evolution of VMS with a new mass-loss recipe that switches from optically-thin O star winds to optically-thick Wolf-Rayet type winds through the model independent transition mass loss.

Keywords. stars: mass loss, stars: winds, outflows, stars: evolution

---

1. Introduction

The interest in very massive stars (VMS) up to 300 $M_{\odot}$ has grown substantially in the last few decades. VMS are defined as stellar objects with masses over 100 $M_{\odot}$ (Vink et al. 2015). They have strong emission line spectral features similar to an emission-line dominated Wolf-Rayet (WR) star of the nitrogen (N) type, but also with left-over hydrogen (H) in their spectra, and are given the WNh spectral type. These stars can be considered as a natural continuation of canonical O stars along the Main sequence (MS) and are likely still H-burning objects (e.g. Massey & Hunter 1998; de Koter et al. 1998).

VMS are often found in young, massive clusters such as the Arches and NGC3603 in our galaxy, and the vast 30 Dor region in the Large Magellanic cloud (LMC) with the central R136 cluster which hosts the most massive stars to date (Crowther et al. 2010). Due to their strong stellar winds these stars are responsible for enormous mechanical and chemical input to the surrounding medium and can significantly influence the evolution of the cluster. In high metallicity (Z) environments such as the Milky Way or the LMC, where winds are extremely strong, VMS may evaporate themselves, largely already on the MS (Vink 2018). If mass-loss rates are low, such as in the early universe or low-Z environments, VMS may produce pair-instability supernovae (PISNe) where the entire star is obliterated leaving no remnants behind, and one such PISN could potentially produce more metals than an entire initial mass function (IMF) below it (Langer 2012).

Due to their high luminosity-to-mass (L/M $\gtrsim 10^4$) ratio, VMS are very close to their so-called Eddington limit ($\Gamma \propto \kappa_{\rm F} L/M \rightarrow 1$). Their evolution is highly uncertain due to the unknown physics in close proximity to this limit of radiative pressure. Despite uncertainties in the mass loss of these stars, both theoretical (Vink et al. 2011) and empirical efforts (Gräfener et al. 2011; Bestenlehner et al. 2014) have hinted towards an enhanced mass loss in comparison to canonical O-type stars. In addition to increased mass loss, objects close to the Eddington limit may be subjected to substantial envelope *inflation*

Table 1. Properties of the transition stars in the Arches and the 30 Dor cluster. The luminosities, terminal velocities, effective temperatures and the surface H are empirically determined (Martins et al. 2008; Bestenlehner et al. 2014). The transition mass-loss rate is obtained from Eq. 3.1. For the transition mass and corresponding electron scattering Γ_e see Sec. 3.

	$L/L_\odot$	v_∞(km/s)	$T_{\rm eff}$(K)	X_s	$\dot{M}_{\rm trans}$	$M/M_\odot$	$\Gamma_{\rm e,trans}$
GAL	$10^{6.06}$	2000	33900	0.7	−5.16	76.9	0.39
LMC	$10^{6.31}$	2550	44400	0.62	−5.0	121.8	0.42

(Ishii et al. 1999; Gräfener et al. 2012) with models predicting very cool temperatures and large radii. While the physical existence of such inflated envelopes is debated, enhanced mass loss could potentially inhibit the formation of such inflated envelopes (Petrovic et al. 2006; Grassitelli et al. 2018).

In this contribution, we derive a new mass loss prescription for VMS using the transition mass-loss rate at two different metallicities and study the effects of enhanced mass-loss on the evolution of stars with initial masses upto 500 $M_\odot$

2. Methodology

The one-dimensional stellar evolution code Modules for Experiments in Stellar Astrophysics (MESA version 12115) is used to compute our grid of stellar models. The models are evolved until the end of core H burning, where the models stop once the central hydrogen mass fraction (X_c) falls below 0.01. The initial mass ranges from 60 to 500 $M_\odot$. We consider two initial metallicities corresponding to the Galaxy ($Z = 0.02$) and the Large Magellanic Cloud ($Z = 0.008$), where the observed transition stars are available to obtain absolute rates.

Convective overshooting beyond the convective regions is treated using the exponential profile parameterized by $f_{\rm ov} = 0.03$. All models begin as solid-body rotators at the ZAMS with $\Omega/\Omega_{crit} = 0.2$. As discussed below, the choice of efficiencies of mixing processes that can influence the core size, such as overshooting and rotation-induced mixing, hardly influences the evolution of VMS.

Theoretical mass-loss prescriptions for OB stars as a function of stellar properties are available (Castor et al. 1975; Vink et al. 2001), and have been previously used to study the evolution of VMS. However these rates quickly start to under-predict the mass loss above the transition point. This has been shown both theoretically (Vink 2006; Vink et al. 2011) and empirically in the 30 Dor cluster (Bestenlehner et al. 2014). The challenge however is to determine the absolute rates of these objects and their mass-loss scaling with different stellar parameters. In this study we use the model-independent transition mass loss in the Arches and the 30 Dor region to calibrate the absolute rates of VMS and further scale these rates with a steeper dependence on the ratio L/M compared to canonical O star scaling.

3. Transition mass loss

Vink & Gräfener (2012) derive a model-independent way to characterize the physical transition from optically-thin winds of O stars to the optically-thick winds of WNh stars, regardless of the assumptions of clumping in the wind. They showed that at this transition the optical depth at the sonic point τ_s crossing unity coincides with the wind efficiency parameter $\eta = \dot{M} v_\infty/(L/c)$ crossing unity, under the assumption of very high Eddington parameter Γ. This condition $\eta = \tau_s = 1$ can be used to obtain a model-independent mass loss, called the *transition mass loss*, given the luminosity and the terminal velocity of the transition objects.

The $\eta = \tau_s = 1$ condition at the transition relies on the assumption that the ratio $(\Gamma - 1)/\Gamma$ is very close to unity or $\Gamma \gg 1$. Vink & Gräfener (2012) tested this assumption of high Γ using a hydro-dynamic model of WR22 from Gräfener & Hamann (2008), by numerically integrating inwards from infinity to the sonic radius to obtain τ_s. They derived a correction factor $f \approx 0.6$ for their transition condition: $\eta = f\tau_s$.

In the Arches cluster, we have a spectral morphological transition from O to the WNh sequence with the transition stars being classified as Of/WNh stars. Using the properties of the transition Of/WNh stars one can derive the transition mass-loss rate using the following formula

$$\dot{M}_{\rm trans} = f \frac{L_{\rm trans}}{v_\infty c} \tag{3.1}$$

Vink & Gräfener (2012) also found that the mass-loss rates of the transition objects in the Arches cluster agrees with the theoretical rates predicted from Vink et al. (2001) for Galactic metallicity. Using this assumption one can derive an 'average' transition mass as well as the electron scattering Eddington parameter at the transition $\Gamma_{\rm e,trans}$. One can perform a similar analysis for the 30 Dor objects in the LMC where we have empirical results of six transition stars (Bestenlehner et al. 2014). The transition properties of the two clusters are provided in Table 1.

We can thus use the transition mass loss as a kink or an anchor point connecting the *optically-thin regime* of canonical O stars below it and the much steeper scaling of *optically-thick regime* of WNh stars above it. In MESA, we implement a mass loss recipe that consists of an optically-thin scaling of $\dot{M} \sim L^{2.194} M^{-1.313}$ from Vink et al. (2001) till the transition point, which then smoothly switches to the optically-thick scaling of $\dot{M} \sim L^{4.77} M^{-3.99}$ derived in Vink et al. (2011). In Sabhahit et al. (2022), we go into more details of the effect of electron number density on the mass-loss rates and the consequent effects on the evolution of VMS.

As for the metallicity dependence of the winds, we implement a $\dot{M} - Z$ dependence that only reflects changes in surface iron (Fe) abundance as originally intended in Vink et al. (2001), with the following scaling $\dot{M} \sim Z^{0.85}$ in the optically-thin regime. The transition mass loss in two different Z environments allows us to derive a mass loss-metallicity scaling of $\dot{M} \sim Z^{0.76}$ at the transition point, which is then applied throughout the optically-thick regime. The terminal velocity v_∞ is also implemented to vary with the metallicity according to Leitherer et al. (1992), $v_\infty \sim Z^{0.13}$.

4. Implications for VMS temperatures and surface abundances

In Fig. 1 we plot a full grid of stellar tracks of VMS models with varying initial mass at Galactic metallicity. At the upper end of the initial mass spectrum ($\gtrsim 200\, M_\odot$) considered here, the VMS begin H-burning with luminosities in excess of $\log(L/L_\odot) = 6.5$. These VMS lie above the transition point and in the optically-thick regime of our mass loss recipe. Owing to the higher absolute mass loss these models quickly evaporate a large fraction of their initial mass, resulting in an overall drop in their luminosity. This is in stark contrast to the MS evolution of canonical massive stars where the luminosity increases as the overall mean molecular weight increases (μ effect) during H burning.

There are qualitative differences in the evolution of VMS model temperatures as well. VMS models with the new mass loss recipe evolve vertically at almost constant temperatures, which is in stark contrast to stars below the transition point where stars tend to expand to cooler temperatures and evolve horizontally during the MS. This reduction in luminosity suppresses both the effects of envelope inflation as well as enhanced mass loss (both being a function of L/M), thus maintaining a balance between the two effects for all initial masses. The drop in the luminosity owing to higher mass loss rates has a self-regulatory action in maintaining constant temperatures throughout the evolution.

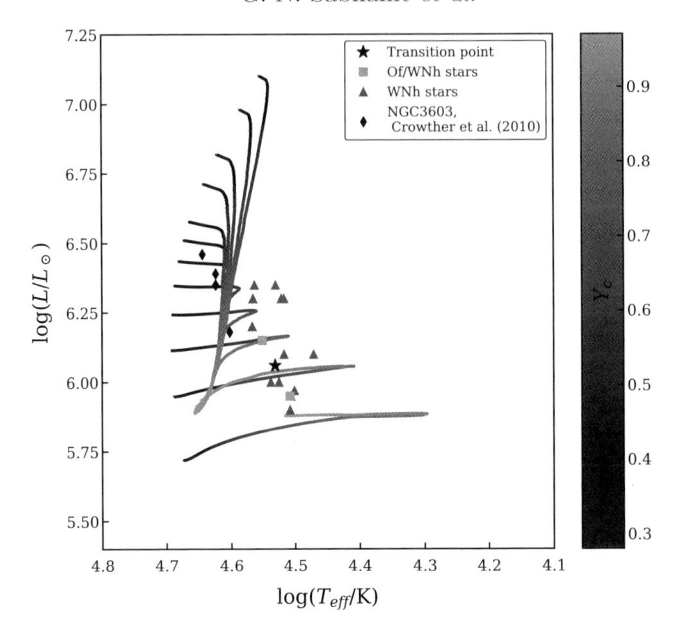

Figure 1. Stellar tracks of VMS models with initial mass ranging from 60 $M_{\odot}$ to 500 $M_{\odot}$. Empirical luminosities and temperatures of VMS in the Galaxy are also plotted (see text).

We over-plot the observed HRD locations of the transition Of/WNh stars as squares and WNh stars as triangles in the Arches cluster from Martins et al. (2008). The WN6h stars in the NGC3603 cluster is also plotted as diamonds from (Crowther et al. 2010). The star symbol marks the 'average' luminosity and the effective temperature of the transition stars obtained by averaging the surface properties of the two Of/WNh stars. The WNh stars in the Arches cluster occupy a narrow band of temperatures in the HRD with $\log(T_{\rm eff}) \approx 4.5 - 4.6$ above $\log(L/L_{\odot}) \approx 6.0$. The vertical alignment of the observed temperatures could just be due to the similar ages of the stars in a cluster. While it is possible to reproduce the observed temperatures with VMS models evolving horizontally in the HRD, their radii and temperatures are highly sensitive to the effect of inflation which grows with L/M. This causes VMS models to have varying temperatures with varying initial mass (for the same age) making it is highly unlikely for models with different initial masses to maintain such constant temperatures long enough to be observed. Vertical evolution would give a natural explanation of the observed constant temperatures of VMS over the entire mass range.

Another interesting property of VMS can be understood by studying the evolution of their surface abundances. The convective core size of a star is determined at the location where the temperature gradient required for radiation diffusion to transport the stellar luminosity outwards, $\nabla_{\rm rad} \sim \chi l/m$ equals the adiabatic temperature gradient $\nabla_{\rm ad}$. Owing to increasing l/m with initial mass, the convective core mass fraction increases with initial mass, reaching values greater than ≈ 0.9 at the beginning of MS for $M_{\rm init} \gtrsim 200\ M_{\odot}$.

A consequence of having an almost fully mixed star is the insensitivity of the evolution of VMS to processes that can affect the core size. Models with $M_{\rm init} \gtrsim 200\ M_{\odot}$ (see thick grey line in Fig. 2) remain fully mixed throughout the MS regardless of their rotational and overshooting inputs. The surface abundances closely maps the central abundances, and can be used as a clock to obtain constraints on the age of these objects (Higgins et al. 2022). The effects of varying overshooting and rotation on the core size becomes negligible and the evolution is completely dominated by mass loss.

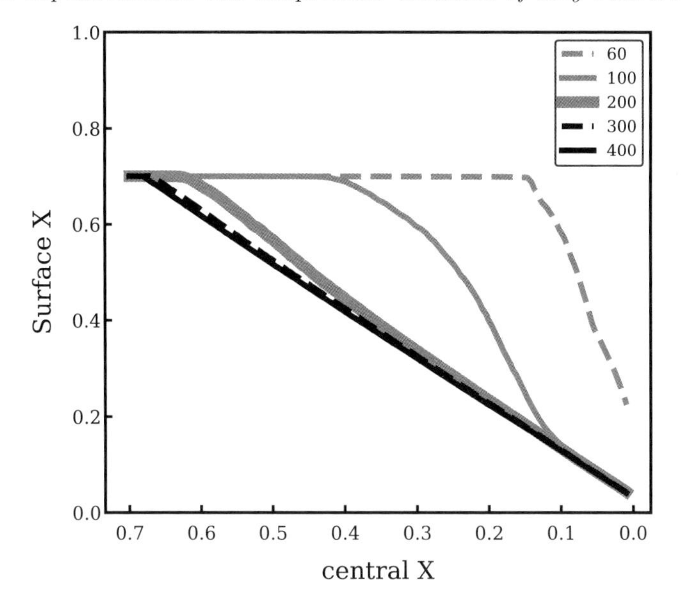

Figure 2. Evolution of surface and central hydrogen mass fraction for selected models in our grid (the initial masses of these models are provided in the legend).

In this study we used the model-independent transition mass-loss rate in two different metallicities to calibrate the absolute rates of VMS, and investigated the effects of a new mass loss recipe with a simple power law scaling on the evolution of stars above 60 $M_{\odot}$. The mass loss scaling instead could be more complicated than a simple power law and could vary with the surface electron density or the temperature. A promising step forward is to understand the mass-loss enhancement at the $\eta = \tau_s = 1$, but now using stellar atmosphere codes such as PoWR that are capable of consistently solving the wind hydro-dynamics to determine wind parameters including mass loss.

References

Bestenlehner, J. M., et al. 2014, A&A, 570, A38
Castor, J. I., Abbott, D. C., & Klein, R. I. 1975, ApJ, 195, 157
Crowther, P. A., Schnurr, O., Hirschi, R., Yusof, N., Parker, R. J., Goodwin, S. P., & Kassim, H. A. 2010, MNRAS, 408, 731
de Koter, A., Heap, S. R., & Hubeny, I. 1998, ApJ, 509, 879
Gräfener, G., & Hamann, W. R. 2008, A&A, 482, 945
Gräfener, G., Owocki, S. P., & Vink, J. S. 2012, A&A, 538, A40
Gräfener, G., Vink, J. S., de Koter, A., & Langer, N. 2011, A&A, 535, A56
Grassitelli, L., Langer, N., Grin, N. J., Mackey, J., Bestenlehner, J. M., & Gräfener, G. 2018, A&A, 614, A86
Higgins, E. R., Vink, J. S., Sabhahit, G. N., & Sander, A. A. C. 2022, MNRAS, 516, 4052
Ishii, M., Ueno, M., & Kato, M. 1999, 51, 417
Langer, N. 2012, Annual Review of Astron and Astrophysis, 50, 107
Leitherer, C., Robert, C., & Drissen, L. 1992, ApJ, 401, 596
Martins, F., Hillier, D. J., Paumard, T., Eisenhauer, F., Ott, T., & Genzel, R. 2008, A&A, 478, 219
Massey, P., & Hunter, D. A. 1998, ApJ, 493, 180
Petrovic, J., Pols, O., & Langer, N. 2006, A&A, 450, 219
Sabhahit, G. N., Vink, J. S., Higgins, E. R., & Sander, A. A. C. 2022, MNRAS, 514, 3736

Vink, J. S. 2006, in Astronomical Society of the Pacific Conference Series, Vol. 353, Stellar Evolution at Low Metallicity: Mass Loss, Explosions, Cosmology, ed. H. J. G. L. M. Lamers, N. Langer, T. Nugis, & K. Annuk, 113

Vink, J. S. 2018, A&A, 615, A119

Vink, J. S., de Koter, A., & Lamers, H. J. G. L. M. 2001, A&A, 369, 574

Vink, J. S., & Gräfener, G. 2012, ApJ Letters, 751, L34

Vink, J. S., Muijres, L. E., Anthonisse, B., de Koter, A., Gräfener, G., & Langer, N. 2011, A&A, 531, A132

Vink, J. S., et al. 2015, Highlights of Astronomy, 16, 51

Winds of Stars and Exoplanets
Proceedings IAU Symposium No. 370, 2023
A. A. Vidotto, L. Fossati & J. S. Vink, eds.
doi:10.1017/S174392132200494X

The Evolution of Atmospheric Escape of Highly Irradiated Gassy Exoplanets

Andrew P. Allan[1], **Aline A. Vidotto**[1] **and Leonardo A. Dos Santos**[2]

[1]Leiden Observatory, Leiden University, Postbus 9513, 2300 RA Leiden, The Netherlands

[2]Space Telescope Science Institute, 3700 San Martin Drive, Baltimore, MD 21218, USA

Abstract.
Atmospheric escape has traditionally been observed using hydrogen Lyman-α transits, but more recent detections utilise the metastable helium triplet lines at 1083nm. Capable of being observed from the ground, this helium signature offers new possibilities for studying atmospheric escape. Such detections are dependent however on the specific high-energy flux received by the planet. Previous studies show that the extreme-UV band both drives atmospheric escape and populates the triplet state, whereas lower energy mid-UV radiation depopulates the state through photoionisations. This is supported observationally, with the majority of planets with 1083nm detections orbiting a K-type star, which emits a favourably high ratio of EUV to mid-UV flux. The goal of our work is understanding how the observability of escaping helium evolves. We couple our one-dimensional hydrodynamic non-isothermal model of atmospheric escape with a ray-tracing technique to achieve this. We consider the evolution of the stellar radiation and the planet's gravitational potential.

Keywords. hydrodynamics

---

1. Introduction

Since the first detections of escaping helium by means of metastable helium triplet 1083nm transit observations (Nortmann et al. 2018; Spake et al. 2018; Allart et al. 2018), there have been over 15 detections as well as constraints set by non-detections. Both theoretically and observationally, K-type host stars have been found to be favourable for producing such detections due to their relatively low mid-UV flux which depopulates the helium triplet state and high EUV flux which populates the state through photoionisations followed by recombinations (Oklopčić 2019; Poppenhaeger 2022). During the lifetime of a planet, the emitted flux and the planetary radii both vary. In Allan et al. (in prep), we study how such evolutionary variations affect the resulting escape and the corresponding helium 1083nm observational signature.

2. Solving helium populations self-consistently vs. post-processingly

Our model (Allan et al. in prep) for hydrodynamic escape is an upgraded version of the hydrogen-only version first presented in Allan & Vidotto (2019) which was based on the model of Murray-Clay et al. (2009). We have also updated the ray-tracing technique for simulating spectroscopic transits used in Allan & Vidotto (2019) to be capable of modelling the helium triplet 1083nm signature. We invite the reader to check Allan et al. (in prep) for a more detailed description of the work. Figure 1 briefly summarises our model for hydrodynamic atmospheric escape. In short, our hydrodynamic model uses a

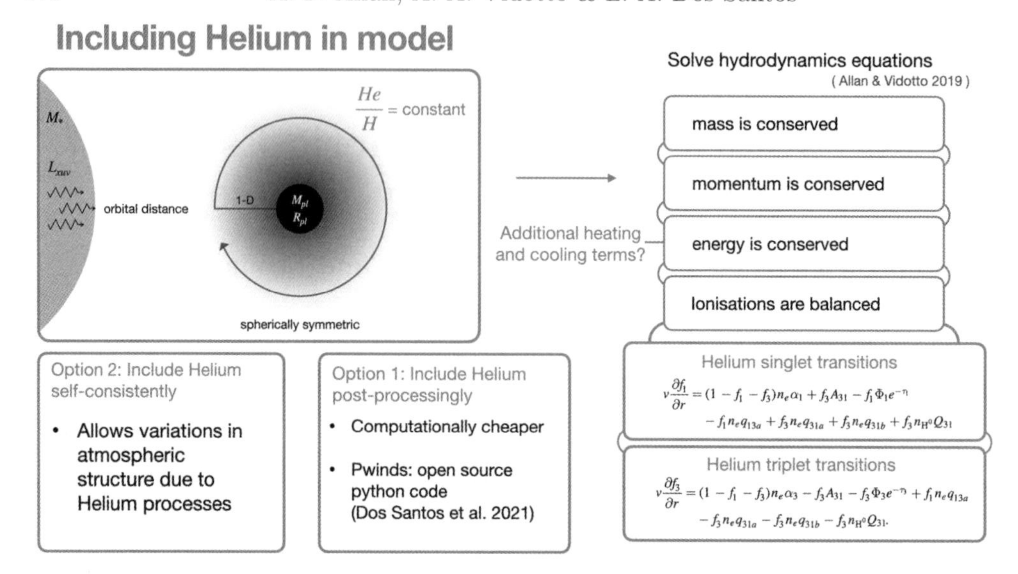

Figure 1. A brief summary of our model which will be presented in greater detail in Allan et al. (in prep). In Allan & Vidotto (2019), we present a similar, hydrogen-only version of this model. The main inputs remain similar, with the addition of the X-ray and EUV1 flux inputs as well the assumed hydrogen / helium fraction.

fluid approximation for the atmosphere, numerically solving equations of fluid dynamics in a co-rotating frame, using a shooting method approach based on the model of Vidotto & Jatenco-Pereira (2006). These four coupled differential equations ensure that mass, momentum and energy are conserved and that ionisations are balanced.
In order to model the helium 1083nm signature, the density of helium in the triplet state must be known. To obtain this, we must solve two additional coupled differential equations which account for transitions into and out of the helium singlet and triplet states. These equations are shown in the bottom right of Figure 1 and are discussed in Oklopčić & Hirata (2018). We approach solving these additional equations in two different ways. One option is to solve for the singlet and triplet populations after already solving the fluid dynamics equations. We refer to this approach as solving the helium populations 'post-processingly'. While the helium / hydrogen fraction is considered in the fluid equations, any heating or cooling processes due to helium must be omitted as these processes require predictions of the helium populations. While not used in the models presented here, `P-winds` (Dos Santos et al. 2022) is an open source python code that is capable of solving the coupled helium population equations, either for a given atmospheric structure or from an iso-thermal Parker wind assumption. An alternative approach is to solve all six equations simultaneously or 'self-consistently'. This allows for the inclusion of heating and cooling processes due to helium in the fluid dynamic equations, potentially affecting the resulting atmospheric structure.

3. The evolution of atmospheric escape

Allan & Vidotto (2019) previously showed that the evolution of atmospheric escape of a close-in planet depends on two important factors:

(a) as the host star evolves, its activity declines due to spin down, resulting in declining fluxes in the X-ray and extreme ultraviolet (XUV) and
(b) as the planet evolves, cooling causes it to contract with time.

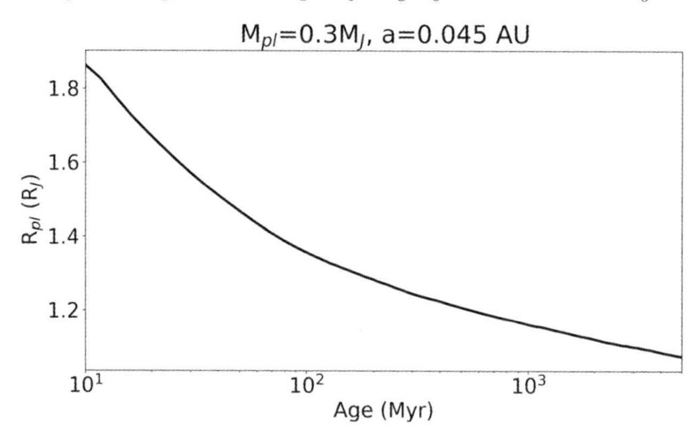

Figure 2. Planetary radius with respect to age for a 0.3-$M_{\rm jup}$ warm-Neptune sized planet orbiting a solar-like star at 0.045 au (Fortney & Nettelmann 2010).

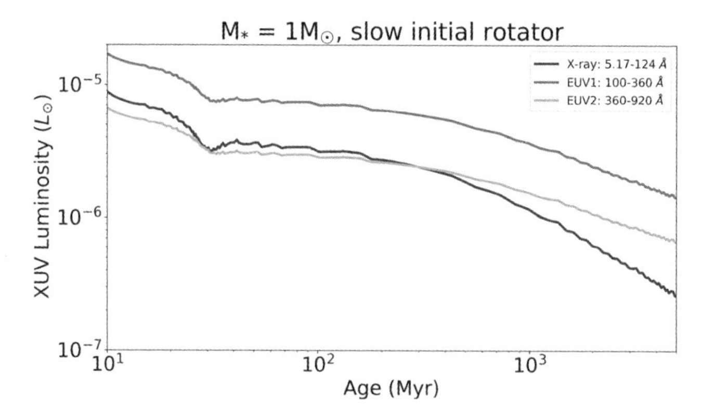

Figure 3. Stellar XUV luminosity in various wavelength bins specified in the legend as a function of stellar age. The predictions are obtained by utilising the model of (Johnstone et al. 2021) and normalising the flux in each wavelength channels until they each reproduce their solar value at the solar age.

The level of atmospheric escape and consequently the observational signatures of escaping hydrogen in the Lyman-α and Hα lines were found to vary strongly with the evolution of the modelled Hot-Jupiter and warm-Neptune planets, with younger planets exhibiting greater escape. This is the result of a favourable combination of higher irradiation fluxes and weaker gravities. Consequently, Lyman-α and H-α absorption are also greater for the younger planets. In a continuation of this work, we now study how the helium 1083nm signature evolves over the lifetime of a planet. We use the same planetary radius as a function of evolution input (Fortney & Nettelmann 2010) as was used Allan & Vidotto (2019), corresponding to a 0.3 $M_{\rm jup}$ gas-giant orbiting a solar-like star at 0.045 au (see Figure 2). For the XUV flux, we look to Johnstone et al. (2021), from which we obtain the evolution of flux emitted in 3 separate wavelength bins corresponding to X-ray (5.17-124 Å), EUV1 (100-360 Å) and EUV2 (360-920 Å) wavelengths (see Figure 3). Following Murray-Clay et al. (2009) and Allan & Vidotto (2019), we approximate that the flux in each of these bins is concentrated on one representative wavelength, X-ray (50 Å), EUV1 (200 Å) and EUV2 (620 Å).

With the evolving inputs of received XUV flux and planetary radius, we run two versions of our model, either post-processingly or self-consistently solving for the helium triplet fractions as explained in the previous section. Figure 4 displays the resulting mass-loss rate as a function of planetary evolution. As found in Allan & Vidotto (2019),

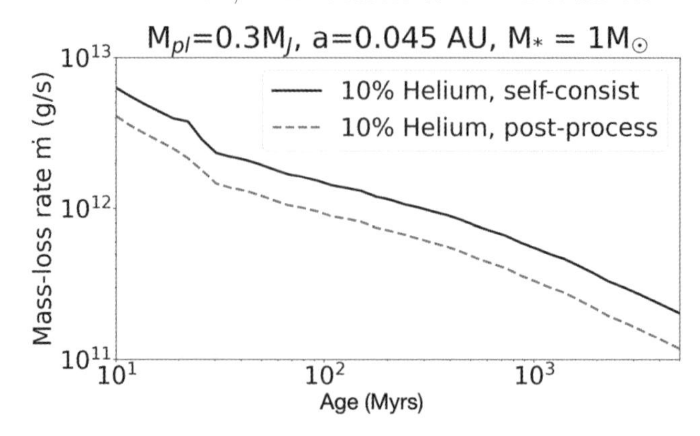

Figure 4. Predicted mass loss rate for a 0.3 $M_{\rm jup}$ gas-giant orbiting a slow initial rotator, solar-like star at 0.045 au. The black solid line corresponds to models which solves helium population equations consistently with the fluid dynamic equations, while the dashed grey line denotes models which solve the helium population equations post-processingly. In both cases a helium to hydrogen fraction of 0.1 was used.

the diminishing XUV flux required to heat the planetary atmosphere combined with the growing gravitational potential due to the shrinking planetary radii leads to the decline of atmospheric escape as the planet evolves. Interestingly, helium heating and cooling processes affect the resulting escape. If the helium populations are solved after the fluid dynamic equations are already solved (shown by the dashed grey line in figure 4), meaning heating and cooling effects due to helium are omitted, then the mass-loss rate is under-predicted. In other words, for the chosen planetary parameters, heating arising from the photo-ionisation of helium can further enhance atmospheric escape. In this modelled case, EUV1 photons photo-ionising singlet state helium are the primary contributor although this will likely vary dependant on the chosen stellar and planetary parameters.

4. The evolution of the Helium 1083nm signature of atmospheric escape

Naturally, the predicted helium triplet 1083nm signature of atmospheric escape weakens with declining escape as the planet ages. This is clearly seen in figure 5 which compares the predicted 1083nm transmission spectra at 10 Myr (left-panel) and 5000 Myr (right-panel). Allan & Vidotto (2019) found that the hydrogen Lyman-α and H-α signatures of atmospheric escape also follow such a trend with evolution. Solving the helium population equations either post-processingly or self-consistently as shown by the solid black and dashed grey spectra in figure 5 respectively, can also impact the resulting 1083nm signature. This is particularly true during younger planetary ages when the higher flux levels enhance heating through hydrogen and helium photo-ionisations.

5. Effects of the assumed helium / hydrogen fraction

An important parameter often featuring in models of hydrodynamic escape which incorporate helium is the fraction of helium / hydrogen in the atmosphere. For simplicity, we assume a constant fraction, both with respect to planetary age and with respect to atmospheric depth. While choosing a helium to hydrogen fraction of 0.1 (black line of Figure 6) or 0.0001 (cyan line) has a negligible effect on the resulting mass-loss rate, this is only true if the helium singlet and triplet populations are solved self-consistently and hence heating due to the photo-ionisation of helium is allowed. Further escape arising from this additional heating is counteracted by the larger gravitational force due to the greater mean molecular weight of helium. Altering the helium / hydrogen fraction

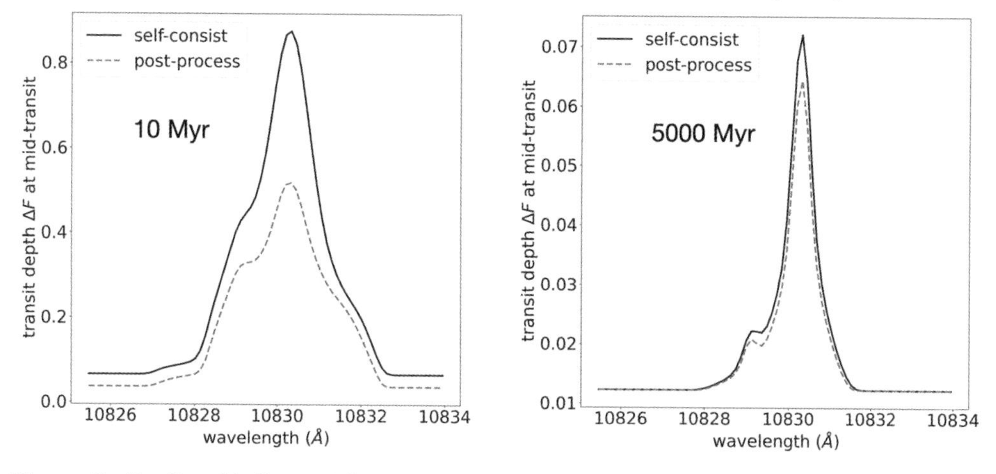

Figure 5. Predicted helium triplet 1083nm transmission spectra for a 0.3 $M_{\rm jup}$ gas-giant orbiting a slow initial rotator, solar-like star at 0.045 au. Note the differing scales of the y-axes. The left (right) panel corresponds to an age of 10 (5000) Myr. Solid black lines display spectra for models in which the helium populations were solved self-consistently whereas the dashed grey spectra were obtained through solving the helium populations post-processingly. In all cases, a helium / hydrogen fraction of 0.1 was used.

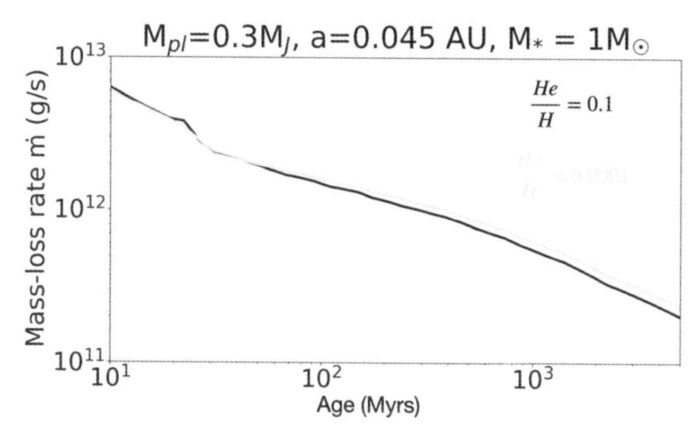

Figure 6. Predicted mass loss rate for a 0.3 $M_{\rm jup}$ gas-giant orbiting a slow initial rotator, solar-like star at 0.045 au. The black line corresponds to models which assume a 0.1 helium to hydrogen fraction, while the cyan line denotes models with a fraction of 0.0001, representing the negligible helium case. In both sets of models the helium populations were solved self-consistently.

while neglecting this heating due to helium (as is the case in a post processing model) would incorrectly exaggerate the dependence of the resulting escape rate on the helium / hydrogen fraction. Despite having only a minor effect on the hydrodynamics of atmospheric escape, the assumed helium / hydrogen fraction remains an important input as it significantly affects the observable metastable helium triplet 1083nm signature.

6. Conclusions

As the planet evolves, atmospheric escape declines due to receiving a weaker XUV flux and its growing gravitational potential. The metastable helium triplet 1083nm signature becomes weakens with diminishing atmospheric escape. This is also true for the hydrogen Lyman-α and H-α signatures. When including helium in the modelled atmosphere, it is important to include additional helium heating and cooling processes which can affect the resulting atmospheric structure and escape. Finally, while the assumed helium to hydrogen fraction has little effect on the resulting escape, it remains important as the 1083nm signature is heavily dependant upon it.

References

Allan, A. & Vidotto, A. A. 2019, MNRAS, 490, 3760
Allan, A. P., Vidotto, A. A., & Dos Santos, L. A. in prep, MNRAS
Allart, R., Bourrier, V., Lovis, C., et al. 2018, Science, 362, 1384
Dos Santos, L. A., Vidotto, A. A., Vissapragada, S., et al. 2022, A&A, 659, A62
Fortney, J. J. & Nettelmann, N. 2010, Space Science Reviews, 152, 423
Johnstone, C. P., Bartel, M., & Güdel, M. 2021, A&A, 649, A96
Murray-Clay, R. A., Chiang, E. I., & Murray, N. 2009, Astrophysical Journal, 693, 23
Nortmann, L., Pallé, E., Salz, M., et al. 2018, Science, 362, 1388
Oklopčić, A. 2019, ApJ, 881, 133
Oklopčić, A. & Hirata, C. M. 2018, ApJ Letters, 855, L11
Poppenhaeger, K. 2022, MNRAS, 512, 1751
Spake, J. J., Sing, D. K., Evans, T. M., et al. 2018, Nature, 557, 68
Vidotto, A. A. & Jatenco-Pereira, V. 2006, ApJ, 639, 416

Winds of Stars and Exoplanets
Proceedings IAU Symposium No. 370, 2023
A. A. Vidotto, L. Fossati & J. S. Vink, eds.
doi:10.1017/S1743921322003696

The future of Jupiter-like planets around Sun-like stars: first steps

T. Konings, R. Baeyens and L. Decin

Institute of Astronomy, KU Leuven - Celestijnenlaan 200D bus 2401, 3001 Leuven, Belgium

Abstract. Planets that orbit low- to intermediate mass main sequence (MS) stars will experience vigorous star-planet interactions when the host star evolves through the giant branches, including the asymptotic giant branch (AGB) phase, due to extreme luminosities and stellar outflows. In this work, we take the first steps towards understanding how a planet's temperature profile and chemical composition is altered when the host star evolves from the MS to the AGB phase. We used a 1D radiative transfer code to compute the temperature-pressure profile and a 1D chemical kinetics code to simulate the disequilibrium chemistry. We consider a Jupiter-like planet around a Solar-type star in two evolutionary stages (MS and AGB planet) by only varying the stellar luminosity. We find that the temperature throughout the AGB planet's atmosphere is increased by several hundreds of Kelvin compared to the MS planet. We also find that CO joins H_2O and CH_4 as a prominent constituent in the AGB planet's atmospheric composition.

Keywords. stars: AGB and post-AGB, Planets and satellites: gaseous planets, Planets and satellites: atmospheres

1. Introduction & Methods

The majority of discovered planets orbit low- to intermediate mass main sequence (MS) stars. Star born with masses between 0.8 and 8 $M_\odot$ evolve through the asymptotic giant branch (AGB) phase towards the end of their lifetime. AGB stars are cool, luminous giants that have developed a stellar outflow with mass loss rates upwards from $10^{-8}\, M_\odot yr^{-1}$ with luminosities around $10^3 - 10^4\, L_\odot$. Spiegel & Madhusudhan (2012) showed that Jovian planets around AGB stars (AGB planets henceforth) at orbits of several AU will receive irradiation levels comparable to hot Jupiters, which will increase the planetary temperatures to over ~ 1000 K. The question then arises how the AGB planet's temperature-pressure profile and chemical composition will adjust to these increased irradiation levels. A higher incident stellar flux will strengthen photochemical processes in the atmosphere, which are known to drive the chemical composition of MS hot Jupiters out of equilibrium (Baeyens et al. 2022). In this work, we asses how the thermal profile and chemical composition of a Jupiter-like planet is affected when it is irradiated by a MS and AGB solar-type star. We consider a $1\, M_J$ gaseous planet at an orbital distance of 5 AU from a $1\, M_\odot$ MS star (MS planet henceforth) and keep these parameters fixed when examining the AGB planet. The stellar luminosity is changed from $1\, L_\odot$ to $3 \times 10^3\, L_\odot$, the effective temperature from 5700 K to 4000 K and the stellar radius from $1\, R_\odot$ to $100\, R_\odot$. We use the one-dimensional radiative transfer package petitCODE (Mollière et al. 2015) to compute a thermal structure under radiative-convective equilibrium, starting from an atmosphere with a solar elemental gas mixture. (for more details, see Sect 2.1 of Baeyens et al. (2021)). We use the one-dimensional chemical kinetics model of Agúndez et al.

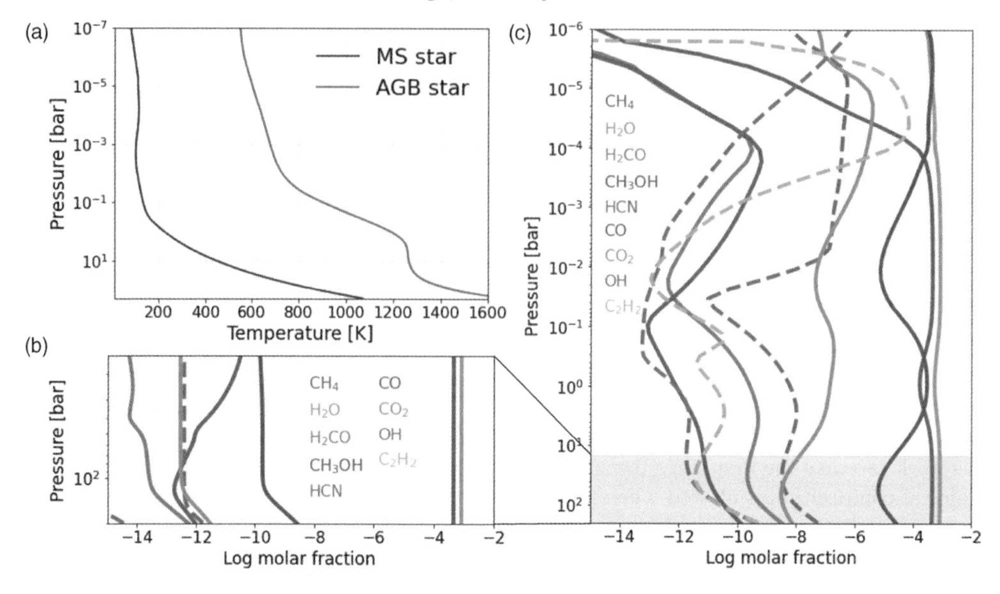

Figure 1. *a)* Temperature-pressure profile of the Jupiter-like planet considered in the work with a MS (blue) and AGB host star (red). *b)* Molar fractions of species in the deep atmosphere of the MS planet. *c)* Molar fractions of species in the atmosphere of the AGB planet. The shaded area indicates the pressure range that was probed for the MS planet.

(2014) with a chemical network of Venot et al. (2020) to compute the molecular composition. To compute the photochemical rates, we use the solar spectral energy distribution (SED, see Baeyens et al. (2022)) and scale it accordingly.

2. Results

Below ~ 1 bar, the MS planet's temperature (Fig. 1a) remains below 200 K, which does not allow the chemistry code to compute molar fractions. Therefore, we constrain the pressure range to the deepest layers in the chemistry model (Fig. 1b). H_2O and CH_4 are the most abundant species, with the latter being the expected dominant carbon-bearing molecule in cooler planets (Lodders & Fegley 2002). In general, we compute very similar quantities of H_2O, CH_4, CO, and CO_2 as Visscher et al. (2010), who modelled Jupiter's atmosphere. The AGB planet exceeds a temperature 500 K throughout the entire simulated pressure range. The distribution of constituents (Fig. 1c) is showing a complex pattern owing to hot temperatures in the middle/deep layers and intense photochemistry in the upper regions. The main difference with the MS planet is the enhancement in CO, which becomes a dominant constituent. The upper layers experience photodissociation as, for example, CH_4 is substantially depleted below ~ 0.1mbar. Finally, we note that we are most likely overestimating the XUV flux of the AGB star by scaling the solar SED with the luminosity of the considered AGB star, as AGB stars generally emit less XUV flux relative to the radiated black body flux (Montez et al. 2017).

References

Agúndez, M., Parmentier, V., Venot, O., Hersant, F., & Selsis, F. 2014, A&A, 564, A73
Baeyens, R., Decin, L., Carone, L., et al. 2021, MNRAS, 505, 5603
Baeyens, R., Konings, T., Venot, O., Carone, L., & Decin, L. 2022, MNRAS, 512, 4877
Lodders, K. & Fegley, B. 2002, Icarus, 155, 393
Mollière, P., van Boekel, R., Dullemond, C., Henning, T., & Mordasini, C. 2015, ApJ, 813, 47

Montez, Rodolfo, J., Ramstedt, S., Kastner, J. H., Vlemmings, W., & Sanchez, E. 2017, ApJ, 841, 33
Spiegel, D. S. & Madhusudhan, N. 2012, ApJ, 756, 132
Venot, O., Cavalié, T., Bounaceur, R., et al. 2020, A&A, 634, A78
Visscher, C., Moses, J. I., & Saslow, S. A. 2010, Icarus, 209, 602

Winds of Stars and Exoplanets
Proceedings IAU Symposium No. 370, 2023
A. A. Vidotto, L. Fossati & J. S. Vink, eds.
doi:10.1017/S1743921322003465

Rapid orbital precession of the eclipsing binary HS Hydrae

A. M. Matekov[1,2] and **A. S. Hojaev**[1]

[1]Ulugh Beg Astronomical Institute of the Uzbekistan Academy of Science, 33 Astronomicheskaya, Tashkent 100052, Uzbekistan

[2]University of the Chinese Academy of Sciences, Yuquan Road 19, Sijingshang Block, Beijing 100049, P.R. China

Abstract. We investigated the evolution of HS Hya system's inclination based on analysis of its light curves in the period 1964–2021. HS Hya is EA type eclipsing binary star, belonging to separate group with changing orbital inclination. We used our recent observations as well as the data from sky surveys.

Keywords. (stars:) binaries: eclipsing, methods: data analysis, techniques: photometric

1. Introduction

HS Hydrae was classified as EA type eclipsing binary star (EBs) based on observations of Strohmeier et al. (1965) and has been studied in detail as a normal binary system by Gyldenkerne et al. (1975). Torres et al. (1997) first suggested the presence of a third body in the system and determined its orbital period as 190 days. Zasche & Paschke (2012), analysing long-term photometric data, found that HS Hya changes its orbital inclination with $P_{\rm nodal} \sim 631$ yr and predicted that eclipses for HS Hya would cease around 2022. Based on analysis of the unique data set for 125 years, created from the DASCH photographic plate database, Davenport et al. (2021) derived HS Hya's precession period as 1194±20 years. According to their estimations, its eclipses will appear again around 2195±3 which is consistent with Vokrouhlický & Zasche (2022).

2. Observation & Data Reduction

The observations were performed during May 17–20, 2020 at the Maidanak observatory (Uzbekistan) using the 60-cm telescope Zeiss-600 equipped with FLI IMG ProLine 1024×1024 CCD, the scale is 0.372″ per pixel. In total 1138 images in Johnson-Cousins R band with 5 second exposure were obtained during these observations. CCD images were processed and analyzed by aperture photometry using the IRAF packages.

3. Discussion & Conclusions

For comprehensive study of the HS Hya orbital inclination evolution we have collected all available photometric data since its discovery. The light curves of Gyldenkerne et al. (1975) and that of based on Hipparcos Perryman et al. (1997) and ASAS Pojmanski (2002) databases are shown in Figure 1. Based on analysis of the data for 1964-2020, we found that the HS Hya orbital inclination has changed $i \geq 17.5$ deg. Light curves based on HS Hya (TIC434479378) data obtained by TESS mission Keivan et al. (2018) and on our data obtained at Maidanak observatory are shown in Figure 2. According our analysis the eclipses in this system will reappear again in 165 years. It is also noteworthy

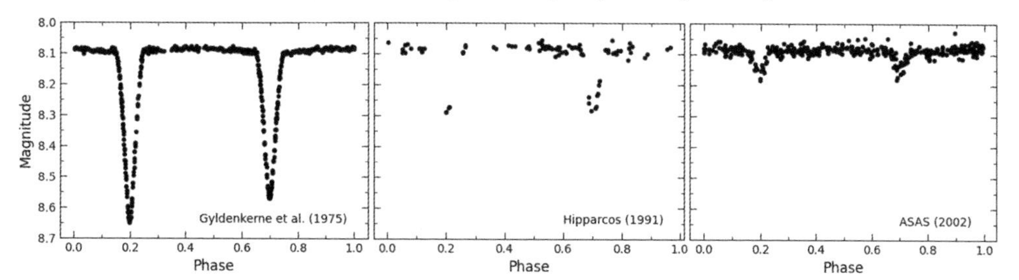

Figure 1. The light curves of HS Hya in the V band. The figure shows clearly the change in the depth of both minima. Three panels have the same scale.

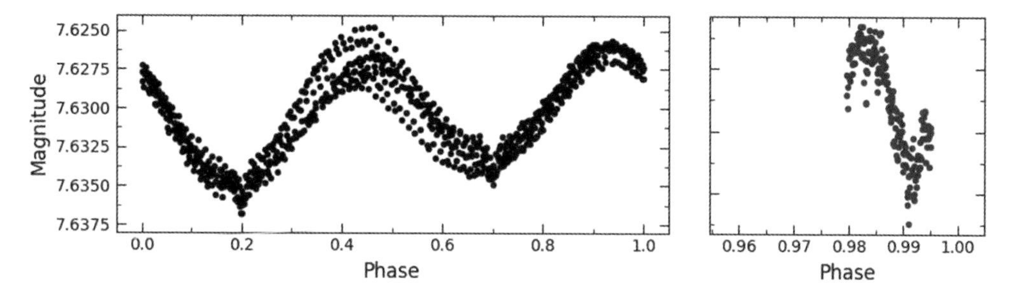

Figure 2. TESS Sector 9 (left)and Maidanak (right) phase-folded light curves for HS Hya: The figure shows amplitude of primary (Phase~0.2) and secondary (Phase~0.7) minima are small.

that according to analysis of the data from TESS Sector 35 (February 9 - March 7, 2021), Davenport et al. (2021) have found no eclipses.

Until our investigation HS Hya belonged to the detached binary systems and classified as EA type EBs. The new analyses of all existed data including our and TESS data revealed that in addition to the eclipses the binary system has superposition of different modes of variability including an ellipsoidal variation out of eclipses. The last may be due to the ellipsoidal shape of at least one of the components of the HS Hya, a thin changing shell around this binary system, and the flows from the Lagrange points of the Roche lobe. Investigations of changes of orbital inclination (nodal motion) effect could allow us to determine the presence of third body around EBs including the black hole. Thus, the study of this phenomenon is very essential for further research in the field. We plan to continue studying this effect in other EBs.

Acknowledgements

We thank the International Grant Uzb-Ind-2021-99 for financial support. This research has made use of the SIMBAD database, operated at CDS, Strasbourg, France, NASA's Astrophysics Data System Bibliographic Services and the Mikulski Archive for Space Telescopes (MAST).

References

Davenport, J. R. A., Windemuth, D., Warmbein, K., et al. 2021, AJ, 162, 189
David Vokrouhlický, and Petr Zasche. 2022, AJ, 163, 94
Gyldenkerne, K., Jorgensen, H. E., & Carstensen, E. 1975, A&A, 42, 303
Keivan, G. Stassun., et al. 2018, AJ, 156, 102

Perryman, M. A. C., Lindegren, L., Kovalevsky, J., et al. 1997, A&A, 323, L49
Pojmanski, G. 2002, AcA, 52, 397
Strohmeier, W., Knigge, R., & Ott, H. 1965, IBVS, 107, 1
Torres, G., Stefanik, R. P., Andersen, J., et al. 1997, AJ, 114, 2764
Zasche, P., & Paschke, A. 2012, A&A, 542, L23

Winds of Stars and Exoplanets
Proceedings IAU Symposium No. 370, 2023
A. A. Vidotto, L. Fossati & J. S. Vink, eds.
doi:10.1017/S174392132200360X

To the dynamics of the two-body problem with variable masses in the presence of reactive forces

A.T. Ibraimova[1] **and M.Zh Minglibayev**[2]

[1]Fesenkov Astrophysical Institute, Almaty, Kazakhstan

[2]Al-Farabi Kazakh National University, Almaty, Kazakhstan

Abstract. We studied the problem of two spherical celestial bodies in the general case when the masses of the bodies change non-isotropically at different rates in the presence of reactive forces. The problem was investigated by methods of perturbation theory based on aperiodic motion along a quasi-conic section, using the equation of perturbed motion in the form of Newton's equations. The problem is described by the variables $a, e, i, \pi, \omega, \lambda$, which are analogs of the corresponding Keplerian elements and the equations of motion in these variables are obtained. Averaging over the mean longitude, we obtained the evolution equations of the two-body problem with variable masses in the presence of reactive forces. The obtained evolution equations have the exact analytic integral $a^3e^4 = a_0^3e_0^4 = const$.

Keywords. stars: mass loss, variable masses, two-body problem, reactive force, osculating elements, perturbation theory.

1. Introduction

Real celestial bodies are nonstationary, their masses, sizes, shapes and structures change during evolution Eggleton (2006). In this connection, we investigated the problem of two spherical celestial bodies in the general case when the masses of the bodies change nonisotropically at different rates in the presence of reactive forces Minglibayev (2012), Minglibayev et al. (2020).

2. Model description

Consider a gravitating system consisting of two celestial bodies with variable masses $m_1 = m_1(t)$, $m_2 = m_2(t)$, $(m_1 \geq m_2)$. Let us assume that the bodies are spherical with spherical distributions of masses. Let us assume that the bodies' masses decrease due to separating particles and increase due to joining (adhering) particles. In this case, the relative velocity of separating particles from the body differs from the relative velocity of joining (sticking) particles to the body. Let's consider the general case when the masses of bodies change non-isotropically at different rates $\dot{m}_1/m_1 \neq \dot{m}_2/m_2$ in the presence of reactive forces.

3. System of differential equations of secular perturbations in osculating elements

Let us consider the two-body problem with variable masses in the presence of reactive forces as the equations of perturbed aperiodic motion along a quasi-conic section in the osculating elements Minglibayev (2012). Then, after averaging over mean longitude, the equations of secular perturbations have following form

$$\dot{a} = \frac{2a^{3/2}m_0}{m\sqrt{\mu_0}}\sqrt{1-e^2}F_\tau(t), \qquad \dot{e} = -\frac{3}{2}\frac{e\sqrt{a(1-e^2)}}{\sqrt{\mu_0}}\frac{m_0}{m}F_\tau(t), \tag{3.1}$$

$$\frac{di}{dt} = -\frac{3}{2}\frac{e\sqrt{a}\cos(\pi-\Omega)}{\sqrt{\mu_0}\sqrt{1-e^2}}\frac{m_0}{m}F_n(t), \quad \dot{\Omega} = -\frac{3}{2}\frac{e\sqrt{a}}{\sqrt{1-e^2}\sqrt{\mu_0}}\frac{m_0}{m}\frac{\sin(\pi-\Omega)}{\sin i}F_n(t), \tag{3.2}$$

$$\dot{\pi} = \frac{m_0}{m}\frac{\sqrt{a(1-e^2)}}{\sqrt{\mu_0}}\left\{F_r(t) - \frac{3a}{2}\frac{d^2}{dt^2}\left(\frac{m_0}{m}\right) - \frac{3e\sin(\pi-\Omega)}{2(1-e^2)}\cdot \mathrm{tg}\frac{i}{2}F_n(t)\right\}, \tag{3.3}$$

$$\dot{\lambda} = n\left(\frac{m}{m_0}\right)^2 - \frac{m_0}{m}\frac{\sqrt{a}}{\sqrt{\mu_0}}\left\{F_r(t)\left(2+e^2\right) - \left(2+3e^2\right)\frac{d^2}{dt^2}\left(\frac{m_0}{m}\right)a\right\} -$$

$$-\frac{3}{2}\frac{m_0}{m}\frac{e\sqrt{a}\sin(\pi-\Omega)}{\sqrt{\mu_0(1-e^2)}}\mathrm{tg}\frac{i}{2}F_n(t) + \frac{\sqrt{a(1-e^2)}}{1+\sqrt{1-e^2}}\frac{m_0}{m}\frac{e^2}{\sqrt{\mu_0}}\left(F_r(t) - \frac{3}{2}\frac{d^2}{dt^2}\left(\frac{m_0}{m}\right)a\right), \tag{3.4}$$

where $m = m(t) = m_1(t) + m_2(t)$, $m_0 = m(t_0) = m_1(t_0) + m_2(t_0) = const$, $\mu_0 = fm_0 = const$, $n = \sqrt{\mu_0}/a^{3/2}$, F_r, F_τ, F_n are the corresponding radial transversal and normal components of the reactive forces in the orbital coordinate system.

4. The first integral of the differential equation system of secular perturbations

Consider together the equation of the analog of the semi-major axis and the analog of eccentricity (3.1). Then, we obtain the following expression

$$3\dot{a}/a = -4\dot{e}/e, \qquad \frac{a^3}{a_0^3} = \frac{e_0^4}{e^4}, \qquad e_0 = e(t_0) = const, \qquad a_0 = a(t_0) = const. \tag{4.1}$$

The resulting integral (4.1) has a simple analytical form and can be used in the analysis of dynamic evolution of binary systems with variable masses. This integral (4.1) can be used to simplify the system of secular differential equations (3.1)–(3.4).

5. Conclusion

The obtained equations of secular perturbation and their first integral can be successfully used in the study of the dynamic evolution of the two-body problem with variable masses in the presence of reactive forces. They are convenient for describing the dynamics of separated two-body systems with variable masses. In the case of close binary systems, tidal forces should also be taken into account. These and other aspects of the two-body problem with variable masses in the presence of reactive forces will be studied in the following papers.

References

Eggleton, P. Evolutionary processes in binary and multiple stars, Cambridge University Press, New York. 2006, 322 p.

Minglibyaev, M. Zh., Dynamics of Gravitating Bodies with Variable Masses and Dimensions. LAP LAMBERT Academic Publishing. Germany. 2012. 224 p. (in Russ.)

Minglibayev, M. Zh., Omarov, Ch. T., Ibraimova, A. T. Reports of the national academy of sciences of the republic of Kazakhstan, 2020, 2, 5

Winds of Stars and Exoplanets
Proceedings IAU Symposium No. 370, 2023
A. A. Vidotto, L. Fossati & J. S. Vink, eds.
doi:10.1017/S1743921322003611

Evolution equations of the multi-planetary problem with variable masses

A.B. Kosherbayeva and M.Zh Minglibayev

Al-Farabi Kazakh National University, Almaty, Kazakhstan

Abstract. We investigated the influence of the variability of the masses of planets and the parent star on the dynamic evolution of n planetary systems, considering that the masses of bodies change isotropically with different rates. The methods of canonical perturbation theory, which developed on the basis of aperiodic motion over a quasi-conical cross section and methods of computer algebra were used. $4n$ evolutionary equations were obtained in analogues of Poincare elements. As an example, the evolutionary equations of the three-planet exosystem $K2-3$ were obtained explicitly, which is a system of 12 linear non-autonomous differential equations. Further, the evolutionary equations will be investigated numerically.

Keywords. celestial mechanics, variable mass, analogues of Poincare elements, multi-planetary system, secular perturbation.

---

1. Introduction

To date, there are more than 5,000 confirmed exoplanets and more than 3,800 planetary systems in the NASA (2022) database. To research exoplanetary systems in the non-stationary stage of their evolution is represented important interest.

2. Problem statment

We considered the problem of $n+1$ bodies with variable masses $m_0 = m_0(t)$ – mass of the parent star S, $m_i = m_i(t)$, – the mass of the planet P_i. The laws of mass are known and given functions of time $m_0 = m_0(t),\ \ m_1 = m_1(t), \ldots,\ \ m_n = m_n(t),\ \ (n \geq 3)$. The masses of spherical symmetric bodies change isotropically with different rates $\dot{m}_0/m_0 \neq \dot{m}_i/m_i, \quad \dot{m}_i/m_i \neq \dot{m}_j/m_j \quad i,j = 1,2,\ldots,n, \quad i \neq j$.

Differential equations of motion of n bodies with isotropically varying masses in a relative coordinate system are given in Minglibaev (2012)

$$\ddot{\vec{r}}_i = -f\frac{(m_0 + m_i)}{r_i^3}\vec{r}_i + f\sum_{j=1}^{n}{}' m_j\left(\frac{\vec{r}_j - \vec{r}_i}{r_{ij}^3} - \frac{\vec{r}_j}{r_j^3}\right), \quad (i,j = 1,2,\ldots,n), \tag{2.1}$$

where $r_{ij} = r_{ji} = \sqrt{(x_j - x_i)^2 + (y_j - y_i)^2 + (z_j - z_i)^2}$ – mutual distances of the center of spherical bodies, f – gravitational constant, $\vec{r}_i(x_i, y_i, z_i)$ – the radius-vector of center of the planet P_i, the sign "stroke" when summing means that $i \neq j$.

For our purposes, analogues of the second system of canonical Poincare elements are preferred

$$\Lambda_i, \quad \lambda_i, \quad \xi_i, \quad \eta_i, \quad p_i, \quad q_i, \tag{2.2}$$

which are introduced on the basis of elements of aperiodic motion over a quasi- conical cross section Minglibaev (2012)

3. Evolutionary equations of n planets with variable masses

The evolutionary equations of n planets with variable masses in dimensionless variables (2.2)-(2.5) in the non-resonant case have the form in our work Prokopenya et al. (2022)

$$\xi_i' = \sum_{s=1}^{i-1} m_s \left(\frac{\Pi_{ii}^{is}}{\Lambda_i}\eta_i + \frac{\Pi_{is}^{is}}{\sqrt{\Lambda_i \Lambda_s}}\eta_s \right) + \sum_{k=i+1}^{n} m_k \left(\frac{\Pi_{kk}^{ik}}{\Lambda_i}\eta_i + \frac{\Pi_{ik}^{ik}}{\sqrt{\Lambda_i \Lambda_k}}\eta_k \right) - \frac{3\gamma_i''}{2\gamma_i}\frac{\Lambda_i^3}{\mu_{i0}^2}\eta_i, \quad (3.1)$$

$$\eta_i' = -\sum_{s=1}^{i-1} m_s \left(\frac{\Pi_{ii}^{is}}{\Lambda_i}\xi_i + \frac{\Pi_{is}^{is}}{\sqrt{\Lambda_i \Lambda_s}}\xi_s \right) - \sum_{k=i+1}^{n} m_k \left(\frac{\Pi_{kk}^{ik}}{\Lambda_i}\xi_i + \frac{\Pi_{ik}^{ik}}{\sqrt{\Lambda_i \Lambda_k}}\xi_k \right) + \frac{3\gamma_i''}{2\gamma_i}\frac{\Lambda_i^3}{\mu_{i0}^2}\xi_i, \quad (3.2)$$

$$p_i' = -\sum_{s=1}^{i-1} m_s B_1^{is} \left(\frac{q_i}{4\Lambda_i} - \frac{q_s}{4\sqrt{\Lambda_i \Lambda_s}} \right) - \sum_{k=i+1}^{n} m_k B_1^{ik} \left(\frac{q_i}{4\Lambda_i} - \frac{q_k}{4\sqrt{\Lambda_i \Lambda_k}} \right), \quad (3.3)$$

$$q_i' = \sum_{s=1}^{i-1} m_s B_1^{is} \left(\frac{p_i}{4\Lambda_i} - \frac{p_s}{4\sqrt{\Lambda_i \Lambda_s}} \right) + \sum_{k=i+1}^{n} m_k B_1^{ik} \left(\frac{p_i}{4\Lambda_i} - \frac{p_k}{4\sqrt{\Lambda_i \Lambda_k}} \right). \quad (3.4)$$

At the same time, the expressions Π_{ii}^{is}, Π_{is}^{is}, Π_{kk}^{ik}, Π_{ik}^{ik} in equations (3.1) -(3.4) and the Laplace coefficients retain their form, but they are already dimensionless quantities. All notations are given in the article Prokopenya et al. (2022).

4. The evolutionary equations of the three-planet exosystem $K2-3$ in explicit form

As an example, the case of $n = 3$ was considered. The evolutionary equations (3.1)–(3.4) for the $K2-3$ exosystem are described by a system of 12 linear non-autonomous differential equations, which are obtained explicitly. The resulting system splits into two subsystems for eccentric and oblique elements. The resulting equations of secular perturbations are difficult, so they will be investigated numerically.

5. Conclusion

The evolutionary equations of a multi-planetary problem with isotropically varying masses at different rates in analogues of osculating Poincare elements were obtained. These evolutionary equations can be used for any n planetary problem with variable masses.The evolutionary equations for the $K2-3$ exosystem were written explicitly. The obtained evolutionary equations will be investigated numerically.

References

NASA Exoplanet Exploration, Last update: September 28, 2022, url: `https://exoplanets.nasa.gov/`

Minglibaev, M. Zh., 2012,Dynamics of Gravitating Bodies with Variable Masses and Dimensions, LAP LAMBERT Academic Publishing, 224

Prokopenya, A. N., Minglibayev, M. Zh., Kosherbaeva, A. B., 2022, Derivation of Evolutionary Equations in the Many-Body Problem with Isotropically Varying Masses Using Computer Algebra, Programming and Computer Software, 48, 2, 107–115, DOI:10.1134/S0361768822020098

Winds of Stars and Exoplanets
Proceedings IAU Symposium No. 370, 2023
A. A. Vidotto, L. Fossati & J. S. Vink, eds.
doi:10.1017/S174392132200388X

Planet migration in accretion discs in binary systems

A.D. Nekrasov[1,2], **S.B. Popov**[1,2] **and V.V. Zhuravlev**[2]

[1]Department of Physics, Lomonosov Moscow State University, 119991, Moscow, Russia

[2]Sternberg Astronomical Institute, 119234, Universitetski pr. 13, Moscow, Russia

Abstract. We model evolution of exoplanets of S-type in close binary systems at the stage when the companion starts to lose mass via a slow stellar wind. At this stage an accretion disc is formed around the planets' host. Detailed structure of such discs is calculated in quasi-stationary and non-stationary approaches. We model migration of planets embedded in these discs.

Keywords. planetary systems, binaries: close, accretion disks

---

1. Introduction

An accretion disc can be formed around a secondary non-evolved star in a binary system when the primary companion leaves the Main sequence and starts to lose mass with an enhanced rate via a slow stellar wind. We analyze accretion disc evolution and planetary migration in such discs around solar-like Main sequence stars in binary systems with evolved companions. As the disc is formed from the stellar wind matter, its properties depend on the mass loss rate by the donor and parameters of the binary. In this study we advance the analysis initiated by Kulikova et al. (2019). We use a numerical model to calculate properties of non-stationary discs (NSD) with a variable mass inflow on the whole disc surface within the Bondi radius during late stages of the primary evolution. Then, the migration path of a single planet embedded in such a non-stationary disc is determined by the migration rate varying in the course of the disc evolution. The case of quasi-stationary discs (QSD) is also modeled for comparison.

2. Model and results

Discs discussed in this note can exist for a comparable period of time (or even longer) than usual protoplanetary discs due to an external source of mass. This leads to similar, but usually more significant migration compared to migration in protoplanetary discs. Our code allows to model discs in the mass range from 10^{-10} up to 10^{-2} $M_\odot$.

We consider binary systems with major semi-axis $a < 100$ AU. Primaries have initial masses, M_1, below 8 $M_\odot$, which guarantees evolution of the star with smooth envelope loss at late stages (i.e., without a supernova explosion) and formation of a white dwarf. The secondary component in each system under consideration is formed as a Sun-like star: $M_2 = M_\odot$ whith $M_2 < M_1$ and evolves slower than the primary. Slight variations of the secondary mass would not change our main conclusions. It is assumed that just one planet is formed around the secondary on a dynamically stable orbit. Thus, any possible interactions in a multi-planetary system are ignored. Properties of the donor are calculated using the code MESA (Paxton et al. 2011).†

† MESA tracks were calculated by A. Andryushin, whom we thank for providing these data.

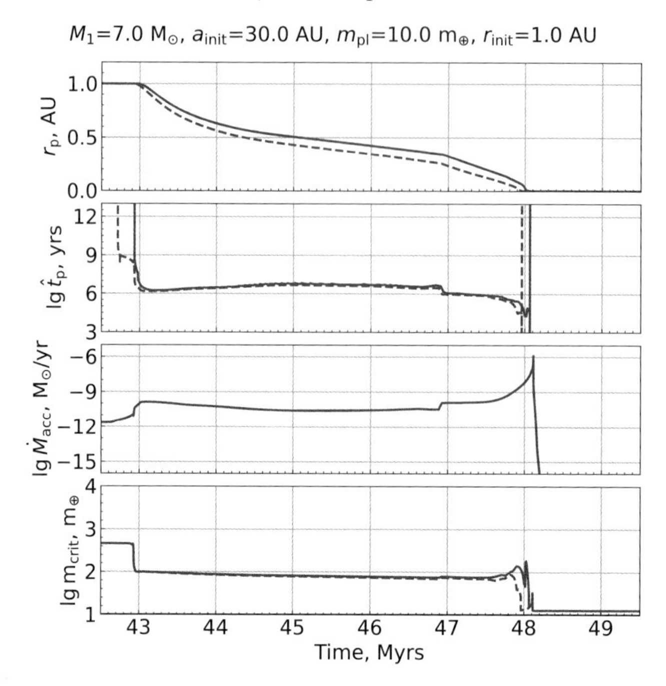

Figure 1. Orbital radius evolution and characteristic migration time for planet with mass $m_{\rm pl} = 10$ m$_\oplus$ and initial orbital radius $a_{\rm pl} = 1$ AU, total accretion rate and critical planet mass for gap opening at radius of the planet for system $M_1 = 7.0$ $M_\odot$, $a = 30$ AU. NSD curves are solid, QSD curves are dashed.

In binary systems with an initial separation $a \lesssim 80 - 100$ AU (we take into account orbital evolution due to mass loss from the donor star) gas giants efficiently migrate in such discs and typically approach short distances from the host star where tidal forces become non-negligible. Neptune-like planets can reach these internal parts of the system in cases when a donor is a relatively massive star (5-8 M$_\odot$) or in binaries with $a \lesssim 20$ AU. An example of our calculations is given in the Figure. Complete set of calculations and detailed description of the model will be published elsewhere.

We conclude that in binaries, mass loss from the primary at late evolutionary stages can significantly modify structure of a planetary system around the non-evolved secondary component, probably resulting in mergers of massive planets with the host star.

Acknowledgements

This study was supported by the Ministry of Science and Higher Education of the Russian Federation grant 075-15-2020-780 (N13.1902.21.0039). The work of A. Nekrasov and V. Zhuravlev was also partly supported by the Foundation for the Advancement of Theoretical Physics and Mathematics 'BASIS'.

References

Kulikova, O,, Popov, S. B., Zhuravlev, V. V., 2019, MNRAS, 487, 3069
Paxton, O., Bildsten, L., Dotter, A., et al., 2011, ApJS, 192, 3

Winds of Stars and Exoplanets
Proceedings IAU Symposium No. 370, 2023
A. A. Vidotto, L. Fossati & J. S. Vink, eds.
doi:10.1017/S1743921322003477

Signatures of wind formation in optical spectra of precursors of planetary nebulae

Kārlis Puķītis and Laimons Začs

Laser Center, Faculty of Physics, Mathematics, and Optometry, University of Latvia, Raiņa bulvāris 19, LV-1586 Rīga, Latvia

Abstract.
Generally it is thought that shaping of planetary nebula from initially spherical envelope of asymptotic giant branch stars into non-spherical morphologies is a consequence of binary interactions. However, post asymptotic giant branch stars HD 235858 and HD 161796 seem to be at odds with this idea and perhaps the non-spherical nebulae surrounding them arose from intrinsic change in the nature of the stellar wind which is poorly understood for this evolutionary phase. Spectroscopic monitoring of these two stars has revealed signatures in the spectra that point to variable outflow. This indicates the prospect of spectroscopic monitoring to advance the knowledge of wind launching mechanism in post asymptotic giant branch stars and other dynamical processes in their extended atmospheres.

Keywords. stars: AGB and post-AGB, stars: atmospheres, stars: winds, outflows, stars: individual (HD 235858, HD 161796)

---

1. Introduction

The role of different processes in shaping of planetary nebula (PN) around post-asymptotic giant branch (post-AGB) stars is not fully understood. It is known that formation of PN begins in AGB stage when intense stellar winds expel outer layers of the star. Virtually all AGB stars are observed to have spherically symmetric wind-created envelopes; however, only around 20% of PN are found with such symmetry with the rest showing mostly elliptical or bipolar morphologies. It is known that the rapid transition to non-spherical morphology and a change in the nature of stellar wind occurs near the end of AGB and beginning of post-AGB phases. Recently, binary interactions have been thought of as the main shapers of PN; however, cases of non-spherical envelopes around unlikely binaries are known. It is possible that an intrinsic change in the nature of stellar wind during the post-AGB phase plays an important role in the formation of the PN. While there is poor knowledge of wind launching mechanism in post-AGB stage, it is more or less understood in AGB stars with the current paradigm being that molecule interaction with shock waves in the extended atmosphere produce dust grains which, by the pressure of stellar radiation, drive the stellar wind.

2. Observations and analysis

We have carried out spectroscopic monitoring of HD 235858 and HD 161796. Former is a carbon rich G type star and the latter - oxygen rich F type star. Both are pulsationally variable post-AGB supergiants in early stage - they are relatively cool and likely to have winds that share similarities with the ones operating in AGB stars. Both have aspherical wind-created envelopes; however, there is no evidence for binarity. It was neither revealed

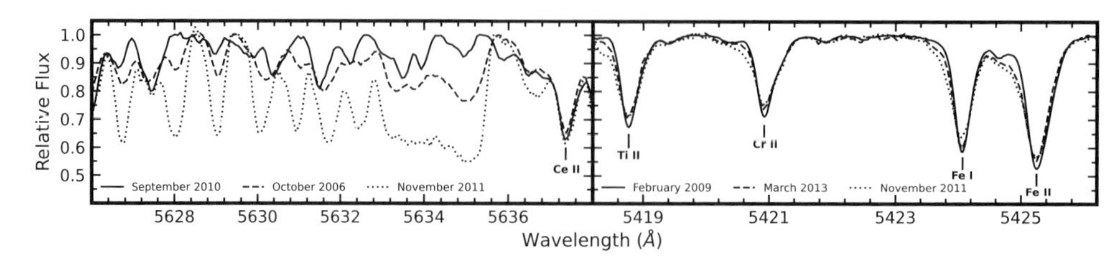

Figure 1. Left panel: intensity changes in the spectra of HD 235858 blueward of 5636 Å (C_2 Swan system bandhead). Right panel: typical variation of absorption lines in HD 161796.

in long term radial velocity monitoring (Hrivnak et al. 2017) nor latest Gaia data seems to support it (Parthasarathy 2022).

For HD 235858 time-domain spectroscopic observations have revealed pulsation induced cool outflow (Začs et al. 2016). This was inferred from intensity variations of blueshifted C_2 Swan system and CN Red system absorption lines, although such molecular features of significant strength are not normally expected in G type stellar spectrum.

In case of HD 161796 spectra acquired at multiple epochs point to an variable outflow of warm matter from the stellar surface (Puķītis et al. 2022). This follows from the specific variation of absorption line profiles - blue wings are variable, most often being extended while red wings remain virtually unchanged. Such variation is seen for both atomic and ionic lines, and for both low- and high-excitation lines of different chemical elements. Additionally, the specific shape of Hα profile in HD 161796 - variable narrow central absorption dip showing emission components which is superposed on a normal broad absorption without any significant variation - as interpreted by Sánchez Contreras et al. (2008), is a consequence of incipient mass loss and therefore support the conclusion of an outflow from HD 161796. For both stars splitting of intense absorption lines can be observed which is a manifestation of shocks which, in turn, are an integral part of wind launching for related stars.

3. Prospects

Spectroscopic monitoring of these and other similar post-AGB stars has a potential to advance the knowledge of stellar wind in this evolutionary phase and to uncover variety of dynamical processes in the atmospheres of such objects. The latter is corroborated by the study of Začs & Puķītis (2021) in which, based on interday variability of molecular and low-excitation metallic lines, it was shown that the star experienced an episode of infall of matter. Recently we have embarked on spectroscopic monitoring of HD 235858 and other similar objects in the near-infrared. Additional spectral lines of different species and excitation energies including bands of the abundant CO molecule will allow to probe dynamic phenomena over larger range of depths of the extended atmospheres of post-AGB stars.

References

Hrivnak, B. J., Van de Steene, G., Van Winckel, H., et al. 2017, ApJ, 846, 96
Parthasarathy, M. 2022, RNAAS, 6, 33
Puķītis, K., Začs, L., & Grankina, A. 2022, ApJ, 928, 29
Sánchez Contreras, C., Sahai, R., Gil de Paz, A., et al. 2008, ApJS, 179, 166
Začs, L., Musaev, F., Kaminsky, B., et al. 2016, ApJ, 816, 3
Začs, L. & Puķītis, K. 2021, ApJ, 920, 17

Author index